STRING & STICKY TAPE EXPERIMENTS

By R.D. Edge

Published by
American Association of Physics Teachers

String & Sticky Tape Experiments

Publications Department
5112 Berwyn Road
College Park, MD 20740, U.S.A.

Cover Art: Designed by Megan Church McGill

ISBN # 0-917853-28-8

String and Sticky Tape Experiments

August 1981

Introduction

We live in a vicarious age, often having our experiences through the medium of television rather than doing the touching and feeling ourselves. It is more enlightening to watch "Nova", or to take apart a motorcycle? Physics is an experimental science, and only by doing "hands-on" experiments--messing about the the equipment--can you get a feel for it. Most physics gear sold to schools is too expensive to allow students to work with it alone, and to have the teacher hovering by is inhibiting. To avoid this problem, the equipment must either be very strong, unbreakable in fact, or so cheap it can be replaced at little cost.

Even in this age of atom bombs and rockets, schools still have difficulty finding experimental equipment for their students. The following experiments were put together to see what could be done with the simplest, least expensive materials. The experiments are mostly in physics, with a few in the psycho-physics of perception included.

The equipment can all be purchased at the nearest store--not even a stop watch is required, nor are you exhorted to "go down to the junk yard and pick up a 2000 volt transformer" as is done in some books requiring economy. In spite of their simple nature, the experiments are quite meaningful, and demonstrate fundamental physics laws in a practical way. They can be done at home, as well as at school.

Equipment

Almost all the experiments involve only common rubber bands (about 3" x 1/16", if available), cellulose tape (the cheapest, clear, half-inch kind), regular paper clips (#1 are best), styrofoam or paper cups, string, drinking straws (plastic, and preferably translucent), glass marbles, paper, blackboard chalk, a foot ruler with a channel down the center, (provided to prevent the pencil rolling of the desk), coins, a pencil and scissors. It is a good idea to collect these simple items and keep them in a box, so that they are easily available.

For some experiments on electricity and magnetism, (see introduction to section 10, 11 and 12) certain additional items prove necessary, namely, aluminum foil (e.g. "Reynolds wrap"), empty aluminum drink cans, small magnets, batteries or power supplies, and copper wire (e.g. #16 insulated magnet wire).

In the "miscellaneous" section, thumb tacks are required, and Mason (preserve) jars and a coat hanger wire for the vacuum experiments.

Please Note: Readers with suggestions or comments should contact Professor Edge directly at the following address: Department of Physics, University of South Carolina, Columbia, SC 29208.

STRING AND STICKY TAPE EXPERIMENTS

Since the ordering of the experiments does not follow the normal textbook pattern, some teachers have suggested that it would be useful to rate the experiments. As a result, the columns on the right in the index provide this facility: Firstly, an asterisk -*- indicates a "fun experiment"--one where the experiment is both amusing and educational.

The experiments can be described as "qualitative", "quantitative" or both. The difficulty of each part is given in the columns, easy at the left, difficult the right. The number in each column rates the fundamental nature of the experiments from 1 to 5:

1. The experiment does not exemplify fundamentals.
5. The experiment demonstrates important fundamental principles of physics.

In addition, it was felt it would be helpful to give some indication of the educational level to which the experiments might appeal. This is given on the right in the list of experiments. An absence of symbols means the experiment will have some interest at all levels. E signifies it would be of most interest at an elementary level, H high school, and U university.

Since the first edition of this book, many of these experiments have been published in the "Physics Teacher". Readers are urges to read this journal (published by the American Association of Physics Teachers) to learn of new ideas.

We have endevored to use the S.I. system in this book whenever possible. However, we feel it is essential for employ familiar units rather than strange ones (e.g. slugs, Newtons) so in many cases we have compromised--e.g. miles per hour for speed.

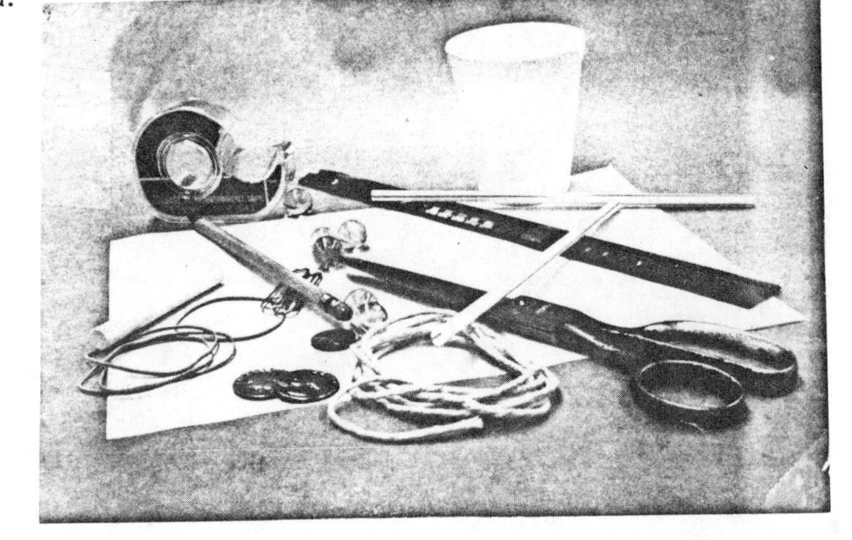

STRING AND STICKY TAPE EXPERIMENTS

List of Experiments

5=Most fundamental
*=Interesting

			QUALITATIVE					QUANTITATIVE					
			Easy				Difficult	Easy				Difficult	Elementary / High School / University
1.	**Mechanics**												
1.01	Measurement and Error								4				HU
1.02	Probability									4			HU
1.03	Vectors										5		HU
1.04	The Rubber Band Balance		3					3					EH
1.05	Elastic Forces		3					3					HU
1.06	Torsion Balance					3							HU
1.07	The Modulus of Rigidity of a Drinking Straw											3	U
1.08	Equal Armed Balance		2					2					EH
1.09	Levers - the Steelyard			3					3				EH
1.10	The Letter Balance	*		3					3		4		HU
1.11	The Microbalance										4		HU
1.12	Stresses in Beams			2						2			EH
1.13	The Triangle of Forces			4					4				HU
1.14	Applications of Couples and Torques							2					
1.15	Center of mass (gravity)			5					5				
1.16	Friction and the Climbing Monkey	*	3					3					HU
1.17	Friction										3		H
1.18	Relative density and the hydrometer				5					5			HU
1.19	Buoyancy - Archimedes principle			5					5				HU
1.20	Velocity		2					2					

5=Most fundamental
*=Interesting

No.	Topic	*	QUALITATIVE Easy				Difficult	QUANTITATIVE Easy				Difficult	Elementary / High School / University
1.21	Acceleration								5				
1.22	Acceleration due to g.		5					5					
1.23	Independence of vertical and horizontal motion		5										
1.24	Two dimensional kinematics using coins					3						3	HU
1.25	Kinetic and Potential Energy							5					
1.26	Inertia (Newton's 1st Law)		3										E
1.27	Action and Reaction	*	5						3				
1.28	Weightlessness	*	5										
1.29	Atwood's machine					3					3		HU
1.30	Collision of spheres	*			4					5			
1.31	The Ballistic Pendulum										4		HU
1.32	Central Forces		4					4					
1.33	Moment of Inertia			4				4					HU
1.34	Moments of inertia of books and straws	*	5							4			HU
1.35	Moments of Inertia					4				4			HU
1.36	Rotational Energy and Momentum					5				5			
1.37	Time and the pendulum		5					5					
1.38	Oscillatations of a spring -- Inertial and Gravitational Mass			5					5				HU
1.39	Coupled and Forced Oscillations	*		5					5				
1.40	Foucault Pendulum	*			3					3			
1.41	Coriolis Effect	*		4					4				
1.42	Law of Equal Areas						2					2	U
1.43	Inertia		3										

5=Most fundamental
*=Interesting

No.	Topic	*	QUALITATIVE Easy	QUALITATIVE			QUALITATIVE Difficult	QUANTITATIVE Easy	QUANTITATIVE			QUANTITATIVE Difficult	Elementary / High School / University
1.44	Car Accelerometer								4				
1.45	The Size of the Sun							2					
1.46	Rolling Uphill or Antigravity	*		3									
1.47	The Monkey Puzzle					4							
1.48	Angular momentum of coins	*	4					4					
1.49	Mobiles and Moments	*	4						4				
1.50	Angular Acceleration - the ball and ruler	*		3									
1.51	Inertia, Momentum, and Blow Football	*	2										
1.52	Quantum Properties								3				
1.53	Brachistochrone		3					3					
1.54	Levitating a Marble - Motion in a circle	*	3					3					
1.55	The Reflex Tester - g	*	3					3					
1.56	Trajectories, drops and the strobe		4					4					
1.57	Spinning Tops		4										
2.	Properties of Matter												
2.01	Stress and Tension in Chalk	*	2										
2.02	Torsional Stress in Chalk	*	3										
2.03	Surface Tension			4									EH
2.04	Surface Tension in a liquid stream	*		4									
2.05	Young's Modulus for Cellophane Tape											4	U

5=Most fundamental
*=Interesting

		QUALITATIVE Easy				QUALITATIVE Difficult	QUANTITATIVE Easy				QUANTITATIVE Difficult	Elementary / High School / University
3. Hydrodynamics												
3.01 Hydrostatics and Hydrodynamics - Hydrostatic Pressure		3						4				
3.02 The Syphon		1										
3.03 Bernoulli Effect		2										
3.04 Bernoulli Principle		2										
3.05 The Atomiser				1								
3.06 Atmospheric Pressure	*	2										
3.07 Vortex Rings	*		3									
3.08 Pressure of the Air	*	2										
3.09 Sailboat log									3			HU
3.10 Bernoulli Effect		2										
3.11 Streamlining of Aerofoils	*			3								
3.12 The Flight of a Baseball or Golf Ball				3								
3.13 Levitating a Dime	*	2										
3.14 Viscosity				4							4	U
3.15 Effects of Pressure	*	3										
4. Heat												
4.01 The Hygrometer			2									EH
4.02 Thermal Expansion of a Soda Straw	*								4			
4.03 The Effect of Heat on a Rubber Band		2										EH
4.04 Heat and Work	*	2									2	
4.05 A Soda-straw Thermometer	*	5					5					

5=Most fundamental
*=Interesting

Experiment		QUALITATIVE Easy	QUALITATIVE	QUALITATIVE Difficult	QUANTITATIVE Easy	QUANTITATIVE	QUANTITATIVE Difficult	Elementary / High School / University
4.06 The Pressure-volume Relationship in a Gas								
4.07 Convection				2				EH
4.08 Thermal Capacity	*				5			
4.09 Cooling a Cup of Coffee	*					5		HU
4.10 Relative Humidity with a Drinking Straw Thermometer			4			4		
4.11 The Mechanical Equivalent of heat	*				4			HU
4.12 Absorption and Emission of Radiation	*	4						EHU

5. Wave Motion

Experiment		QUALITATIVE Easy	QUALITATIVE	QUALITATIVE Difficult	QUANTITATIVE Easy	QUANTITATIVE	QUANTITATIVE Difficult	Elementary / High School / University
5.01 Longitudinal Waves	*	5						
5.02 Transverse Waves	**	5						
5.03 Twin Slit Diffraction	*	4				4		HU
5.04 Longitudinal Waves - Crova's Disc	*	4				4		
5.05 Young's Slits With Paper Waves	*	4				4		
5.06 Adding Oscillations		4						EHU

6. Light

Experiment		QUALITATIVE Easy	QUALITATIVE	QUALITATIVE Difficult	QUANTITATIVE Easy	QUANTITATIVE	QUANTITATIVE Difficult	Elementary / High School / University
6.01 Reflection of Light					4			
6.02 Refractive Index of Water	*					5		
6.03 Optics - Positive and Negative Lenses			5				4	
6.04 Real Images			5				4	
6.05 Refraction of Particles			3					HU
6.06 The Phase amplitude diagram for a 12 slit interferometer			2					HU

5=Most fundamental
*=Interesting

	*	QUALITATIVE Easy				Difficult	QUANTITATIVE Easy				Difficult	Elementary High school University
6.07 Diffraction												EH
6.08 Pinhole Camera				3								
6.09 Triboluminescence		2										
7. Sound												
7.01 The Rubber Band Guitar	*			3					3			
7.02 The Way the Tension and Length of a Plucked String Affect its Pitch.										4		HU
7.03 Waves in Pipes			4					4				
7.04 The Reed	*			3					3			
7.05 Combination Tones					2					2		
7.06 Resonance in Sound		3							3			EH
7.07 The String Telephone			2									E
7.08 Acoustics of Rooms--Reverberation Time	*	4							4			
7.09 Velocity of Sound							5					
7.10 A Paper Wave to Explain the Doppler Shift												HU
8. Atomic and Nuclear Physics												
8.01 Molecular Size	*		4						4			
8.02 Nuclear Cross-section and the Size of a Penny	*						5					
8.03 Radioactive Decay - An Analogy							4					HU
9. Solid State and Crystallography												
9.01 The Packing Together of Atoms in Metals												

5=Most fundamental
*=Interesting

		*	QUALITATIVE: Easy				Difficult	QUANTITATIVE: Easy				Difficult	Elementary / High School / University
10.	Electrostatics												
10.01	The Electroscope	*	5					5					HU
10.02	The Versorium		4										EH
10.03	The Electrophorus		4										HU
10.04	The Faraday Ice Pail Experiment							4					HU
10.05	Capacitance							4					
11.	Magnetism												
11.01	Why is the North Pole South?		4										
11.02	Making a Magnet, and Using it as a Compass	*	4										
11.03	Lines of Force Around a Magnet			4									
11.04	Tangent Magnetometer, and Magnetic Strength of a Bar Magnet										5		HU
11.05	The Force Law Near a Bar Magnet										5		HU
11.06	Strength of Bar Magnets		2										
11.07	The Concentration of Field in a Magnet		3										
11.08	The Dip Circle											3	
11.09	What is Magnetic?	*	2										E
11.10	Penetration of Magnetism		3										EH
11.11	The Mysterious Magnet	*	5										HU

5=Most fundamental
*=Interesting

No.	Topic	*	Qualitative: Easy				Qualitative: Difficult	Quantitative: Easy				Quantitative: Difficult	Elementary / High School / University
12.	**Current Electricity**												
12.01	The Magnetic Field Around a Coil of Wire										5		
12.02	The Tangent Galvanometer	*									5		HU
12.03	Electrical Resistance, (series and parallel)					4						4	HU
12.04	Magnetic Induction											4	U
12.05	Mutual Inductance											4	U
12.06	Electrolysis				3								
12.07	Forces between Parallel Conductors			4									
12.08	The Current Balance											4	HU
12.09	A Current Balance Using a Magnet										4		HU
12.10	A Simple Method of Measuring Alternating Current, and the Transformer											4	HU
12.11	The Simplest Electric Motor	*		5					2				
13.	**Psychophysics**												
13.01	Benham's Top	*	2										
12.02	Physical Perception - Optical Illusions	*	2										
13.03	Moiré Patterns	*	2										
13.04	Persistence of Vision	*			2								
13.05	The Ames Window	*		2									
14.	**Physics Games**												
14.01	Pirates' Treasure Game (Vectors)	*	4										EH
14.02	The Three-meter Dash (Kinematics)		4										
14.03	The Knee-bend Game (Energy and Power)	*	5										

No.	Title	5=Most fundamental *=Interesting	QUALITATIVE: Easy				QUALITATIVE: Difficult	QUANTITATIVE: Easy				QUANTITATIVE: Difficult	Elementary / High School / University
14.04	The Wave Game	*	5										
14.05	The Kinetic Theory Game		4										
14.06	The Nuclear Reaction Game		3										
14.07	The Collision Game		4										EH
14.08	The Electron-in-a-wire Game		3										EH
14.09	The Close Packing Game		4										EH
14.10	Lifting the Body - Force		4										
14.11	Reaction Kinematics	**						5					HU
14.12	Statics-People Pyramids							4					EH
14.13	Stand Up-Vector Forces		4										EH
14.14	Hunker Hawser		4										E
14.15	Conclusion-Indroducing Physics to the Youngest Students												E

15. Miscellaneous

No.	Title	5=Most fundamental *=Interesting	QUALITATIVE: Easy				QUALITATIVE: Difficult	QUANTITATIVE: Easy				QUANTITATIVE: Difficult	Elementary / High School / University
15.01	Reaction - The Rocket and Balloon					2							
15.02	The Platonic Solids	*		2									
15.03	A String and Sticky Tape Digital Computer							4					HU
15.04	Four Dimensional Tic-Tac-Toe							4					HU
15.05	Simple Vacuum Experiments		4										

Experiment 1.01 Measurement and Error

We will start our experiments by measuring accuracy. Lord Kelvin said:

"I often say that when you can measure what you are speaking about, and express it in numbers, you known something about it; but when you cannot express it in numbers your knowledge is of a meagre and unsatisfactory kind; it may be the beginning of knowledge, but you have scarcely, in your thoughts, advanced to the stage of Science, whatever the matter may be."

Many schools avoid the analysis of errors - yet, in fact, it is often the most important part of the experiment - so it is dealt with here first.

Materials: paper or styrofoam cup, ruler.

Procedure: Physics is concerned with measurement. The most fundamental quantities we measure are mass, length and time. The most important feature of such measurements is their accuracy. No measurement can be perfectly accurate. For example, we can measure the speed of light to better than one part in 1,000,000*, but the age of the Universe (20 x 10^9 years) we only know to about a factor of two. Lay a ruler along the line below. Do not put zero against one end of the line, but put the ruler down along the line at random, paying no regard to where the ends are. (If you have no ruler, cut out the scale on page 4)

* 2.997925 ± .000003 x 10^8 m/sec

____

Read each end of the ruler to .01 cm and put down the measurements in the table on the following page. Repeat nine times, displacing the ruler each time.

Do all the measurements agree? The difference between the separate measurements is called the error of the results. All measurements have some degree of error in them. The best value for the length is the average, which you get by adding all the results together and dividing by the number of measurements. If you add together all the differences between the average and each of the results obtained, taking all of the differences as positive, we have a measure of error. Divide this sum by the number of measurements and multiply by 1.25 (obtained from the theory of errors) to get the "standard error" for each individual measurement, which estimates how close an individual measurement will come to the mean. There is an even chance a single measurement will lie closer to the mean, or farther away from the mean, than .67 of the standard error. A class obtained an average of 10.23 cm with a standard error of .03 cm for a line similar to that above. Is your standard error close to this?

A typical example of the standard error for five measurements is shown overpage; the experimenter must have had good eyesight!

Left end cm	Right end cm	Differences	Difference From Average	Squares of Difference
2.00	12.25	10.25	0	0
2.69	12.95	10.26	.01	.0001
3.32	12.58	10.26	.01	.0001
4.34	14.58	10.24	.01	.0001
4.25	14.48	10.23	.02	.0006

Sum = 51.24 | Sum of Differences = .05 | Sum of Squares = .0007

Average = 10.25 | Average = 0.00014

Standard error = 1.25 x .05/5 = .012. | Square Root = Standard Error = .012

DO NOT USE THIS COLUMN UNLESS CALCULATING ROOT MEAN SQUARE

	Left End cm	Right End cm	Right End -Left End	Difference from average	Squares of Differences
1					
2					
3					
4					
5					
6					
7					
8					
9					
	Sum =			Sum of Differences	Sum of Squares

Divide sum by 9 = Average =

Divide by 9 = Mean Square

Take Square Root = Standard Error

Divide sum of differences by 9 and multiply by 1.25 (1.25/9 = .139) to get the Standard Error =

The mean itself is not the true value, and if we divide the standard error or standard deviation as it is sometimes called, by the square root of the number of measurements, we obtain the standard deviation of the mean, which is what you see written after physical results behind the ± sign e.g.

5.4 inches ± .1 inch

which means, a length is 5.4 inches, accurate to .1 inch - it could be as much as 5.5 inches or as small as 5.3 inches, but 5.4 inches is the best value. Divide your standard error by 3. You will now have the standard deviation of the mean since you obtained 9 results. Look at the picture of a meter dial (p.4). What does the pointer read? After guessing, use a ruler to check up - and show you are probably wrong. Optical illusions frequently mislead. Accurate measurement is the only way to go, and tells you how accurate any result is.

Random errors such as those above occur through the statistical fluctuations of measurements, as with throwing dice (unloaded). Systematic errors are predominantly of one sign, e.g. drift of the zero of a meter--and they affect each observation equally--as with loaded dice. If they can be determined, they can be eliminated.

The result of an observation should never be expressed to more figures than the error would suggest. For example, whereas 5.4 ± 0.1 inch is meaningful, 5.4278 ± 0.1 inch is not. The number of digits, two in the example given, is called the number of significant figures.

The conventional way of finding the standard error (or standard deviation s.d.) is to take the root mean square deviation. This is more tedious, but gives a better estimate, and should be done if you use a calculator that takes square roots. Simply square each number individually in the last column (the difference from average), add them together and divide by the number of measurements to get the mean square deviation. Take the square root of this to give the standard error for a single measurement, which we found before. The values found using the two methods will differ, but not by much.

The Vernier Scale

It is difficult to estimate a fraction of a millimeter on a ruler by eye. In order to facilitate this a Vernier scale is used. Such a scale, enlarged, is shown below.

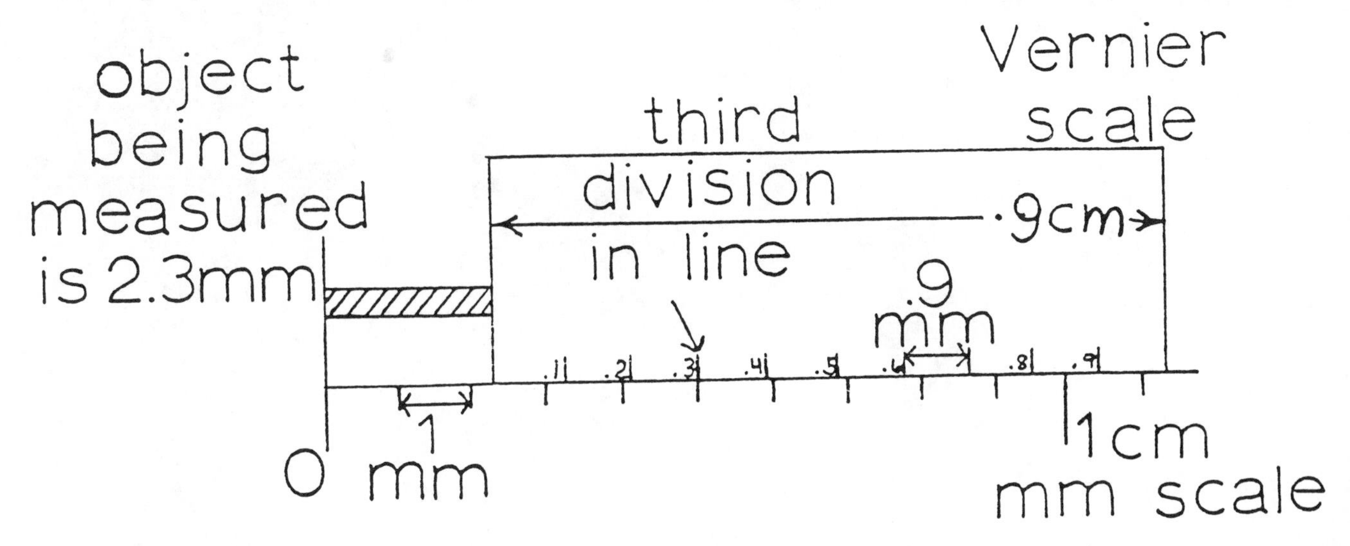

You will see, that if the lower scale is in millimeters, the upper scale has ten divisions to nine millimeters -- i.e. each division is 9/10mm. This means that, on moving from one division to the next on the upper scale we move along by one millimeter, but drop back also by 1/10mm. So, if the zeroth upper division is .3mm from the nearest division of the bottom scale, the first division will be .2, the second .1 and the third division will be exactly over the lower division--so the upper and lower scales are aligned. It is easy to see this, and hence to state that the length shown is 2.3mm. Now, cut out the scale below, and use it to measure the length of the line given above, by laying the zero of the millimeter scale at the zero of the line, and sliding the cut out vernier along, as shown in the picture below.

METRIC 1 2 3 4 5 6 7 8 9 10 11 12 13 14 15 16 17

Vernier scale to be cut out

METRIC 1 2 3 4 5 6 7 8 9 10 11 12 13 14 15 16 17

This line is 125.4 mm long.

Remove the scale completely from the line, then reset it, and repeat the measurements five times. Do your measurements agree? Are they better than if you try to measure the length of the line using a millimeter ruler, estimating fractions of a millimeter, as you did previously?

Still greater accuracy can be obtained by using the same vernier to read both ends of the line, since there is an "end correction" necessary because there is an error which arises by measuring one end of the line in one way, and the other the other. If we lay the ruler along the line, slide the vernier to one end and read it, then to the other and read that, both ends are read in the same way, and the end effect cancels out. For example, if both ends were read .1mm too large, the difference would still be the same in spite of the incorrect absolute value.

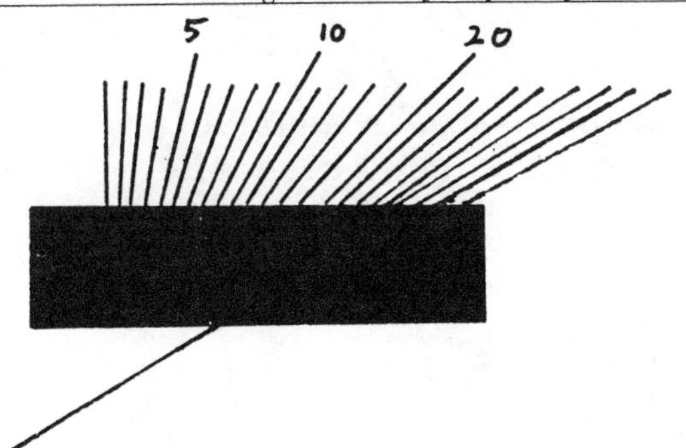

AREA

Length is a fundamental dimension. We can derive areas from length, for example

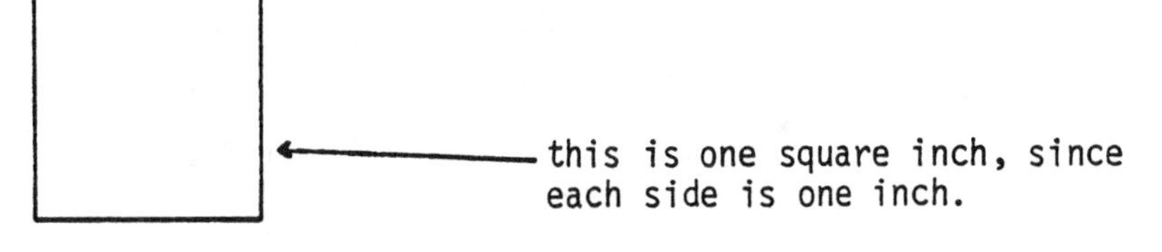

this is one square inch, since each side is one inch.

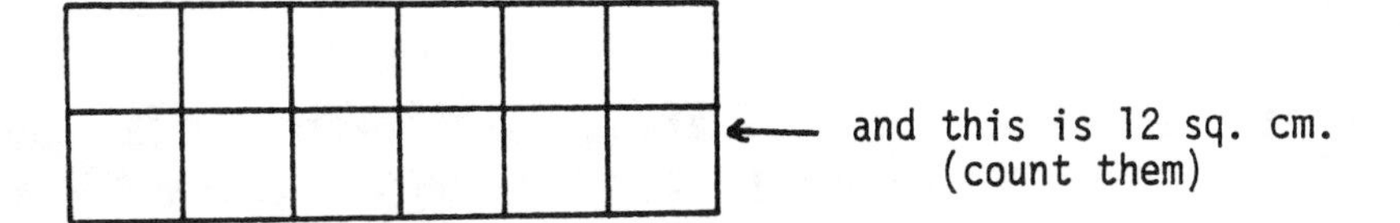

and this is 12 sq. cm. (count them)

Add up the squares to find out how many square centimeters the cross-hatched area below is. Measure it to a fraction of a square and put down the results. Repeat twice.

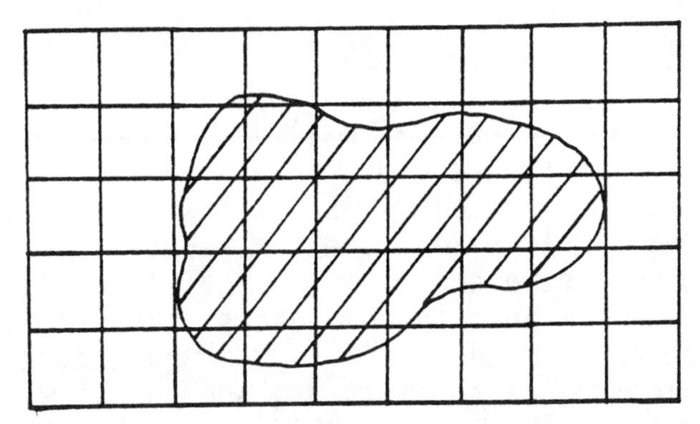

Do you get the result to the same fraction of a square each time? What is the error? Is the percentage error of the area measured greater than the length measured previously? Why?

Vernier Caliper

Although it is measured in square centimeters, an area can be any shape. Are the two areas below the same?

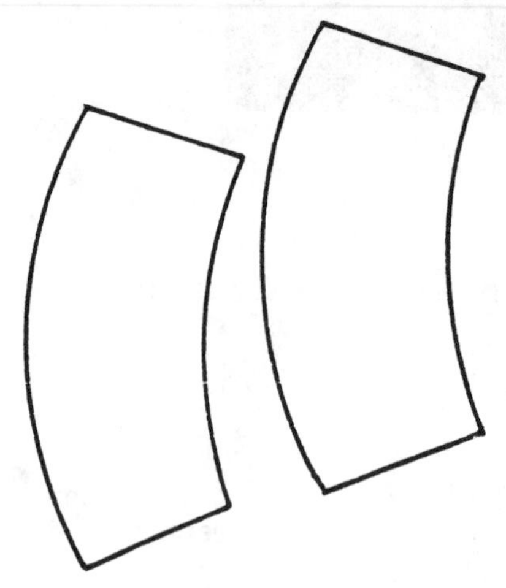

They are, although an optical illusion makes one seem larger -- only measurement can tell.

We see we can add squares to get area, but looking back, you will see that the area 6 cm by 2 cm gives 12 square cm. Area is length times length -- so a square 5 cm by 5 cm totals 25 cm^2.

VOLUME

Now consider volume. On page 9 is a drawing with instructions how to fold it into a cube. If you look, each side is 5 cm by 5 cm, so the number of squares on each side is

$$5 \times 5 = 25$$

So each layer of the cube has 25 small 1 cm cubes in it, of volume 1cc and since there are 5 layers, this amounts to 125 cubic centimeters. So the volume is 5 x 5 x 5 = 125 -- three dimensions of length multiplied together. But again, the volume does not have to be a cube or rectangle. Fill the little cube with water and empty it into the Dixie cup. Has the volume changed when the water pours into the Dixie cup? Make a mark on the side of the cup to which the water fills it, and repeat with another cubeful. This will be 125 + 125 = 250 cc and you have a 250 cc graduated vessel. Since 1 cc of water weighs 1 gm, your cup will have a mass very nearly 250 gms, and you can use it as a calibrated mass as well as volume. You could weigh your marbles against this to find their mass. The whole point is, using just the sheet of paper, and a little water, we end up with standards of mass and volume.

The volume of a paper or styrofoam cup is $\pi\frac{h}{3}(r_1{}^2 + r_1r_2 + r_2{}^2)$ where r_1 and r_2 are the inside radii of top and bottom and h is the length. Most styrofoam cups have a volume of about 200cc when filled level with the brim. Some cups are labeled on the bottom 6 oz. or 180cc, and hold this much when filled just below the brim.

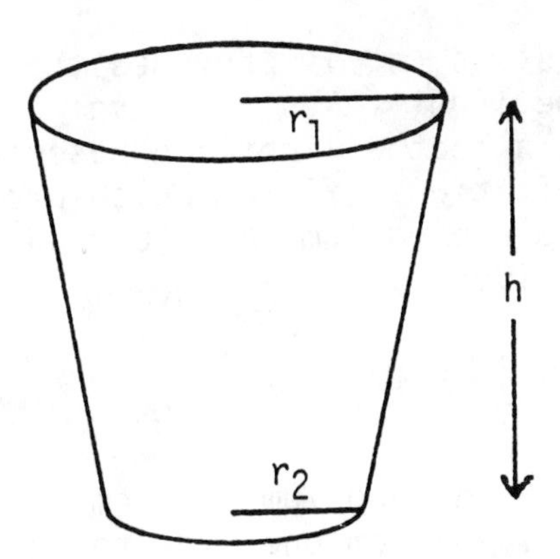

A sheet of paper 8½" square makes a box which holds 180 cc. Fill and empty the box rapidly - unless you use wax paper, the water quickly makes it soggy.

Masses - The following table was obtained by the Bureau of the Mint. Note before 1964, when clad coins were introduced, the mass of a half dollar was twice that of a quarter, which was 2.5 times that of a dime, whose mass was half that of a nickel A dollar bill has a mass close to 1 gm.

Penny	before 1942	3.110 g
	1943	2.700 g
	1944-1982	3.110 g
	1982	2.500 g
Nickel		5.000 g
Dime	before 1964	2.500 g
	1965-date	2.268 g
Quarter	before 1964	6.250 g
	1965-date	5.670 g
	Except the silver-clad Washington Bicentennial	5.750 g
Half-dollar	before 1964	12.500 g
	1965-1970	11.500 g
	1971-1974	11.340 g
	1977-date	11.340 g
	Silver-clad Kennedy Bicentennial	11.500 g
	Cupruo-nickel-clad Kennedy Bicentennial	11.340 g

The marbles used in our experiments vary in mass from 4.8 to 5.8 gm. Marbles taken from one packet rarely vary by more than 10% in mass. An average of 5.4 gm is probably a good choice.

Now that we know how to measure length and mass, how do we measure time? We can do so in terms of length through some standard derived unit such as velocity $(\frac{\text{length}}{\text{time}})$ or acceleration $(\frac{\text{length}}{\text{time}^2})$.

TIME

Light has a constant velocity in all systems, so it provides the best standard -- but it is incoveniently fast -- the time light takes to travel a foot is 10^{-9} seconds, a nanosecond, so we could call a foot a light - nanosecond. The most convenient standard is the earth's gravitational field, which provides an acceleration, g, of 981 cm/sec². A simple pendulum has a period (the time for the pendulum to swing to and fro) of T given by

$$T = 2\pi\sqrt{\frac{L}{g}}$$

where L is the length of the pendulum. So a pendulum 99.4 cm long takes one second for half a complete oscillation, i.e. from one side to the other. One's ear is very sensitive to frequency - so using a rhythmic song ("onward Christian Soldiers" has been used by photographers who must time in the dark) one can time quite accurately.

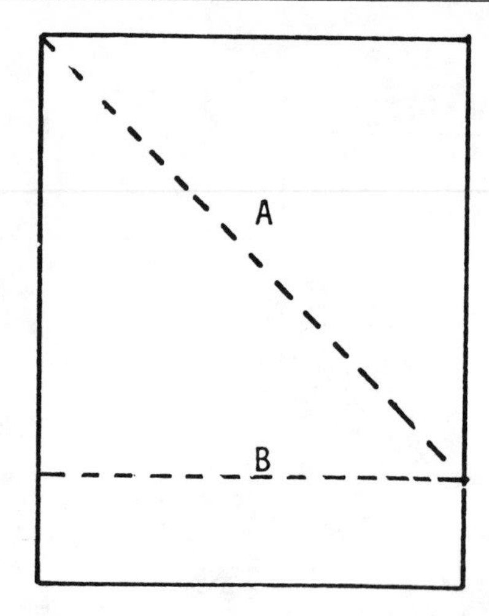

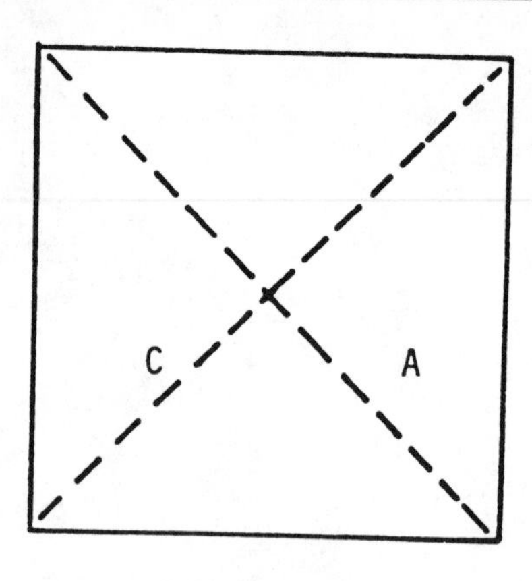

1. Fold along line A. Cut along line B.

2. Open up paper and fold along line C.

3. Open up paper, turn it over and fold along line D. Then unfold the paper. Now turn the paper over again. Bring the two X marks together and then bring the two Y marks together. Your paper should now look like the two diagrams below.

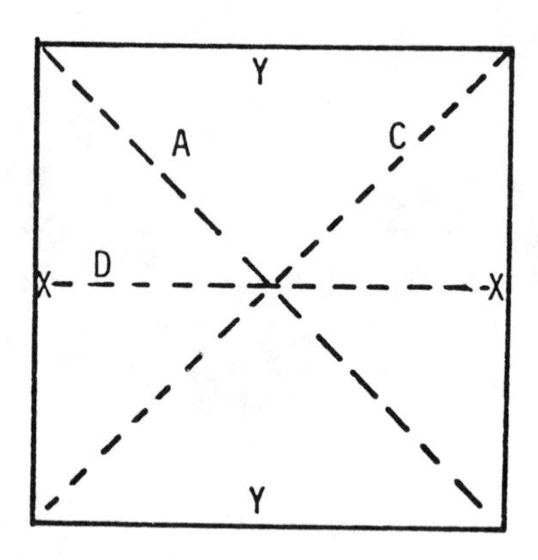

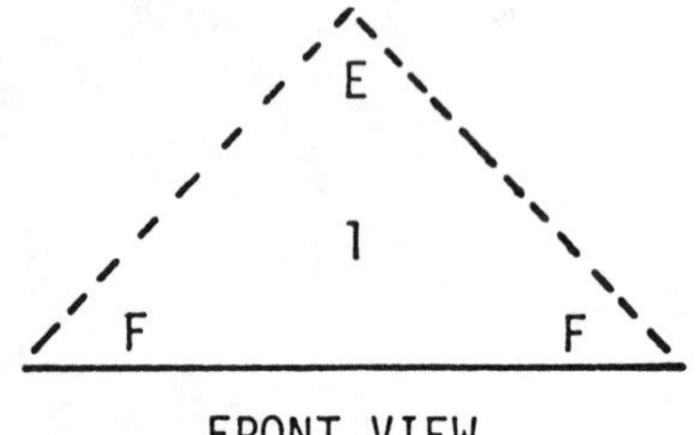

FRONT VIEW

4. You now have two symmetrical sides, sides 1 and 2, as shown in the side view diagram. THE NEXT PAGE OF INSTRUCTIONS PERTAIN TO SIDE 1. AFTER COMPLETING EACH INSTRUCTION TURN THE PAPER OVER AND DUPLICATE THE INSTRUCTION FOR SIDE 2.

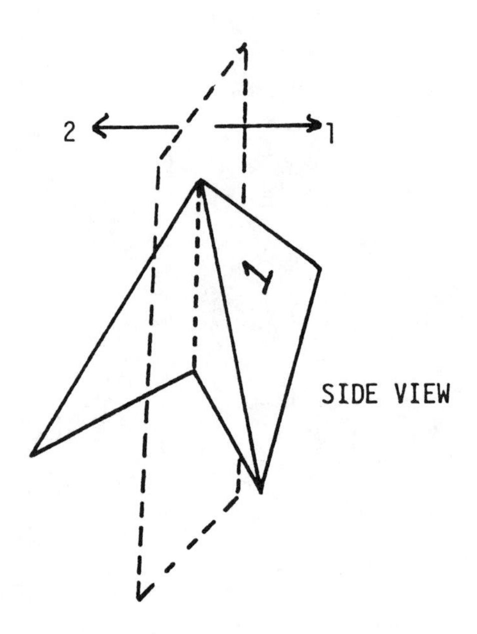

SIDE VIEW

REMEMBER TO DUPLICATE THE INSTRUCTIONS FOR SIDE 2.

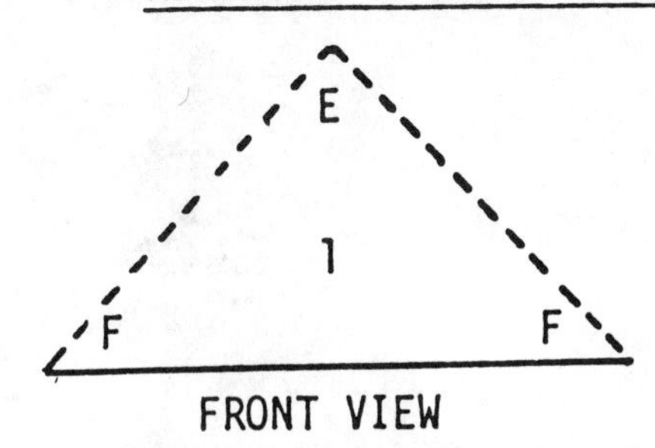

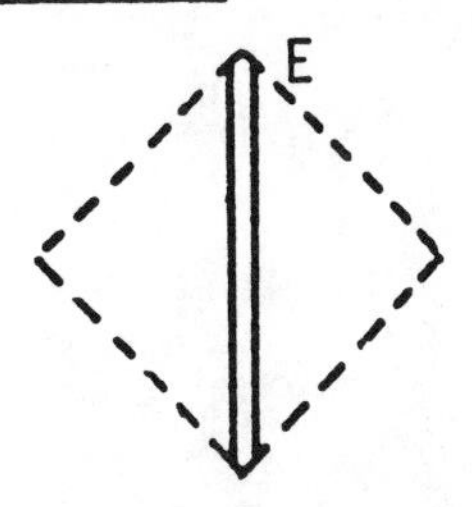

5. Fold points F to point E..............Your paper should look like this.

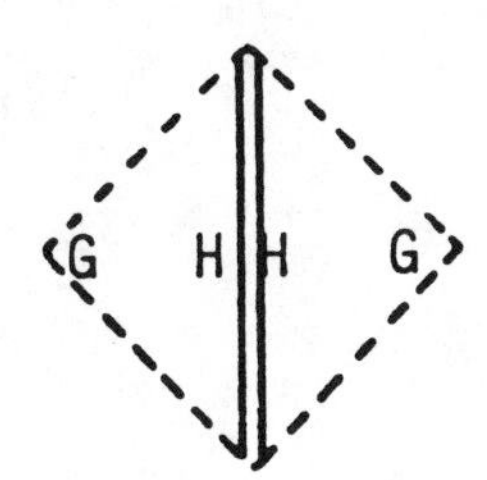

6. Fold points G to H....................Your paper should look like this.

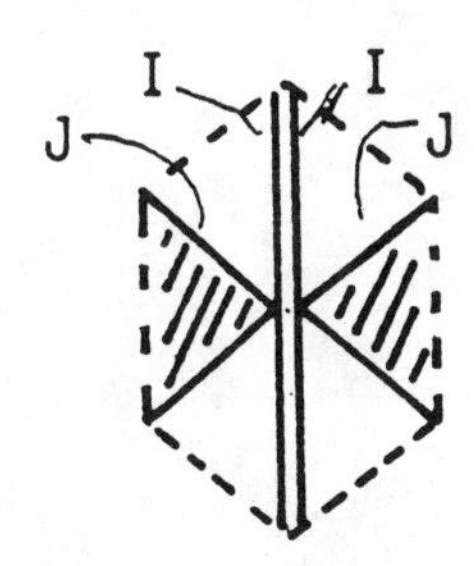

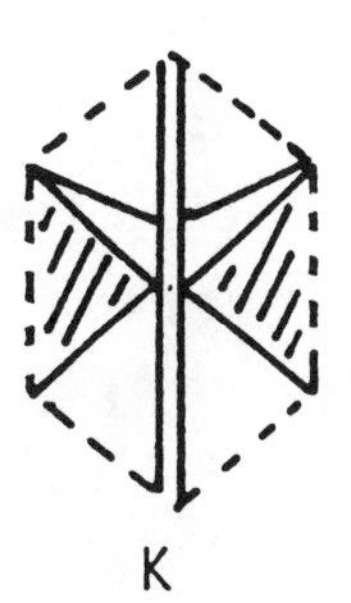

7. fold leafs marked I into slots marked J. Your paper should look like diagram 8.

8. Blow into K to make a cube.

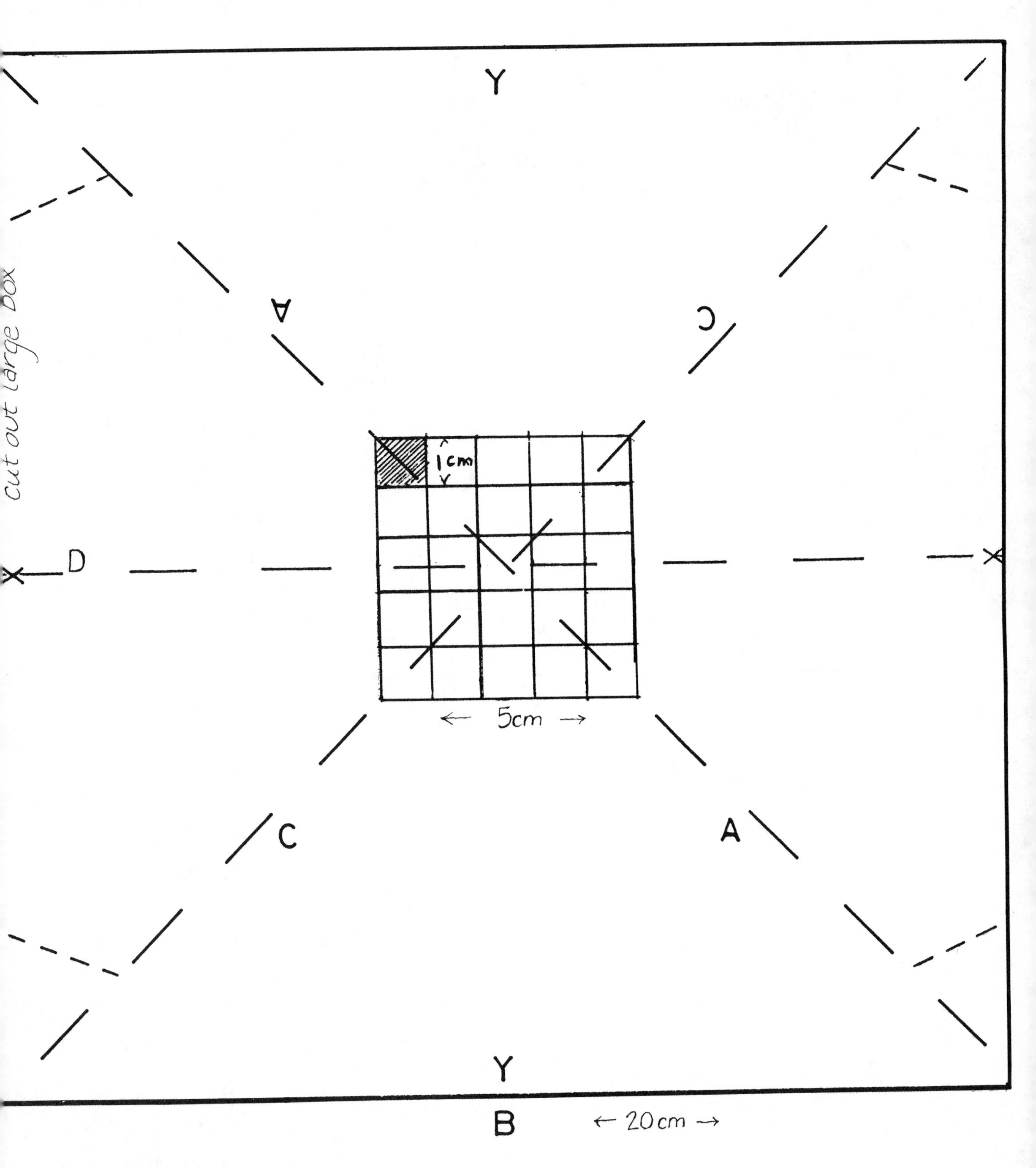

A BOX MADE FROM THIS SHEET HOLDS 125 cc

EXPERIMENT 1.02. Probability and the normal curve of error. It is important to know not merely the average error, but also the distribution.

Materials: Eight coins

Procedure: Unless you have a double headed penny, each coin has two different sides. If you toss one coin, what is the chance, or probability it will come down heads?

Now toss two coins together. Is there the same chance they will come down both heads, both tails, or one head and one tail?

Both heads: the chance of one head is 1/2, so the chance of two is 1/2 x 1/2 = 1/4. HH - one chance in four or a probability of 1/4.

Head and tail - not 1/4, but 1/2, since HT is the same as TH, i.e. the order doesn't matter.

Both tails (1/4)

In practice, you rarely get exactly one quarter of the throws to be both heads, but the more tosses you make the closer it gets to the expected value.

Now, toss eight coins together. Count the number of heads and put an X in the corresponding column of the table below. Repeat this a large number of times, making an X in the next square above if there is already an X in the column, thus

X
X

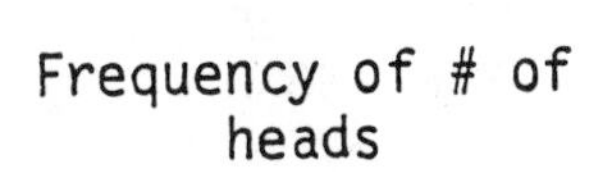

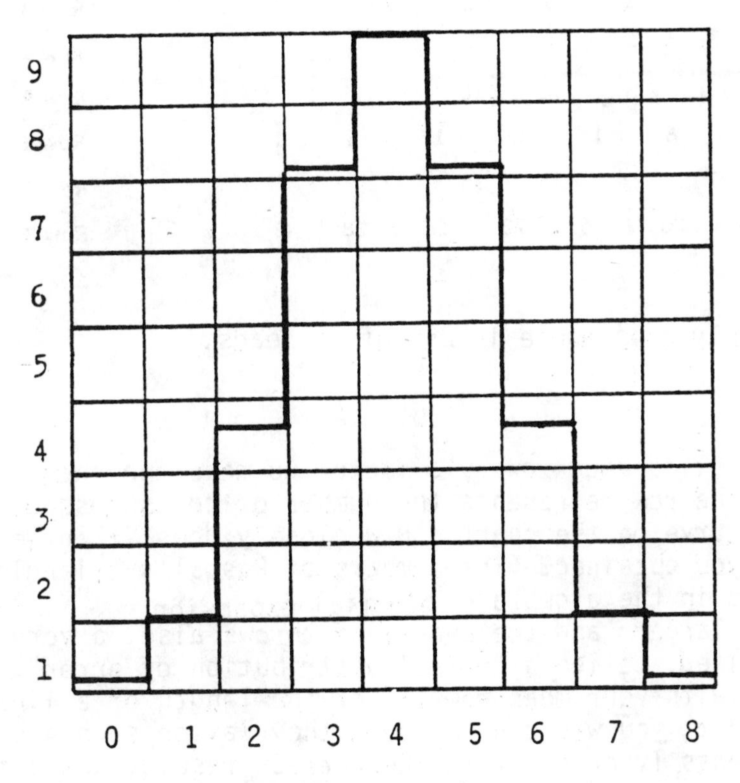

Notice how the number of heads peaks up in the middle, giving a bell shaped curve. How can this be predicted? The famous French Mathematician-Philosopher Blaise Pascal first demonstrated how this could be done in "Pascal's Triangle". This is a way of automatically counting up how many ways a different combination of heads and tails can occur. For example, in the case of two coins, Pascal's triangle looks like:

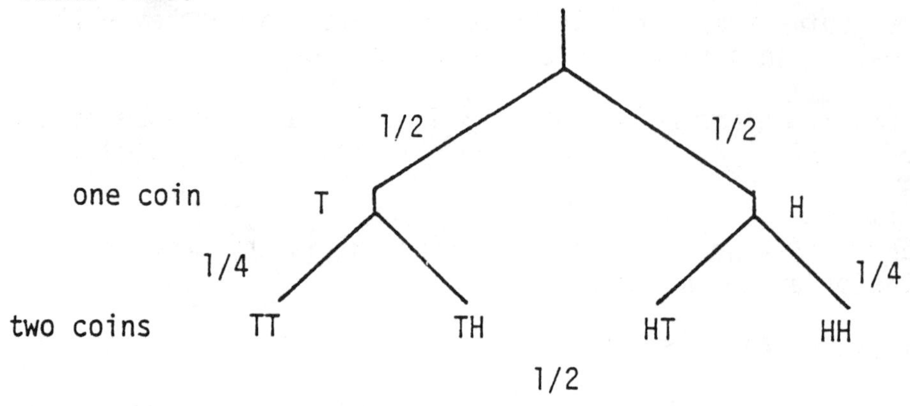

For eight coins, it looks like:

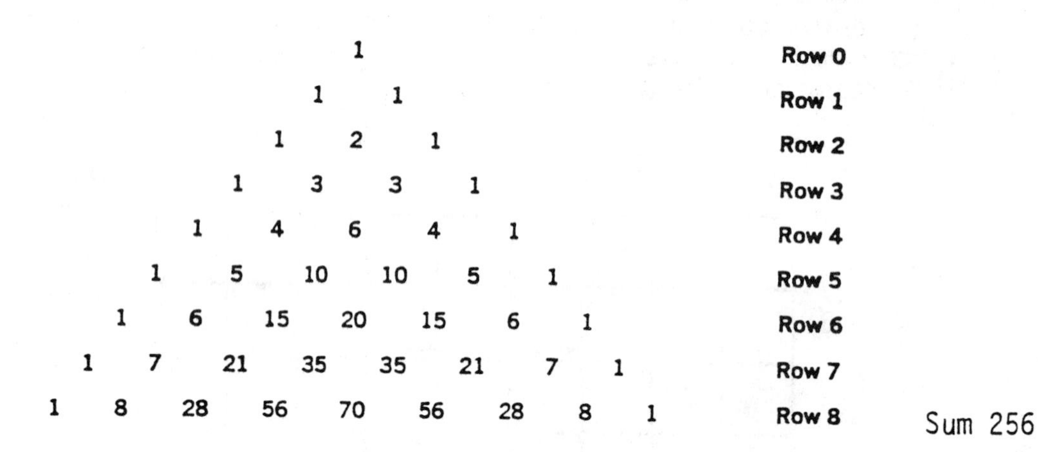

so there is only one chance in 256 of no heads.

Here, we multiplied each row by a factor to make the ends of each row equal to 1. The row represents the number of coins tossed. We have plotted this curve on the chart. How closely does it agree with the distribution you obtained? The numbers of Pascal's triangle are also those obtained in the algebraic binomial expansion As the number of coins gets larger, and the number of throws also, a very smooth curve is obtained, called a "normal distribution of error". If you had continued with your measurements of the length of a line in the first experiment, you would have found they lay on such a curve, with most measurements lying close to the correct result, and a few far away.

More advanced students note that the binomial expansion, whose terms give the number of heads and tails for the eighth row, takes the form

$$(1+1)^8 = 1 + 8 + \frac{8x7}{1x2} \ldots$$

Generally

$$(1 + 1)^N = 1 + N + \frac{N(N-1)}{1x2} + \frac{N(N-1)(N-2)}{1x2x3}$$

where there are N rows (or N coins to be tossed).

The sum of the number of trials is $2^8 = 256$ for the eighth row (in general 2^N). If we let N become very large, the distribution takes the form that the number of times we get $\frac{N}{2}$ - n heads and $\frac{N}{2}$ + n tails is given by

$$S = 2^N \sqrt{\frac{2}{\pi N}}\, e^{(-\frac{2n^2}{N})}$$

where the standard deviation $\sigma^2 = \Sigma \frac{sn^2}{N} = N$.

This approximation is good for large N, but very poor for small N. For example, in the case where N = 8, it gives 13, 20, 27, 36, 34, 27, 20, 13 which does not agree with the previous results, however as N becomes greater than 100, the distribution sharpens very rapidly, and the approximation becomes very much better for very large N. A distribution $\frac{1}{\sigma}\sqrt{\frac{2}{\pi}}\, e^{-\frac{2x^2}{\sigma^2}} \Delta X$ where $\sigma^2 = \Sigma \frac{(X-\bar{X})^2}{N}$ (not necessarily N) is called a Gaussian distribution, or "normal curve of error", and represents the probability an event is between X and X + ΔX.

Experiment 1.03. Vectors

Materials - Pencil, ruler, scissors

Procedure - PART 1: Distance

Quantities such as mass are called "scalars." They can be defined by a simple number, telling us how much we have - if we have 2 kg, we have twice as much as 1 kg. However, some quantities, such as length, require a direction as well to specify them. For example, if we say we walked a mile, we might well be asked where, or in which direction - East, West, North or South. We might have walked half a mile north, and half a mile south, and arrived back at the same spot - or walked one mile north - so clearly we must specify direction together with distance. Direction must be carefully specified too. - For example -

A man walked one mile south, one mile west, and one mile north, arriving back at the place he started, where he saw a bear. What color was it? Answer - White. The only place you can walk one mile south, west and north, and arrive back where you started is the north pole, so it must be a polar bear. But note, this could only happen on a sphere, such as the earth - it would be impossible on a plane.

Now, scalars add arithmetically, but vectors do not - if we walk one mile, and then another mile, we may be two miles from where we started, walking in the same direction, or back at the start, if in opposite directions.

The easiest way to talk about vectors is to draw a diagram. For example, if we let one inch equal one mile, our two walks might look as follows

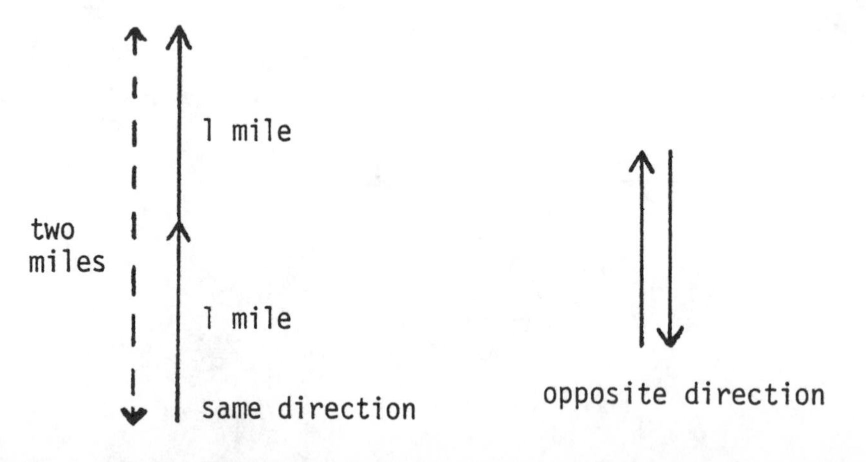

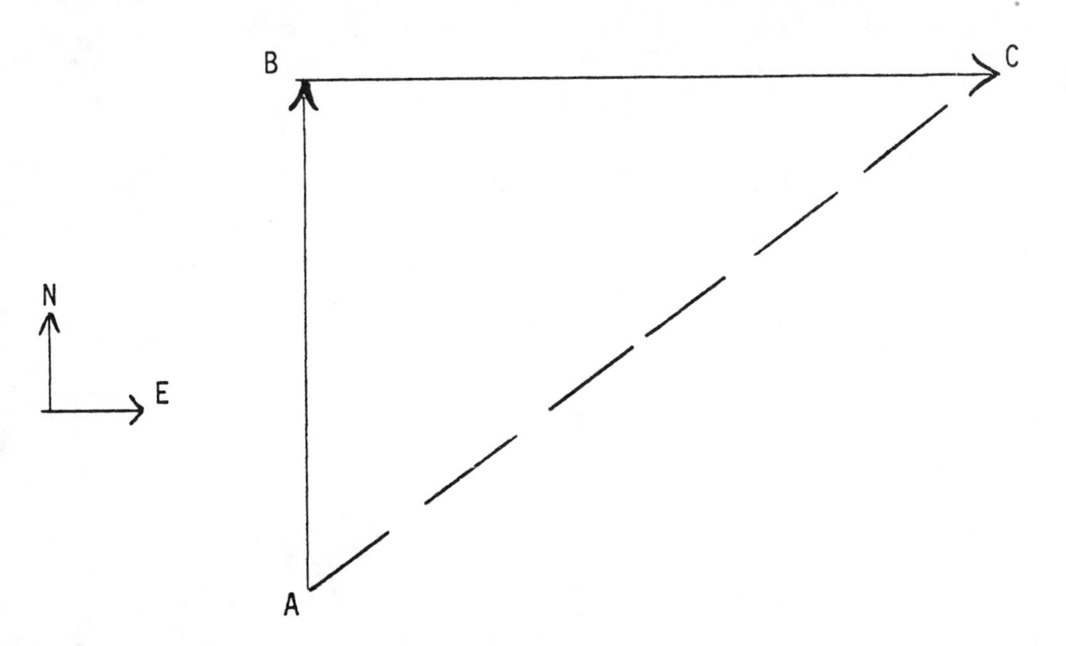

Now, suppose you walk three miles north, and four miles east. We measure three inches north AB on the paper, as shown, and four inches east BC. Then the distance from where you started is AC. Measure it. How far are you from where you started? You are five miles, which is not the arithmetic sum of three plus four. Now try this one out yourself. You walk two miles north and two miles west. Draw it below, but use a scale where 1 cm represents 1 mile. How far are you from the place where you started? The answer should be about 2.8 cm.

Unless we are walking N, S, E or W, it is generally easier to talk about direction in degrees. Cut out the protractor printed below, which is graduated in degrees.

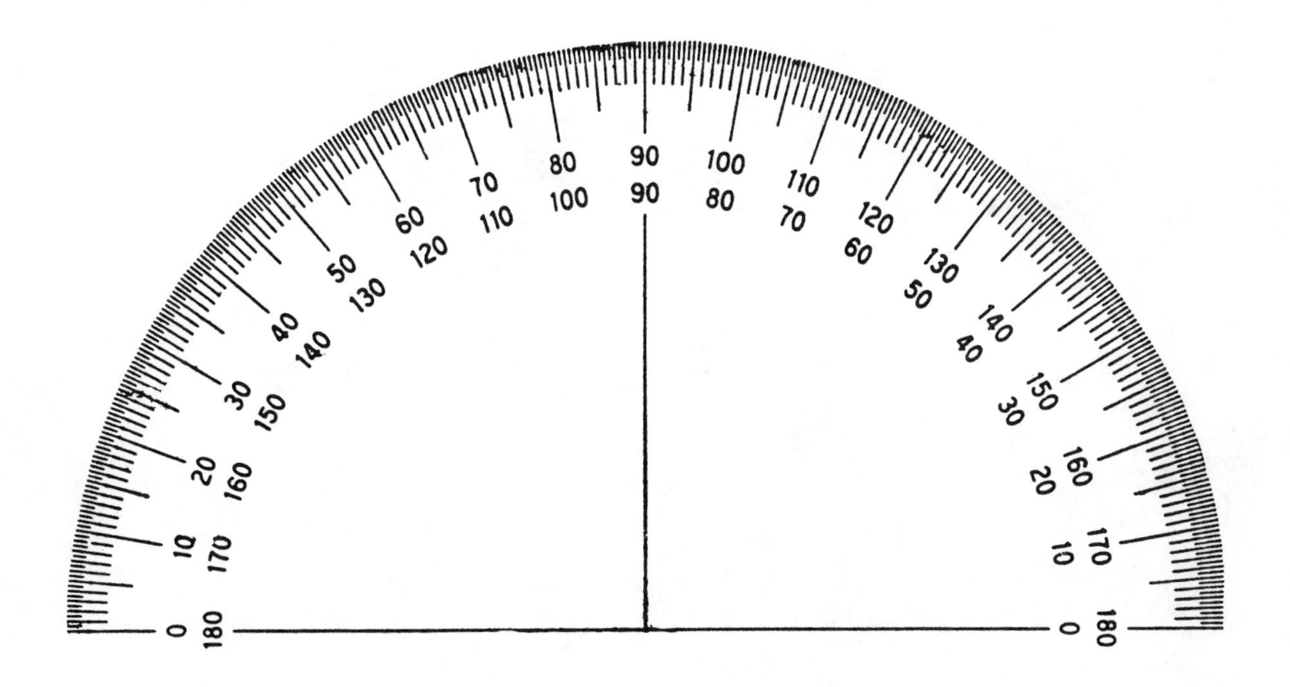

Now, suppose we walk one mile, turn 120^{0} to the left, walk another mile, turn 120^{0} left, and walk a third mile. We draw a map or scale diagram of this as shown below - and you will see you are back at the origin.

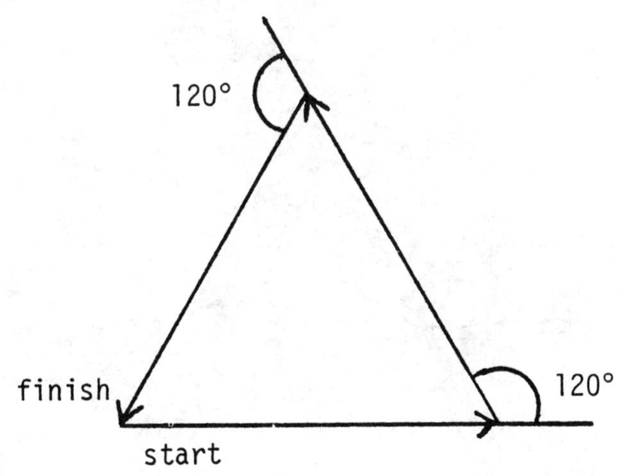

On the following page is a "sectional map" which airmen use to navigate. The scale of miles is given along the top of the chart, and may be cut off for measurement. If you start from Owens, the airport marked at the top of the map (33°59' 81°),and fly due south 22 statute miles, then due west 40 miles, at which airport will you be?

Notice there is a circle of angles marked around the Columbia VOR (which is a radio beacon used for navigation). Suppose you fly over the VOR, head to 249° and travel 39 statute miles; which airport will you be over?

If you fly 138° from north, starting at Columbia Metropolitan Airport, and travel 39 miles, which private airport will you be over?

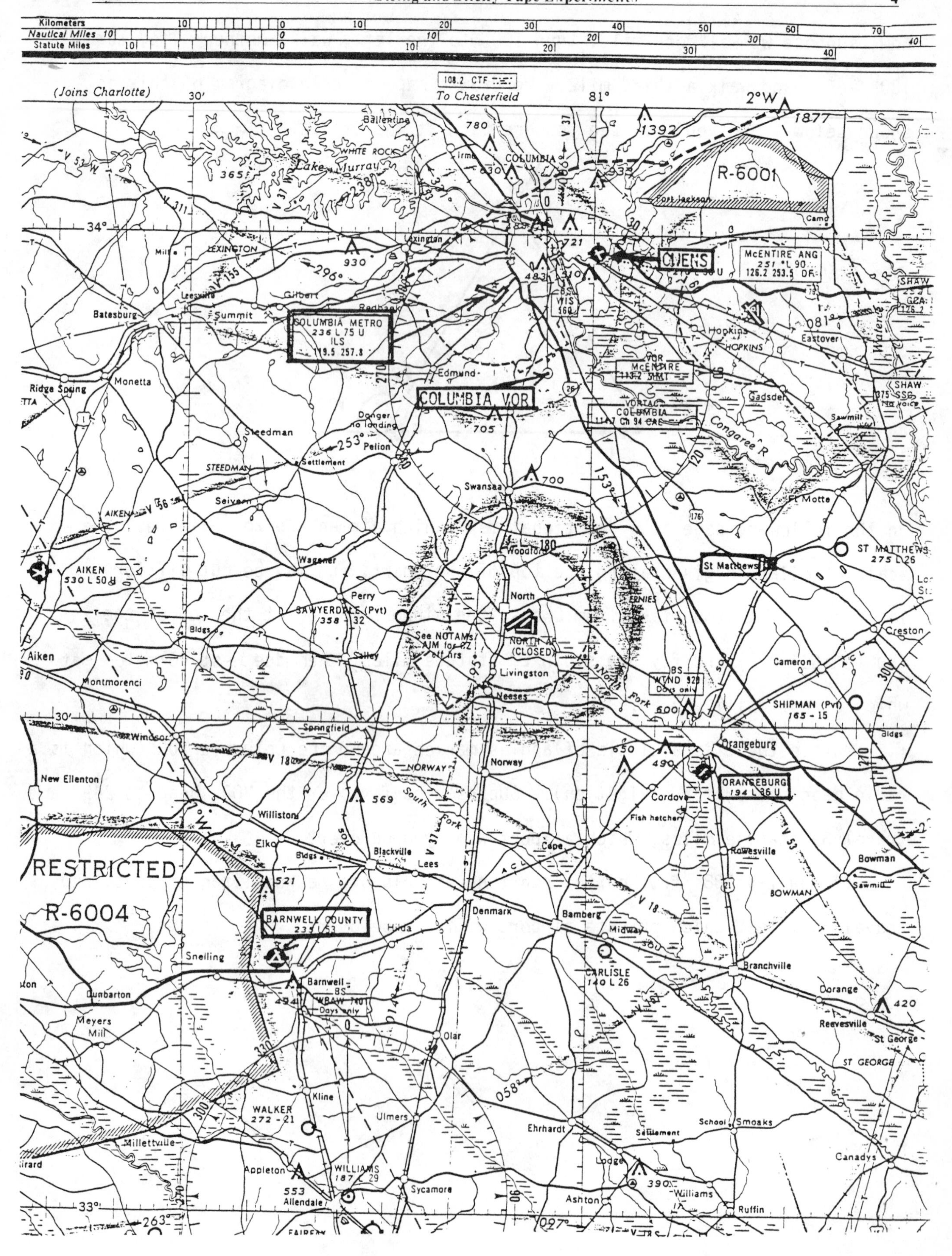
Kilometers
Nautical Miles
Statute Miles
(Joins Charlotte)
To Chesterfield
COLUMBIA METRO
COLUMBIA VOR
R-6001
Fort Jackson
Lake Murray
WHITE ROCK
Ballentine
Irmo
COLUMBIA
LEXINGTON
Lexington
Gilbert
Leesville
Batesburg
Summit
Monetta
Ridge Spring
Edmund
Steedman
STEEDMAN
Pelion
Swansea
Seivern
AIKEN
AIKEN
530 L 50 H
Aiken
Wagener
Perry
SAWYERDALE (Pvt)
Woodford
North
NORTH AF (CLOSED)
See NOTAMs/AIM for PZ eff hrs
Salley
Livingston
Neeses
Montmorenci
Windsor
Springfield
NORWAY
Norway
New Ellenton
Williston
Elko
Blackville
Lees
RESTRICTED
R-6004
BARNWELL COUNTY
Snelling
Barnwell
Dunbarton
Meyers Mill
Denmark
Hilda
Olar
Kline
WALKER
272 - 21
Ulmers
Millettville
Appleton
WILLIAMS
Allendale
Sycamore
Ehrhardt
Lodge
Ashton
Williams
Ruffin
Smoaks
School
Settlement
Canadys
Bamberg
Midway
CARLISLE
Cape
Cordova
Fish hatchery
Orangeburg
ORANGEBURG
194 L 35 U
Rowesville
Branchville
BOWMAN
Bowman
Dorange
Reevesville
St George
ST GEORGE
Cameron
Creston
SHIPMAN (Pvt)
165 - 15
St Matthews
ST MATTHEWS
275 L 26
Ft Motte
Congaree R
Gadsden
Hopkins
HOPKINS
Eastover
McENTIRE ANG
SHAW
OWENS
Danger no landing
Settlement
South Fork
North Fork
Bldgs
Sawmill
Days only
RESTRICTED

PART 2: Velocity

Not only displacements are vectors, but anything associated with distance, such as velocity, or acceleration. Since force is mass times acceleration, force is also a vector quantity. This we can see more intuitively, since the direction an object moves depends on the direction we push it--the direction the force acts.

To see how velocities add, let us take the example of an airplane flying between two points A and B.

It might seem reasonable, once in the air from A, to point the airplane at B - but this might be very bad, for if there is a side wind, as shown below, a plane would not end at B, but somewhere else.

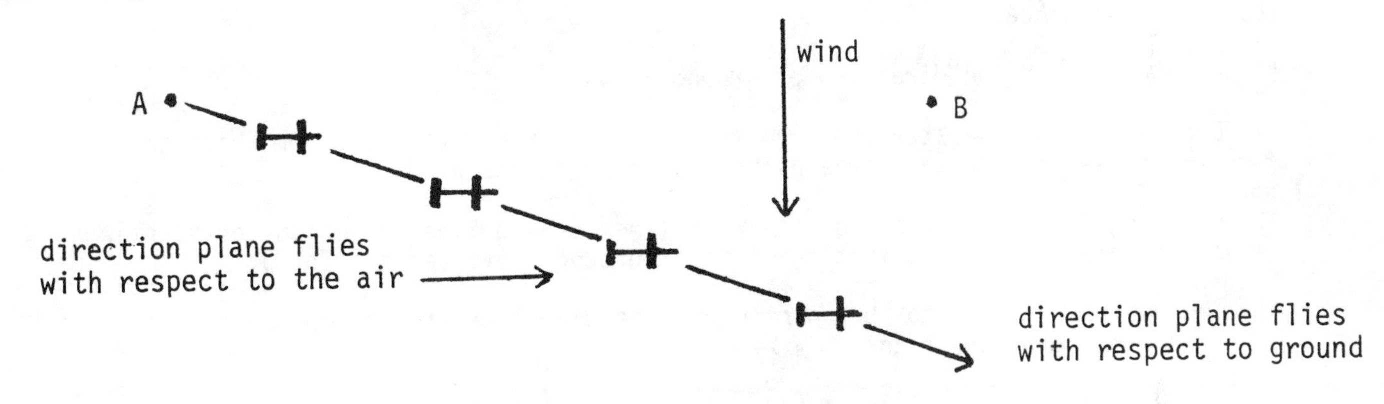

We must arrange that the velocity of the plane plus the velocity of the air takes it to B. For example, if there is a twenty mile an hour cross wind, as shown above, and the plane flies at sixty miles an hour, we make a scale drawing in which one inch represents ten miles an hour. First we draw the cross wind

2"

Then we draw a line in the direction we wish to travel

Then, we mark off a distance from C equal to the planes velocity, that just meets the direction of travel

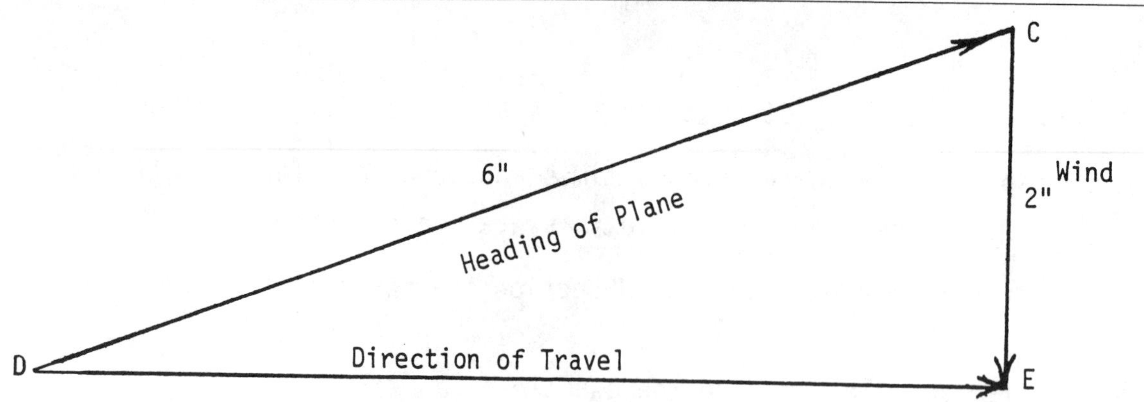

As with distance, the vector sum of the two velocities is the direction we wish to travel. Measure the length D E. This represents the velocity with respect to the ground and is 55 mph.

To summarize: If you know the crosswind, the direction of the plane on the ground, and its airspeed, proceed as follows:

1) Draw a line in the direction of ground travel.

2) Draw a second line starting from the first, in the direction from which the wind comes, of length representing the speed of the wind.

3) Starting from the far end of the wind line, draw a line of length representing the airplane speed, to meet the line of ground travel at a point p.

4) The distance from p to the point where the wind line starts represents the ground speed of the plane.

5) The direction of the line joining the ground line and the wind line represents the plane's heading in the air.

Now try this example, using 1 cm = 10 mph.

Wind from the south of 40 mph
Plane's airspeed is 100 mph
Direction of travel with respect to ground - east
What is the plane's heading and ground speed?

A simple example arises if you are given the plane's direction in the air, airspeed, and the windspeed. Then you draw a line of length and direction the speed and heading of the plane, from the end of which draw a line of length and direction the windspeed, and the third side of the triangle is the ground speed and direction of the plane on the ground.

If a plane flies at sixty miles an hour into a headwind of sixty miles an hour, what is its velocity with respect to the ground? Well, it wouldn't travel very far, I can assure you.

Try this example. A plane points north, airspeed 100 mph, with a crosswind from the west at 40 mph. What is its groundspeed and direction of travel (draw a scale diagram, one cm equals 10 mph, and measure the angle from north with the protractor). Do not confuse the velocity triangle with distances on the ground. First, draw the velocity triangle, then use the computed groundspeed and direction to calculate where the plane is at a given time.

ADVANCED CALCULATION

To give a practical example in vectors, here are some questions based on the regular examination for the private pilot administered by the F.A.A.

You are supplied with a map, and the operating limitations of the airplane.

You are assumed to be a private pilot making a cross-country flight from Barnwell County airport at Barnwell, South Carolina to Columbia Metropolitan Airport at Columbia, South Carolina. The first leg of the flight is from Barnwell Airport to Orangeburg Airport at Orangeburg, S. C., where you plan to land. The second leg is from Orangeburg Airport to Columbia Airport via the small town of St. Matthews and Columbia VOR (used for radio navigation).

The above mentioned airports and VOR stations can be located by means of the following latitude and longitude coordinates.

	latitude		longitude
Barnwell County Airport	$33^{0}15'$	-	$18^{0}23'$
Orangeburg Airport	$33^{0}27'$	-	$80^{o}51'$
Columbia VOR	$33^{0}52'$	-	$81^{0}03'$
Columbia Metropolitan Airport	$33^{0}57'$	-	$81^{0}07'$

Lines of constant latitude are horizontal, and longitude vertical on the map. The degrees of longitude are marked along the top of the map, and latitude along the left side of the map.

Clouds and weather, over South Carolina generally ceiling - 20,000 ft.

Winds aloft forecast

5,000 ft.	10 knots	at 100^{0} to North
10,000 ft.	15 knots	at 150^{0}
15,000 ft.	25 knots	at 180^{0}

(Winds are always given in knots from a particular direction, which is a nautical mile per hour. Nautical miles are marked on the scale along the top of the map - 1 knot = 1.1515 mph.)

1. You are cleared to runway 9 of Barnwell County Airport (runway heading, i.e. the direction the plane faces, is in tens of degrees from north, so runway 9 is 90°). What is your ground speed just after takeoff (at 50 ft. altitude) if the gross weight of the aircraft is 2,000 lb. with a head wind of 15 knots (use the takeoff data to find indicated air speed, i.a.s., the speed of the plane with respect to the air).
2. Immediately after takeoff, you head for Orangeburg. What is the bearing (direction from North) of Orangeburg from Barnwell?
3. If you fly your aircraft at an altitude of 10,000 ft., what heading should you have to arrive at Orangeburg, at an air speed of 108 mph?
4. How far is Orangeburg from Barnwell?
5. If you leave Barnwell at 12:00 noon, at what time will you arrive in Orangeburg?
6. Should you refuel in Orangeburg to arrive in Columbia? Show your calculation. (The plane had 42 gal. of fuel and consumes 6 gal./hour.)

I.A.S. (Indicated Air Speed) for take off

Gross wt.	I.A.S.
1700	75 mph
2000	79
2350	84

This is the air speed read in the plane before it can lift off the ground carrying the weight shown.

7. What is the bearing of St. Matthews from Orangeburg?
8. You fly the rail track to St . Matthews. At what angle to the rails should you fly, if your air speed is now 90 mph? (wind speed and direction 10 knots at 100°)
9. You turn over St. Matthews and make for the Columbia VOR 33° 52, 81° 3 '. What is your heading, and groundspeed, if the airspeed is 90 mph?
10. Flying directly to Columbia Airport, how many minutes would it take from the VOR?
11. The ground wind at Columbia is now 10 mph from the northwest. You make a downwind approach to runway 29. What angle to the runway should you fly if your airspeed is 80 mph?
12. What is the total time for the trip, exclusive of the time on the ground?

Experiment 1.04. The Rubber Band Balance (Hooke's Law) and Stored Energy

Materials needed: ruler, several marbles, rubber band, two paper clips, cup

Procedure: Support the rubber band by a bent paper clip, as shown from the top of the ruler. Add marbles, one by one, to a cup, attached to the other end by another paper clip. Measure the length of the rubber band each time in centimeters, putting 0 cm at the top.

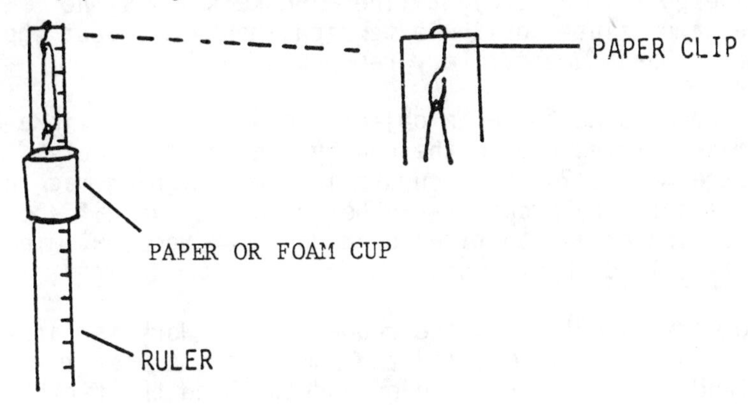

Put the length for each marble on the table.
An example is given below.

EXAMPLE

number of marbles	reading on ruler	difference from 0 marbles
0	15.1	0
5	15.6	0.5
10	16.4	1.1
15	18.0	1.6

number of marbles	reading on ruler	difference from 0 marbles reading
0		
5		
10		
15		
20		
25		
30		
35		
40		
45		
50		

Plot the difference (which is the extension of the rubber band) on the graph over, where a sample plot is also given. Are your results similar to the sample? Can you draw a straight line through your points? What does this show?

Qualitatively: The rubber band stretches uniformly with the increase in weight, which is the force of gravity on the board (mass x acceleration due to gravity). i.e. the extension is proportional to the number of marbles, and the plotted points lie on a straight line. This proportionality is known as Hooke's law. However, as the load increases, the extension becomes greater than predicted by this law, and the points diverge from a straight line. Hooke's law is no longer obeyed. Do you get the same values for the extension when you repeat the experiment? Probably not--rubber bands are not very reproducible.

Quantitavely: We can use the rubber band to weigh objects, now we have calibrated it with a graph--see how many marbles are in the cup by reading the scale. Does this agree with counting the marbles? How accurate is your weighing machine? Do not be too surprised if it is not very accurate--rubber bands behave differently after they have been stretched. The effect that the band does not recover immediately after being stretched is called "hysteresis".

Stored Energy: The marbles do work in extending the rubber band. Work is force times distance. However, since the force is mg, and the number of marbles increases as the rubber band extends, the work done to extend the band the first centimeter is much less than that required to extend it the fifth. This work is stored in the form of potential energy of the rubber band. We would like to plot the energy stored against the extension of the rubber band. The force required to extend the rubber band a small distance is reasonably constant--the work done will be the force times this extension.

In our plot of number of marbles against the extension, the force times a small extension will be proportional to the area under the curve which lies above this extension (see figures) since this is the number of marbles times the extension.

The Number of little squares for each increase of a centimeter is plotted in figure below. Do the same for your curve. Now, since there are 4 little squares for 1 cm and 1 marble, and one marble weighs approximately 5 gm, so the work represented by these squares is .005 x 9.81 Joules, and one little square represents .0123 joules. Hence, we can replace the scale of number of little squares by one of Joules, as shown.

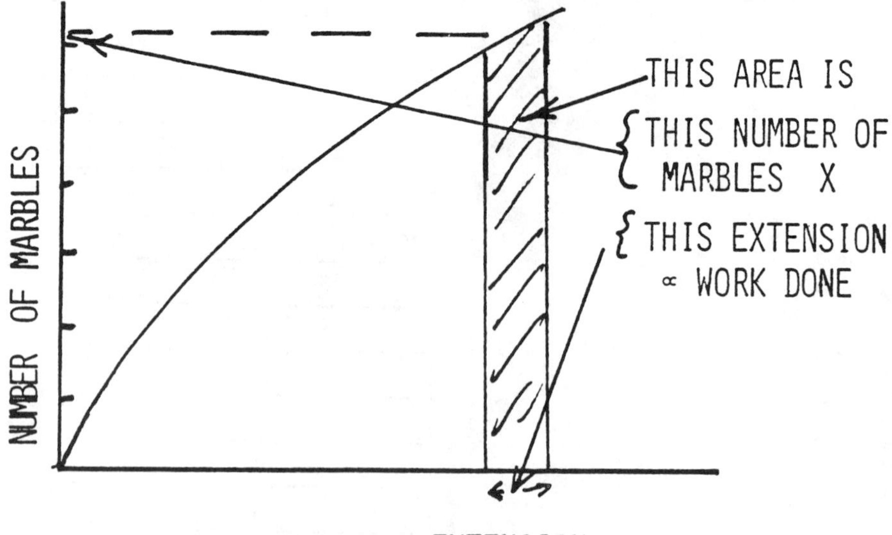

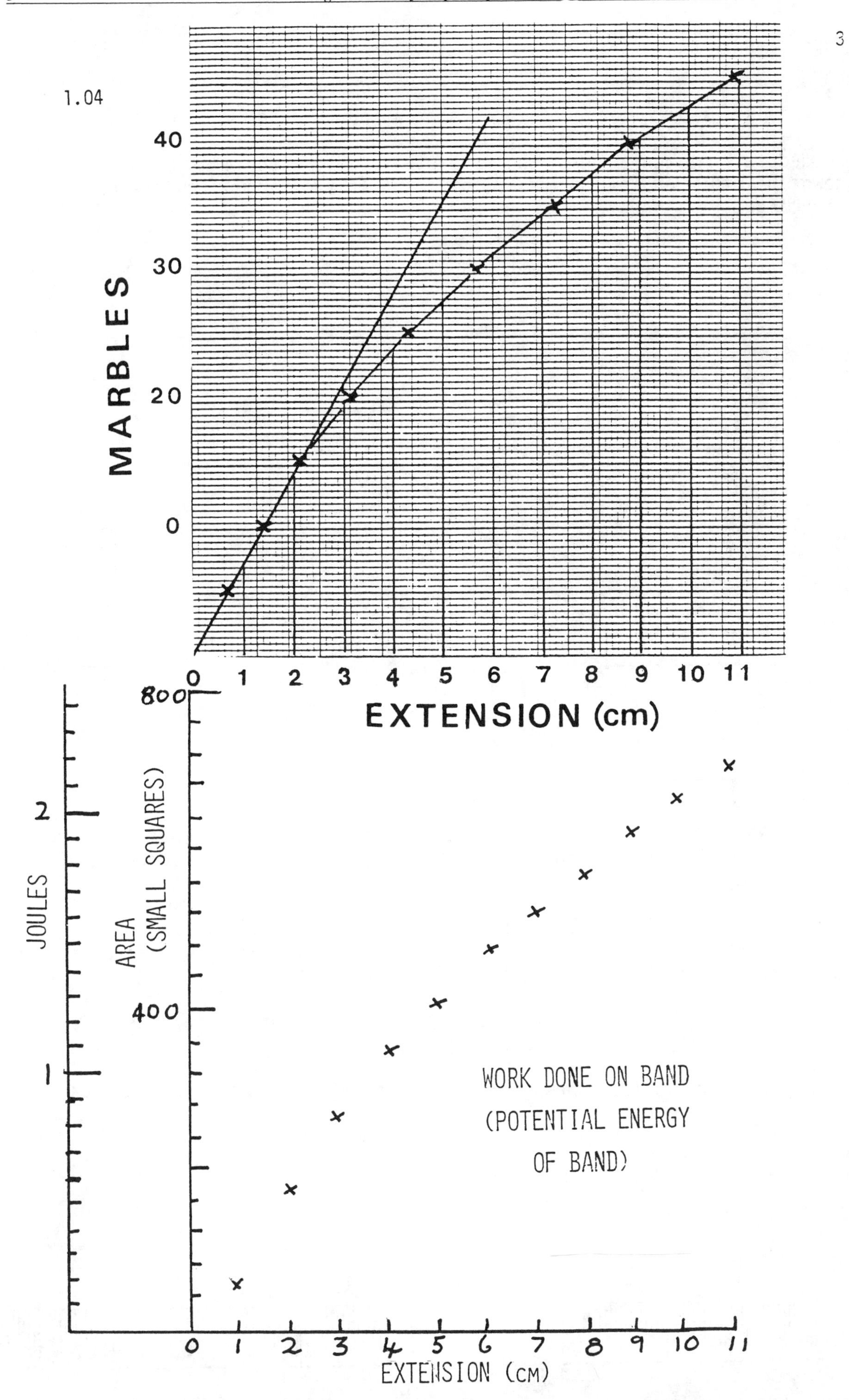
1.04
MARBLES
40
30
20
0
0 1 2 3 4 5 6 7 8 9 10 11
EXTENSION (cm)
JOULES
2
1
AREA
(SMALL SQUARES)
800
400
WORK DONE ON BAND
(POTENTIAL ENERGY
OF BAND)
0 1 2 3 4 5 6 7 8 9 10 11
EXTENSION (CM)

Experiment 1.05. Elastic Forces

Materials: ruler, piece of card, book support, paper cup, marbles, pencil.

Procedure: Attach a paper cup by a string to the end of the ruler. Hold down the other end of the ruler, by its last two inches, to the book support very tightly as shown.

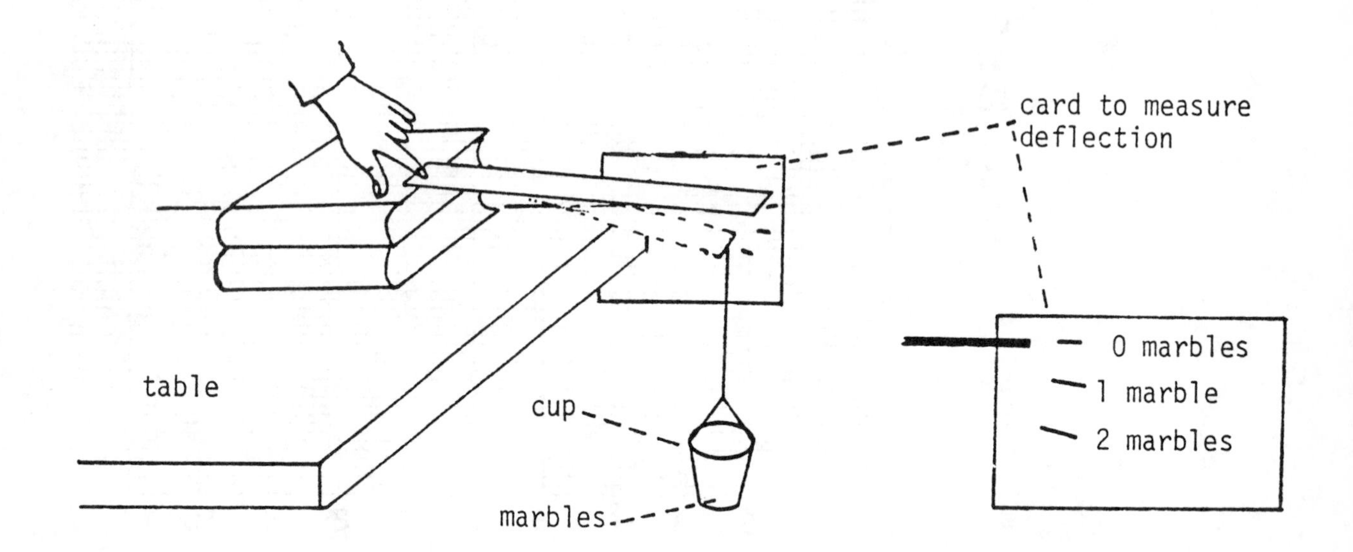

Now using a piece of card, mark on the card the deflection of the end of the ruler with different number of marbles in the cup. A second ruler can be used instead of the card, or the scale below.

Qualitative Questions: Does the ruler go back to its original position when the marbles are removed? What would happen if you kept on adding marbles? Does the deflection go up or down with the number of marbles?

Quantitative: Plot the deflection against the number of marbles. Can you draw a straight line through all the points?

Experiment 1.32 shows how gravitational mass, measured in this experiment, is related to inertial mass.

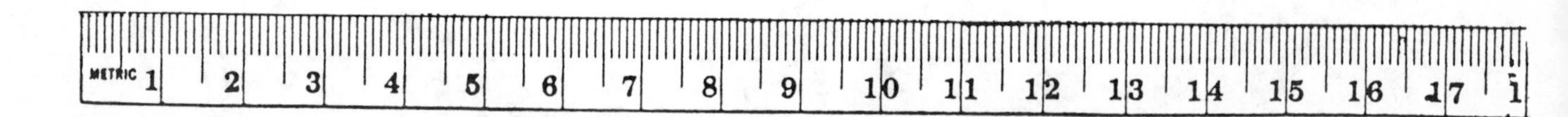

What does this show?
A force for which the deflection is proportional to the magnitude of the force is called an "elastic force". The "spring constant" is the constant of proportionality between the forces and the deflection. If we use the unit of mass as 1 marble, what is k?
force = k x deflection
g = gravitational constant = 9.81 m/sec^2

$$k = \frac{g \times \text{number of marbles} \times \text{mass of marbles}}{\text{deflection}}$$

k generally turns out to be about 39 N/m. You can go on to Experiment 1.38 for another method of finding k.

Number of Marbles	Deflection
1	
2	
3	
4	
5	
6	
7	
8	
9	
10	
11	
12	
13	
14	
15	
16	
17	
18	
19	
20	
21	
22	
23	
24	

Experiment 1.06 Torsion Balance

Materials needed: Two drinking straws, marbles, dixie cups, string, paper clip.

Procedure: - Bend about two inches of a drinking straw through a right angle, and tape it so that about three inches overhang the table, as shown.

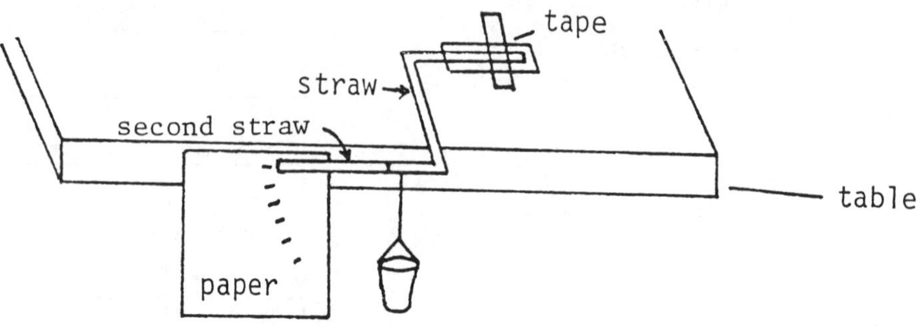

Bend at right angles, so the end lies along the edge of the table, cut the end of the second straw, and slip it in the first. Tape a piece of paper against the table edge. Hang the cup from the end of the straw as shown. Add marbles, one by one, and mark the paper where the end of the straw lies. You can now tell how many marbles are in the bucket by reading the scale you have made.

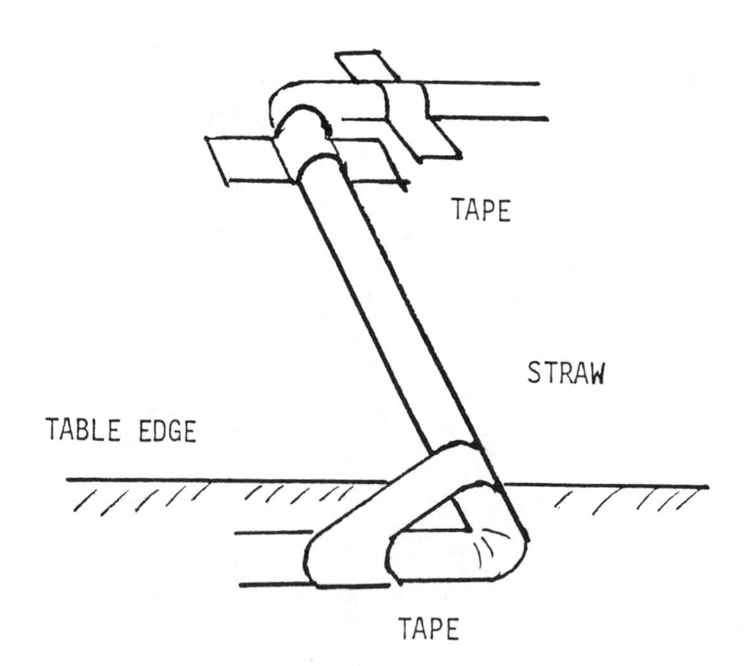

EXPERIMENT 1.07. The Modulus of Rigidity of a Drinking Straw.

Materials needed: Three straws, marbles, cup, string, sticky tape.

Procedure: We wish to measure the torque (moment of the force) which produces a given rotation of the drinking straw.

Attach the straw to the table, as in experiment 6, but now stick two other straws as shown, to measure the rotation over a fixed length, say 10 cm of the straw.

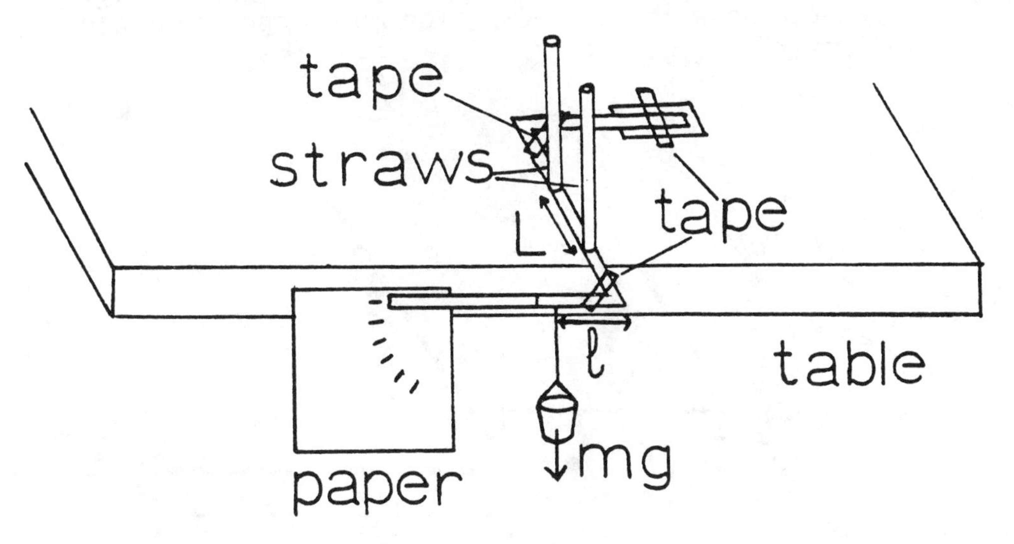

The mass of the marbles should be known. The information we need is:

1. the mass of marbles in the cup m
2. C, the couple rotating the straw, $mg\ell$, where ℓ is shown in the figure.
3. The angle rotation of the more distant straw, θ_1 found by marking a card placed behind it, before and after adding weights to the cup. The torsion constant for the straw, k, is given by couple = k x angle of rotation.
4. The angle of rotation of the nearer straw, θ_2, found similarly

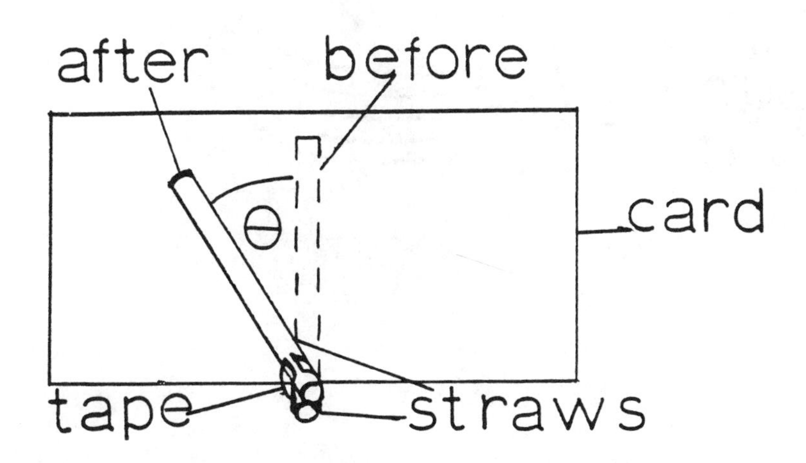

5. The length between the marker straws, L. The "torsional rigidity" can now be found which is defined as

$$\frac{CL}{\theta_2-\theta_1}$$

6. The diameter of the straw 2r must now be found, using a millimeter ruler. The straw can be slit, as shown, for more accurate measurement of the circumference, $2\pi r$.

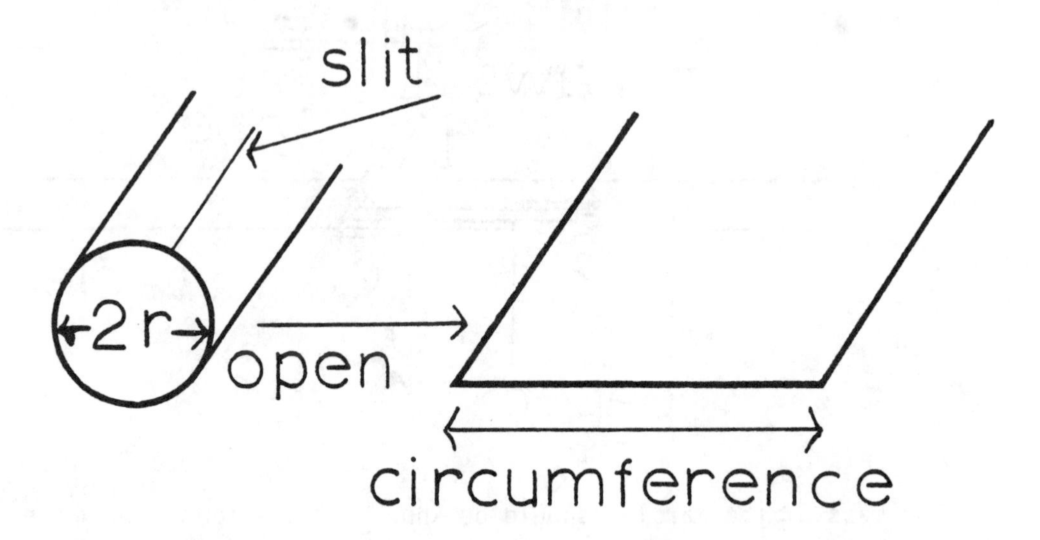

7. The thickness of the straw material must be found, by superposing several layers, and using a millimeter ruler

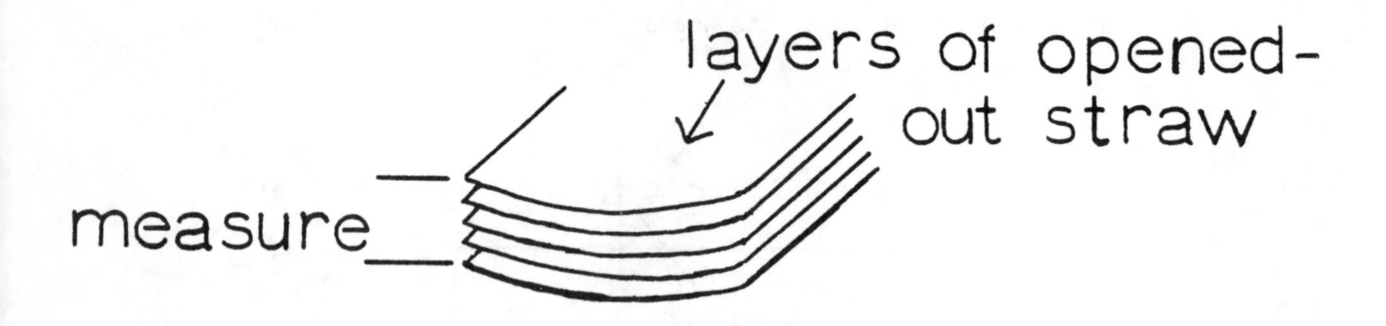

Now the modulus of rigidity of the straw can be found.

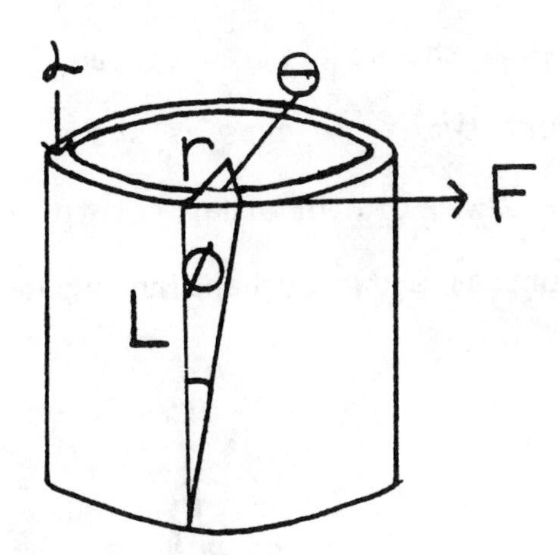

The cross-sectional area of the straw is α, at a distance r from the axis of the straw, which is fixed at one end a distance L from α. The external couple C twists α through an angle θ. If the tangential force on α is F, the couple is given by

$$C = Fr$$

If the angle of shear is ϕ

$$r\theta = L\phi$$

The modulus of rigidity n is given by

$$\frac{\text{tangential stress}}{\text{angular deformation } \phi} = \frac{F/\alpha}{\phi} = \frac{CL}{r^2\theta\alpha}$$

Since the area of cross section of the straw is $2\pi rt$ where t is the thickness of the straw,

$$n = \frac{CL}{2\pi r^3 t} = \frac{mg\ell L}{2\pi r^3 t}$$

Hence the modulus of rigidity of polyethylene straws may be determined. The value should be approximately: 10" dynes/cm^2.

Experiment 1.08. Equal armed balance

Materials needed: - Two paper cups, soda straw, marbles, paper clips.

Procedure: - Poke a hole through the soda straw at each end, and push a paper clip through, to support two cups as shown. Either poke a hole midway between the other two, or, in order not to reduce the strength of the straw, make a support as shown by bending a paper clip.

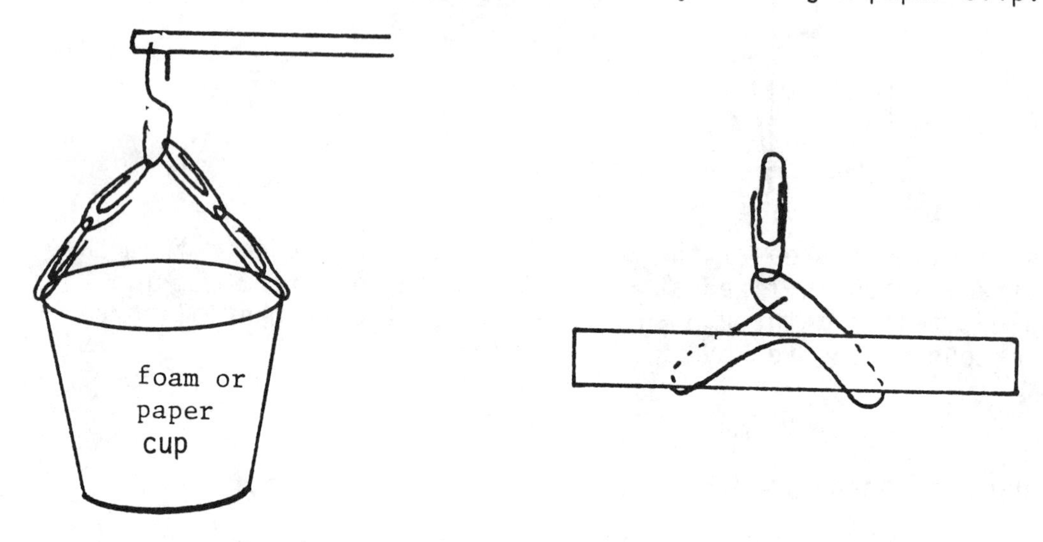

Now you have an equal-armed balance. Try weighing marbles against one another. Change the marbles from one cup to the other, keeping the number in each cup constant.
What does it show?

Qualitative: Even with the same number of marbles, the balance is never exact, but you can tell approximately when balance occurs.

Quantitative: You can get a closer balance by adding paper clips to one cup or the other of the balance. How many paper clips balance one marble? When you have two marbles in each pan, how many additional paper clips are required to balance? Do different marbles weight the same? How big is the difference ?

Experiment 1.09. Levers - the steelyard.

Materials - soda straw, cup, paper clips, string and marbles.

Procedure: - Poke two holes through the straw with the paper clips, one about 1/2 inch from the end, the other about 1 1/2 inches from the end. Attach a marble to a paper clip with sticky tape, or by bending the clip as shown. Now, support the cup by the paper clip close to the end, and slide the marble along the arm until it balances. Then add a marble to the cup, and balance again. Do this for several marbles and mark on the straw where the balance marble is using a marking pen or piece of sticky tape. You can use the steelyard to count marbles.

What does this show?

Qualitatively: - You can lift several marbles by one marble if you move it far enough away from the pivot or fulcrum. If you had a long enough (and strong enough) straw, you could lift the whole world with one marble.

Quantitatively: - Notice that the distance you must move the balance marble every time you add another marble to the cup is always the same. This confirms the relation: -

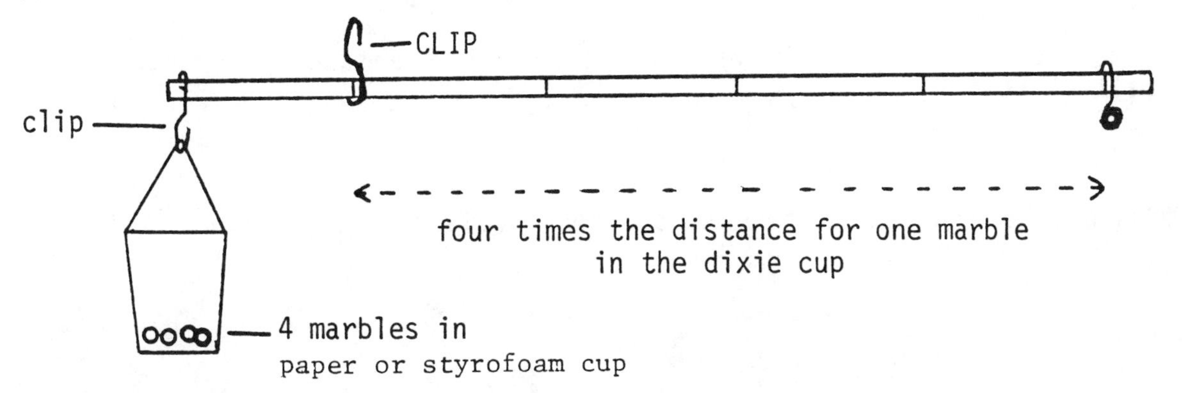

force (1) x distance to axis = force (2) x distance to axis for equilibrium.

The steelyard was used to weigh objects in the days of the Greeks and Romans. It is still used today for some accurate weighing machines.

A very simple balance may be made by placing a coin on the end of a ruler, and pushing it over the edge of the table until it overbalances.

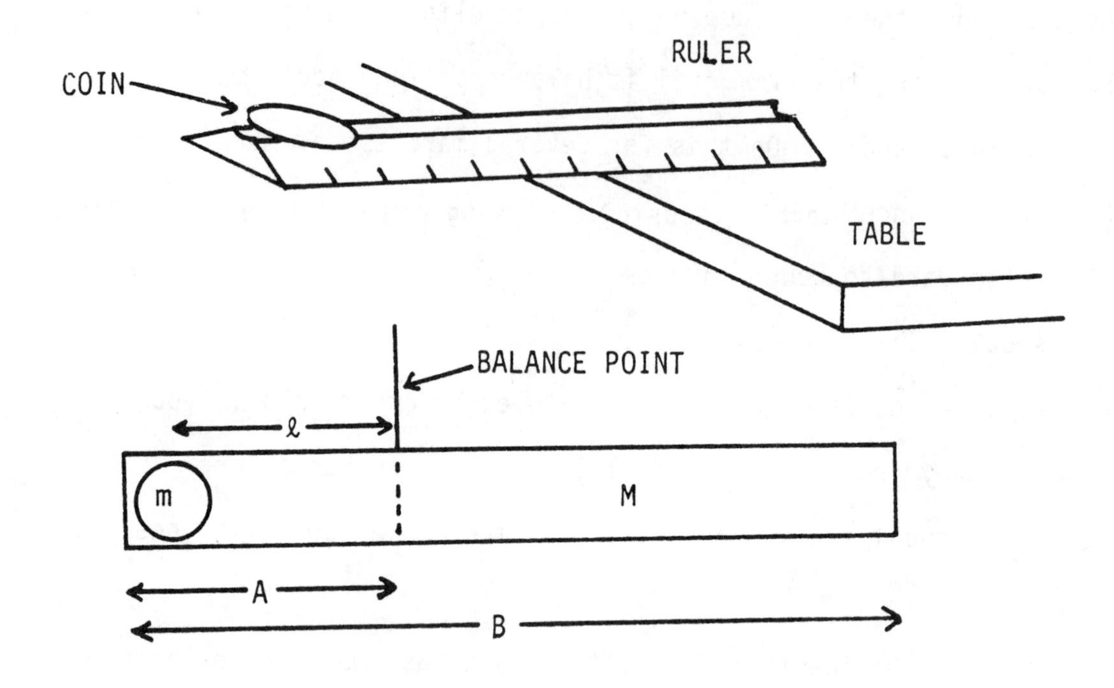

If the mass of the ruler is M, and of the coin m, then taking moments about the balance point

$$mg\ell = Mg\ (B/2 - A)$$

Find the mass of the ruler from the masses of coins given in experiment 1.01.

Experiment 1.10 The letter balance

The letter balance is convenient for weighing small objects. Cut out the diagram shown , folding one side over the other, to provide more strength. Stick one penny on each side of the paper, in the marked spaces, if you wish to use the gram calibration, or two pennies on each side (four pennies in all) for the ounces. Push a paper clip through the hole marked W, and hang the letter to be weighed from it. Another paper clip provides the pivot P as shown in the diagram, and a straightened paper clip, B in the diagram, hangs vertically to provide the pointer.

The letter balance is calibrated by taking moments about the pivot P. The line joining the center of mass of the balance, complete with pennies and paper clip, to the pivot must be at 90° to the line joining the pivot to the point W from which the weight hangs.

If the distances are as shown in the diagram, at equilibrium

$$mgb \cos \theta = Mga \sin \theta$$

where M is the mass of the balance, and m is our unknown. Hence

$$m = \frac{Ma}{b} \tan \theta$$

Since a penny weighs 3 gms, the scale may be drawn directly on the card. This has been done for the two scales shown.

Stick three pennies on the balance in place of the two or four, and calibrate it yourself by hanging dollar bills (each weighing 1 gm) or other objects from the weighing clip, and making a mark where the pointer hangs with no mass, or one or two dollar bills.

Check whether the tangent of the angle you measure is proportional to the mass of the object being weighed.

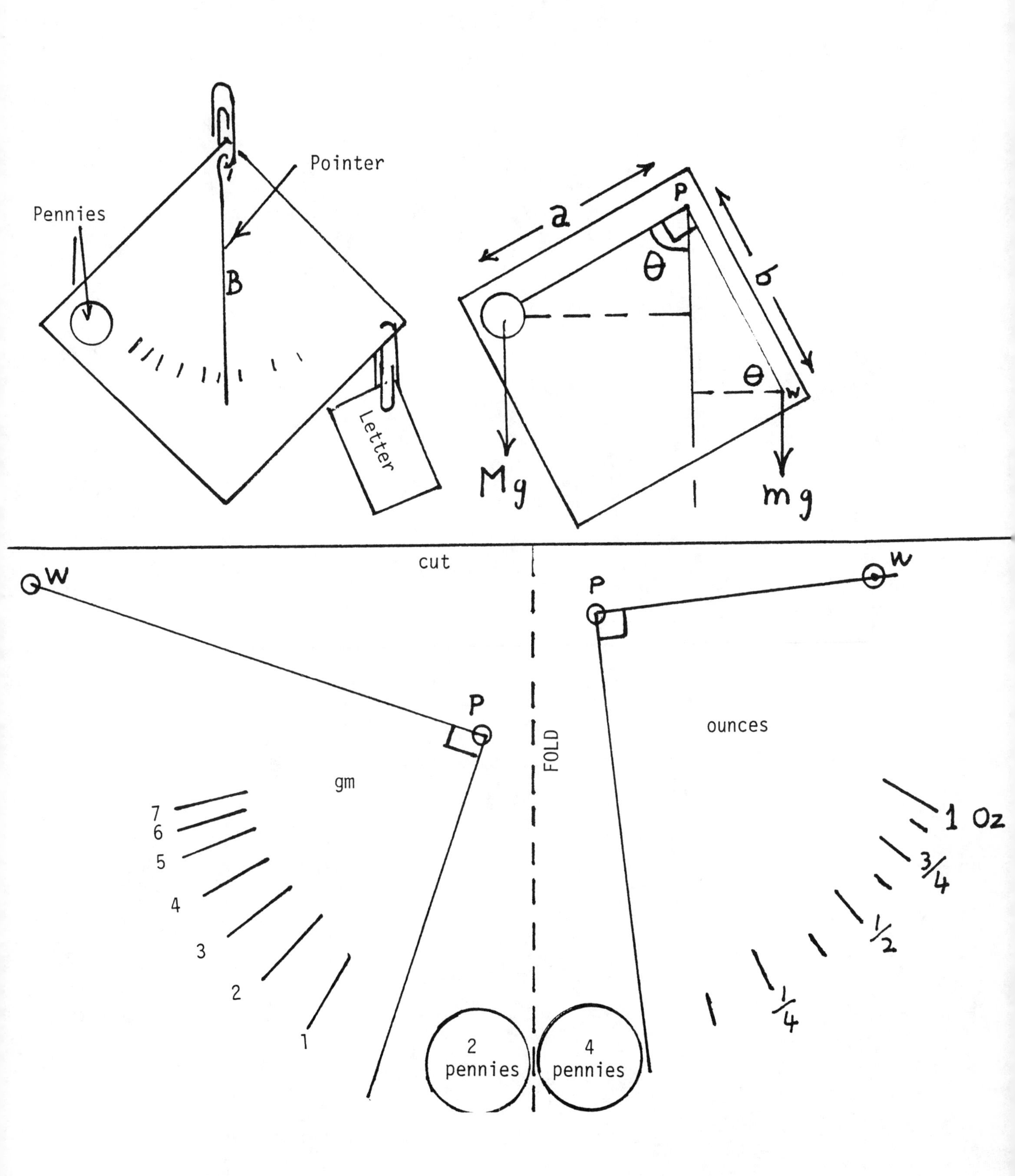
Pointer
Pennies
B
Letter
a
P
b
θ
θ
w
Mg
mg
cut
W
P
P
w
FOLD
ounces
gm
7
6
5
4
3
2
1
1 Oz
3/4
1/2
1/4
2
pennies
4
pennies

Experiment 1.11. The microbalance

Materials: - Soda straw, paper clips, sticky tape, paper.

Procedure: - Only in sensitivity does this differ from the previous experiment. With this experiment, it is easily possible to weigh a human hair.

Take a soda straw and push the narrow part of two paper clips into one end. Balance it roughly on another straw, and poke a hole at the point of balance with a pair of scissors or a needle.

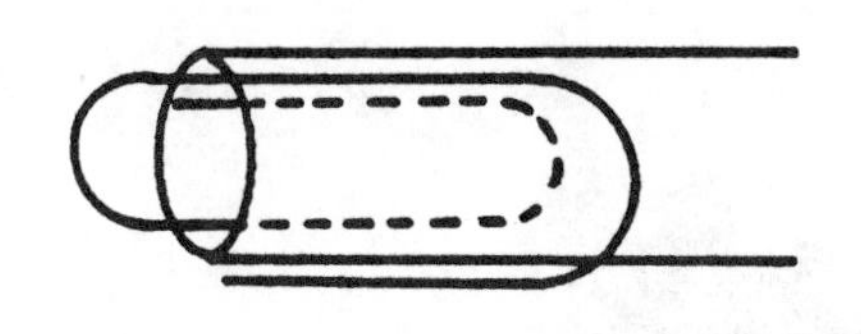

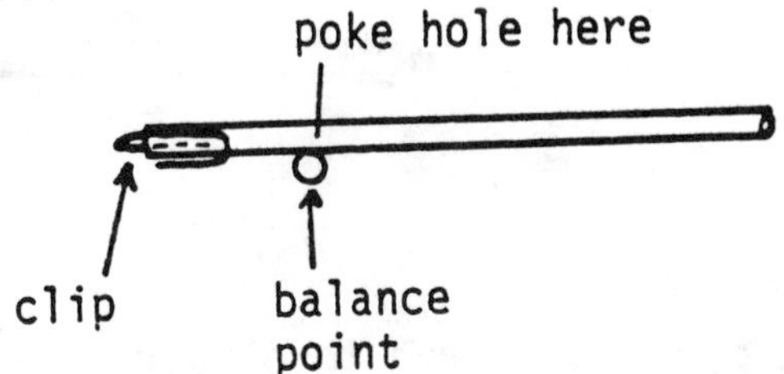

Unfold one end of a paper clip and push it through the hole, as shown.

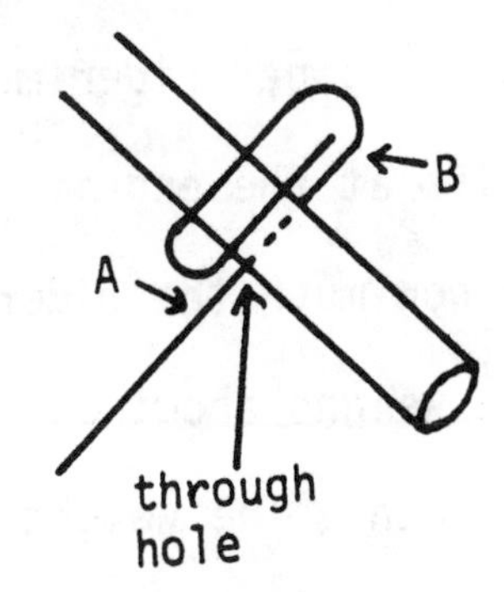

Make sure the region from A to B remains perfectly straight.

The balance must now be placed on frictionless bearings. The best we can do is to use two more paper clips. Tape these, as shown

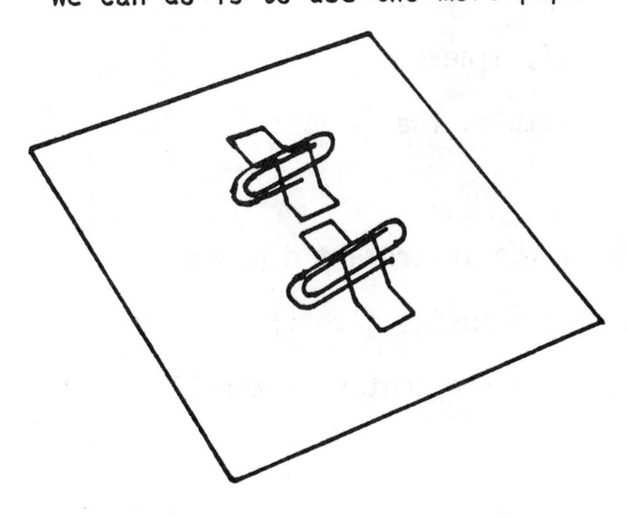

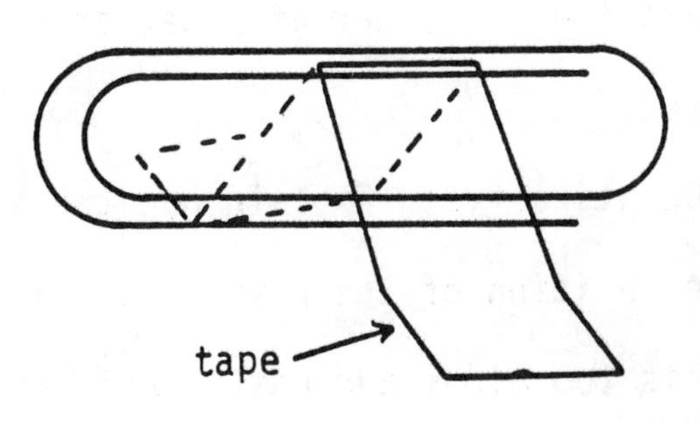

to a sheet of paper, or better card. The tape must pass between the wire of the clip, so the metal of the balance pivot rests on the metal clip. Both are hard, and if the pivot clip rolls on the fixed clip, there is very little friction.

Now comes the tricky part - to balance the straw exactly. To do this, cut a small slip of paper (about 1 cm by .5 cm) and fold it in a V, resting it on the straw.

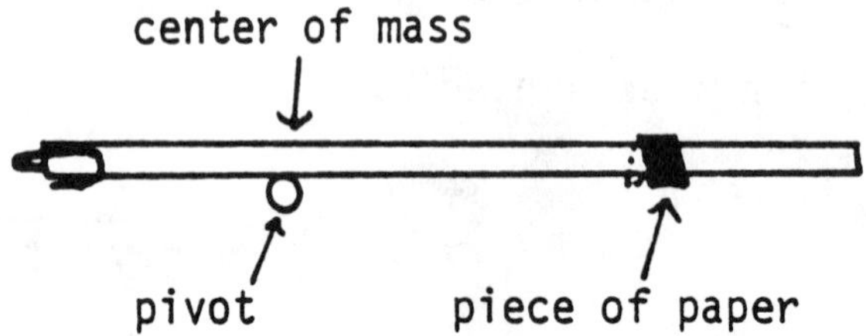

Move it back and forth to obtain a balance. Tap the paper the balance rests on to free the friction locking the pivots. Now, place a hair over the end of the straw and see how much the paper rider must be moved to obtain a balance. Adjust the paper clip at the end if the range is too small. To calibrate the balance, see how much the rider must be moved to balance a small US postage stamp, which weighs about 60 mg. You can cut the stamp into a quarter if it weighs too much. The weight is again proportional to the distance the rider has to move to balance it. Don't forget to tap the base to free the pivots.

Qualitative: Weigh various light objects - a hair, a stamp, a small piece of string, a length of cotton, a seed, a small leaf, a needle.

Put them in order of mass, or weight - for example, the hair is lightest.

Quantitative: Explain how the sensitivity of the instrument depends on the position of the pivot. Why is the balance too insensitive if the pivot is too high, and unstable if the pivot is below the center of mass?

Experiment 1.12. Stresses in beams

Materials needed - several plastic drinking straws, sticky tape, string marbles, foam or paper cup.

Procedure: Hang a cup from a straw by string and a paper clip as shown

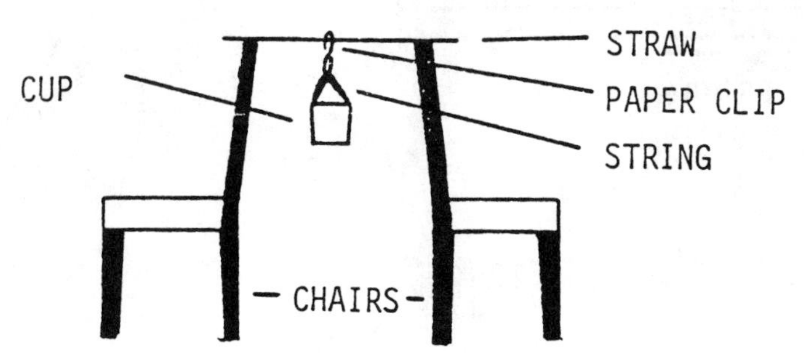

Support the straw as close to each end as possible. Add marbles to the cup hung from the middle, until it gives way. Do the same with the cup hung about a quarter from one end. Why do you think it needs more marbles close to the end? You will need a lot of marbles!

Now, take three straws, fasten the ends of two straws together, and fasten the outer ends using tape to the other straw, as shown. Tape a piece of string round the top, fastening it with tape to be sure, loop it round the lower straw, and again, see how many marbles break this girder. You probably won't have enough marbles, and it will take a book or two, hung from it to break it.

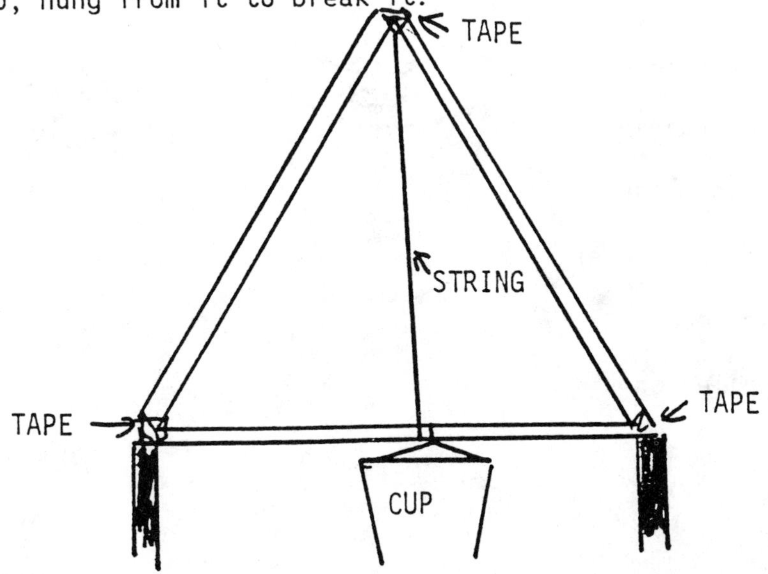

Qualitatively - a narrow girder, like the straw, breaks easily - a large girder distributes the force and can withstand a heavy load.

Quantitatively - the bending torque at the center is very large

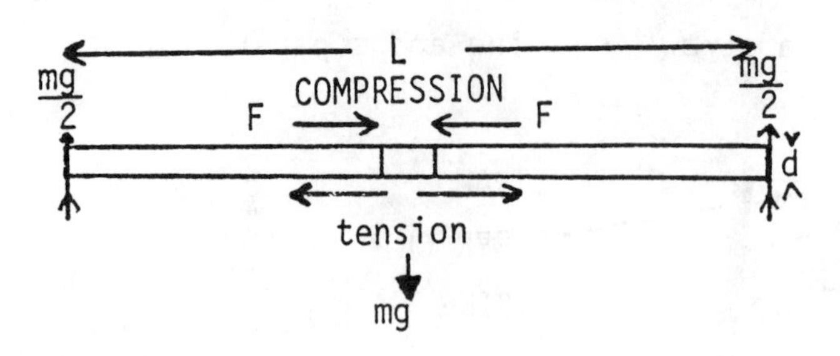

torque = $\frac{mgL}{4}$ force at top $F = \frac{mgL}{4d}$

With the weight a quarter from the end, the torque becomes $\frac{3mg}{4} \times \frac{L}{4}$ i.e. $\frac{3}{4}$ of what it is in the center.

For the girder.

$MG/2 \text{ SIN } \theta = 0.6 \text{ MG if } \theta = 60^0$

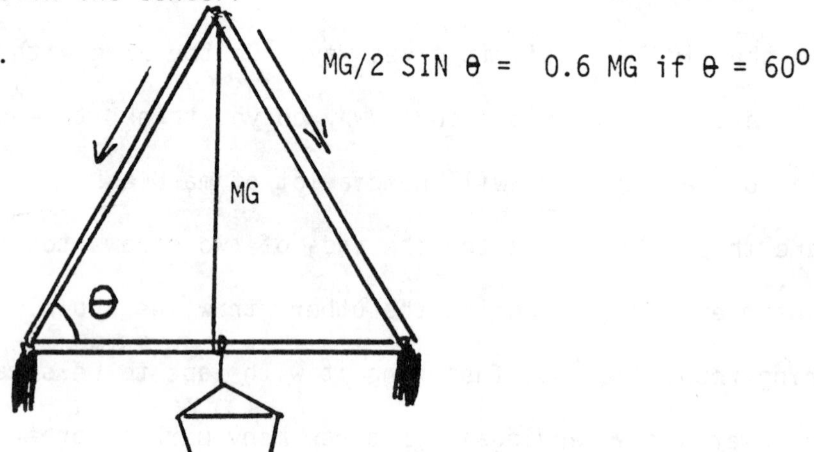

<u>A simple example of the strength of corrugation.</u>

Place a piece of paper the size of a dollar bill over a cup. A quarter placed on the paper will fall into the cup. If the paper is corrugated, however, with half inch pleats as shown, many quarters (or pennies) may be piled on without it breaking.

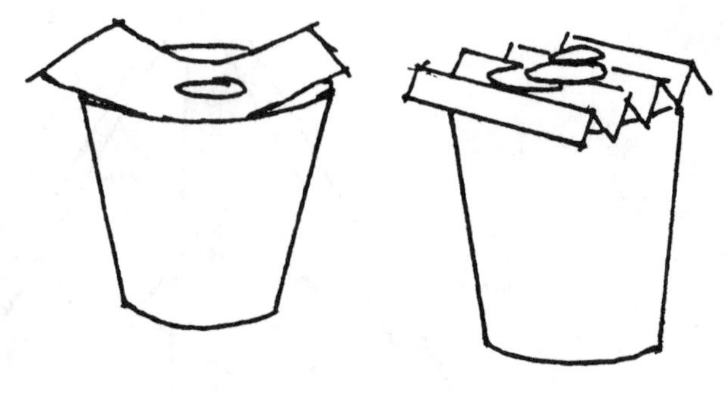

Experiment 1.13 The triangle of forces.

Materials needes: - three rubber bands (as much alike as possible), three paper clips, sheet of paper, pencil, string.

Procedure: Fasten the three rubber bands together as tightly as possible, as shown, using a piece of string. Now, hold two down on a sheet of paper, and stretch the third, so that they spread out under tension. Mark the point they are joined, and the ends of the rubber bands, on the sheet of paper.

Qualitative: The rubber bands are under tension, that is, they are exerting a force on one another, but they are in equilibrium - so the forces balance one another in order that there is no motion.

Quantitative - Subtract the unstretched length of each band from the stretched length.

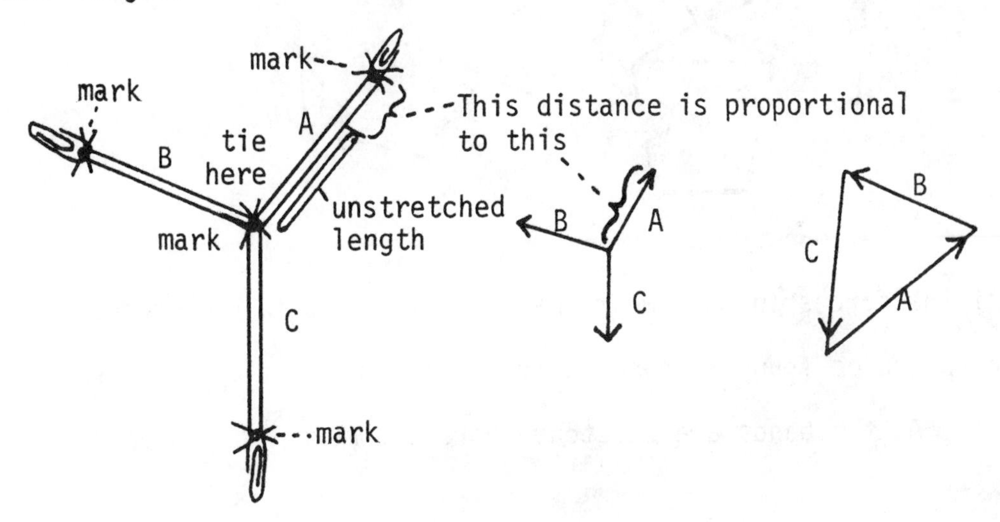

Make a diagram of the forces. You can represent each force by a length, and since the extension of each band is proportional to the force extending it, that length can be the extension itself. Now, the three forces, placed head to tail, should form a triangle, whose sides are parallel to the forces, as shown in the figure. This is the triangle of forces, determining the magnitude and direction of the forces for equilibrium.

Unfortunately, because it is rare to get three comparable rubber bands, about 20% agreement is all that can be expected of this experiment. However, you can use one band as standard. Leave one band loose, and extend the other by pulling it against the standard, as shown. The extension of the standard

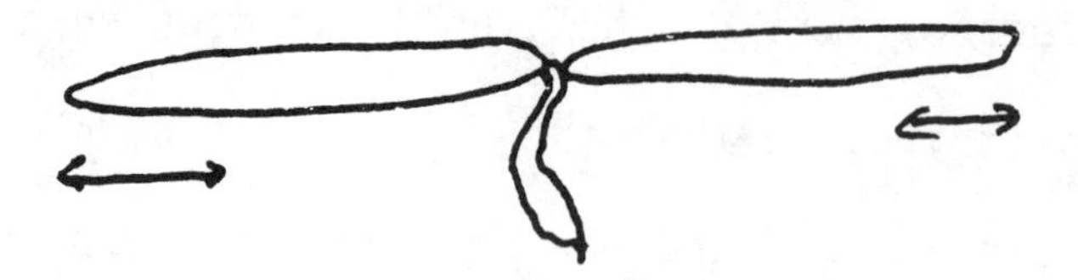

for the extension of the other band in the experiment can then easily be found and used in the experiment.

Another interesting experiment is to hang a cup of marbles from two rubber bands, as shown.

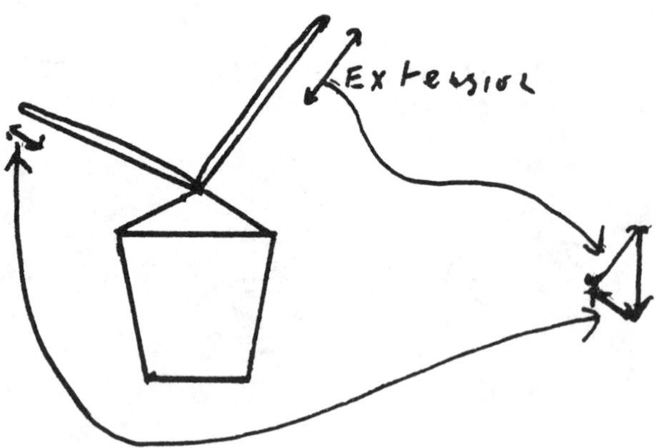

The resultant of the forces in the two rubber bands is found. The force of the marbles in the cup can be found by hanging from one band. Is this the same as the resultant? As the bands are stretched more and more, what happens to the cup?

Another problem with this experiment is that the rubber bands do not obey Hooke's law--hence large extensions should be avoided.

Experiment 1.14 Application of couples and torque.

Materials: - Two foam or paper cups, sticky tape, and string

Instructions: Attach the bottoms of the cups together as shown.

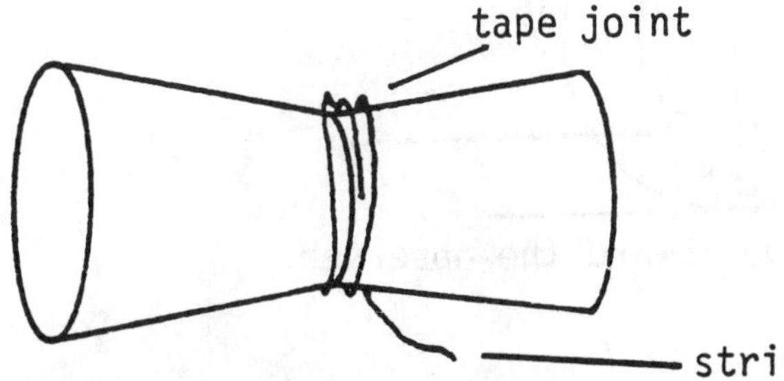

The string is taped to the joint and wrapped around it several times. Now, place the cups on the floor horizontally, hold the string level with the floor, and pull. Which way do the cups move?

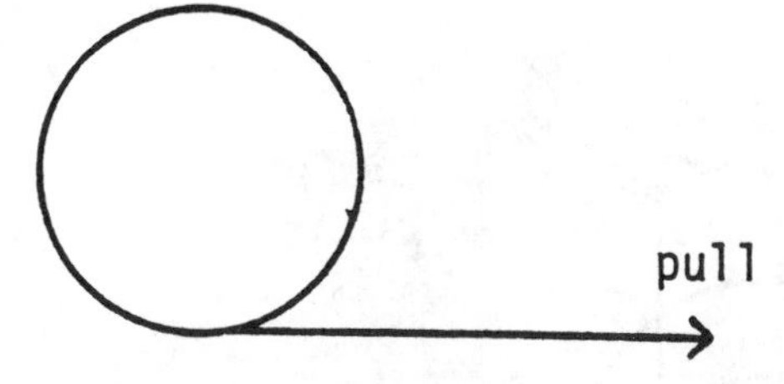

Hold the string at an angle of about 80° and again pull.

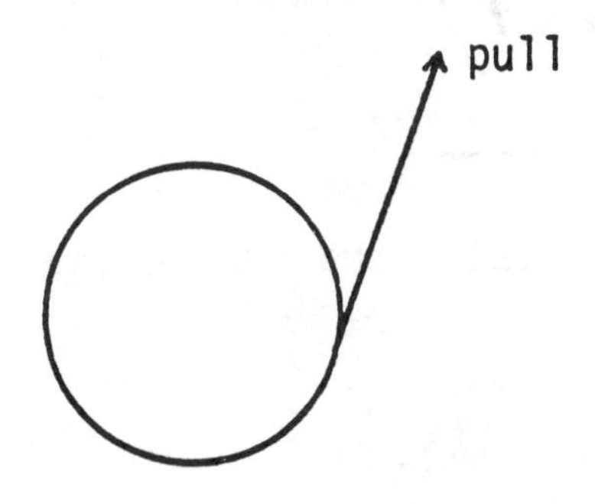

Quantitative: A couple consists of two parallel forces, equal and opposite, which tends to rotate a body, as shown.

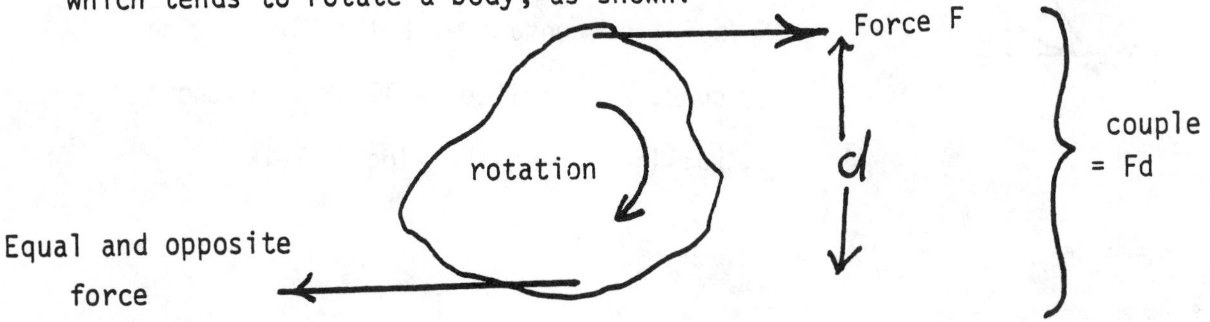

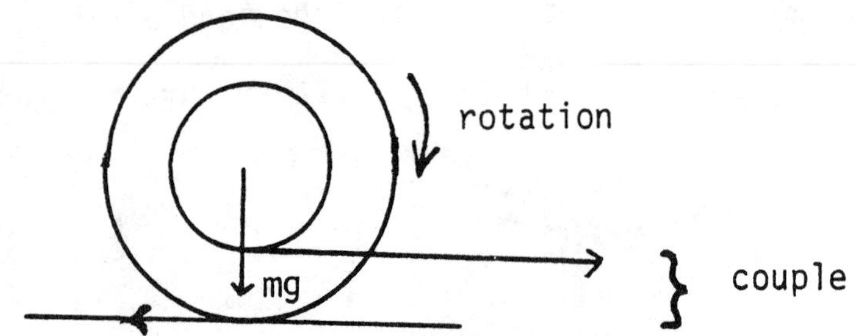

The couple acts to roll the cup toward the observer.

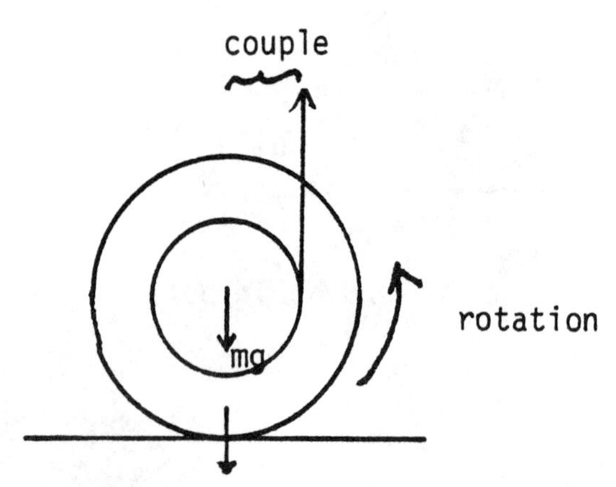

In the above position, the couple rotates the cup in the opposite sense.

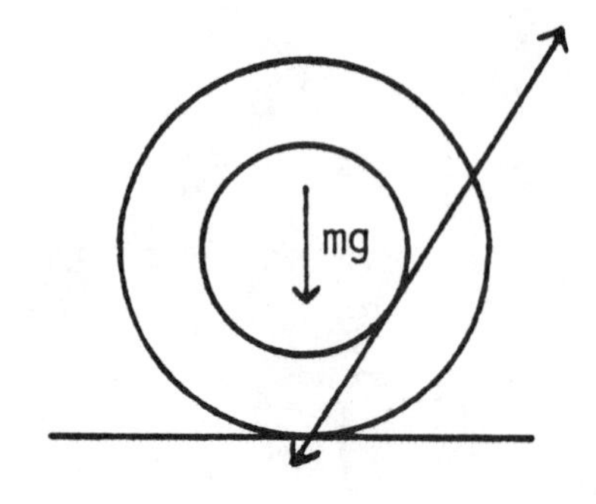

In the position shown, no rotation is possible because the direction of the string passes through the point of contact with the floor. The cups, when pulled, will skid, without rolling, along the floor.

Experiment 1.15. Center of mass (gravity)

Materials: scissors, paper, pencil, string, paper clip, marble, cellotape

Procedure:

When an object is supported from a frictionless pivot, the center of mass lies directly below the pivot. Cut out the shapes over (and any others you would like. Open up a paper clip, tie one end of a piece of fine cord to the clip, and cellotape a marble to the other. Punch a hole, as shown, through one of the points in the cutout and hang it vertically. Mark where the string lies against the paper.

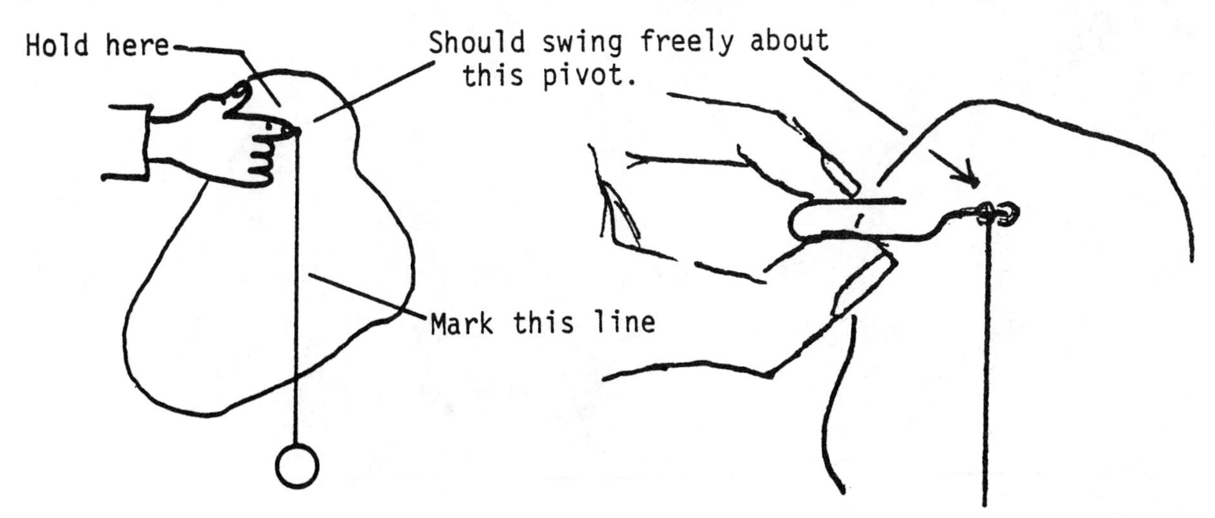

Repeat about another point of suspension. Where the lines cross is the center of mass. (C of M. or C of G.)

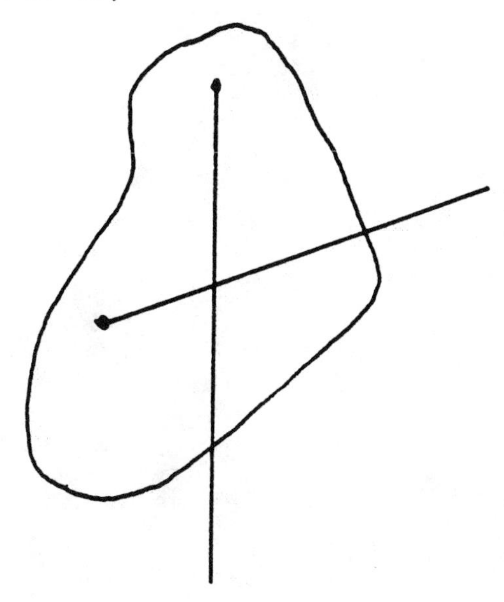

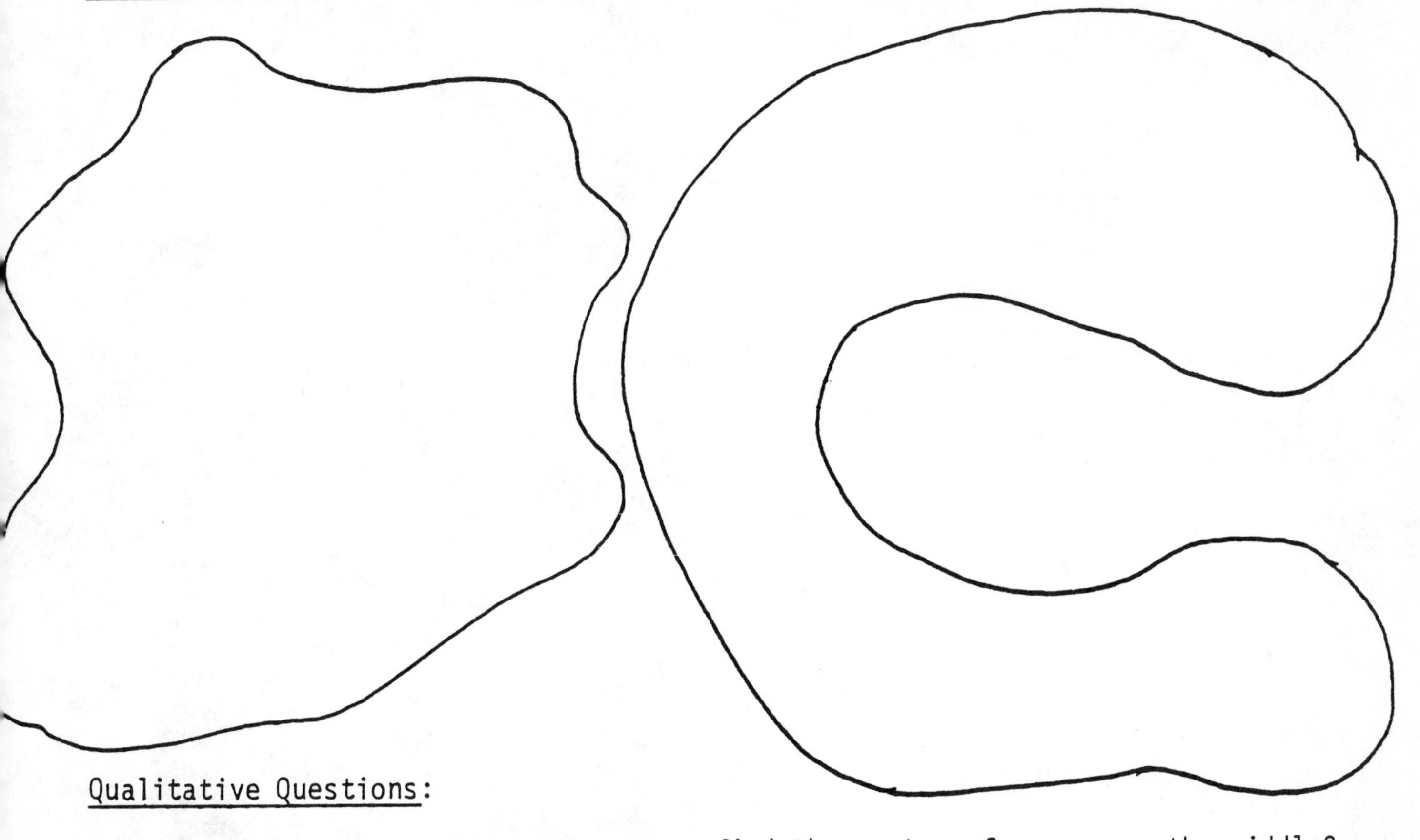

Qualitative Questions:

Object A: Why would you expect to find the center of mass near the middle?

Object B: Can the center of mass lie outside the object?

Quantitative Questions: Calculate where the center of mass of object C should be. Cut out C and check whether you are right. Then cut out the map of the US and find out where its center of mass is.

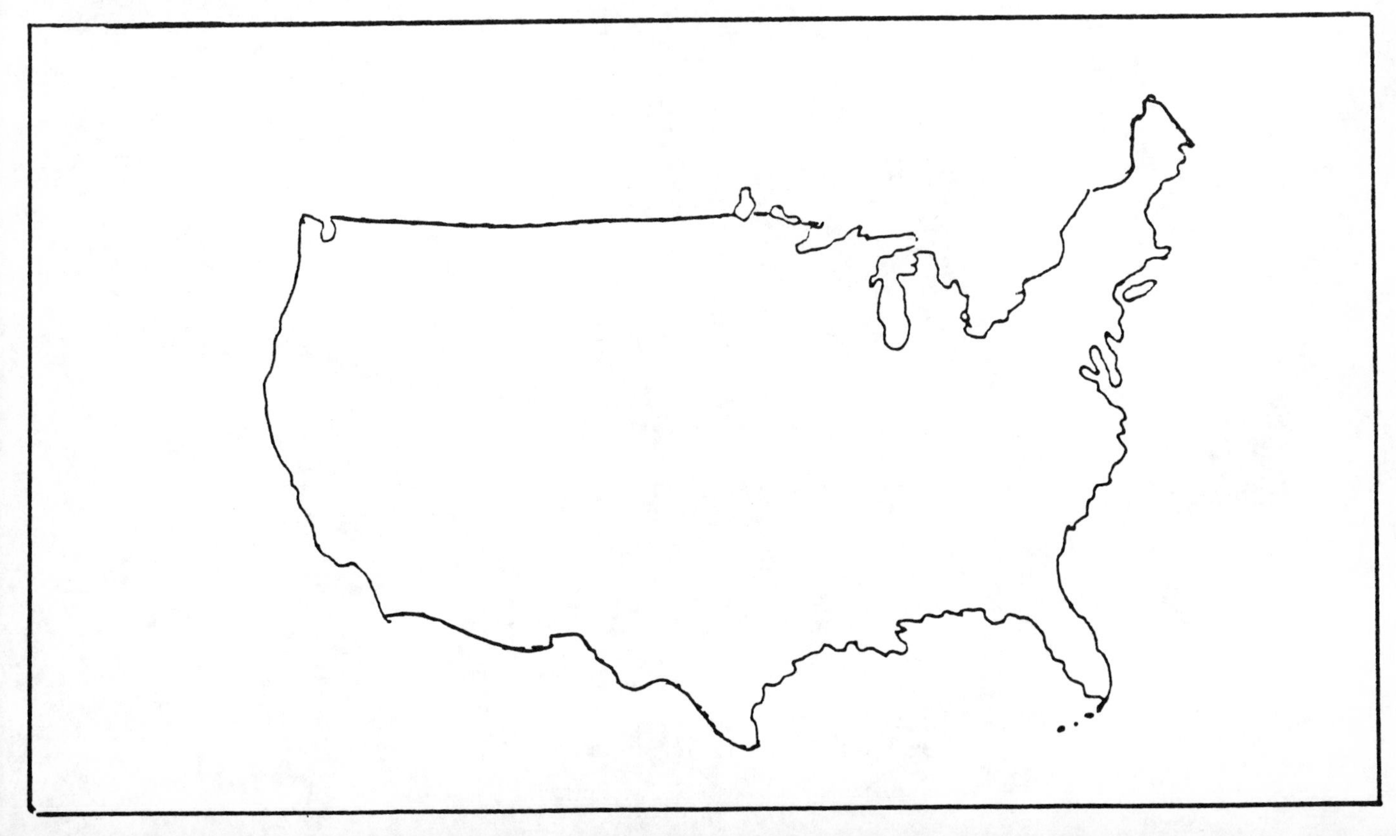

Why is the center of mass of a body so important? Stand with your back to the wall and try to touch your toes. You fall over because your center of mass lies ahead of your toes.

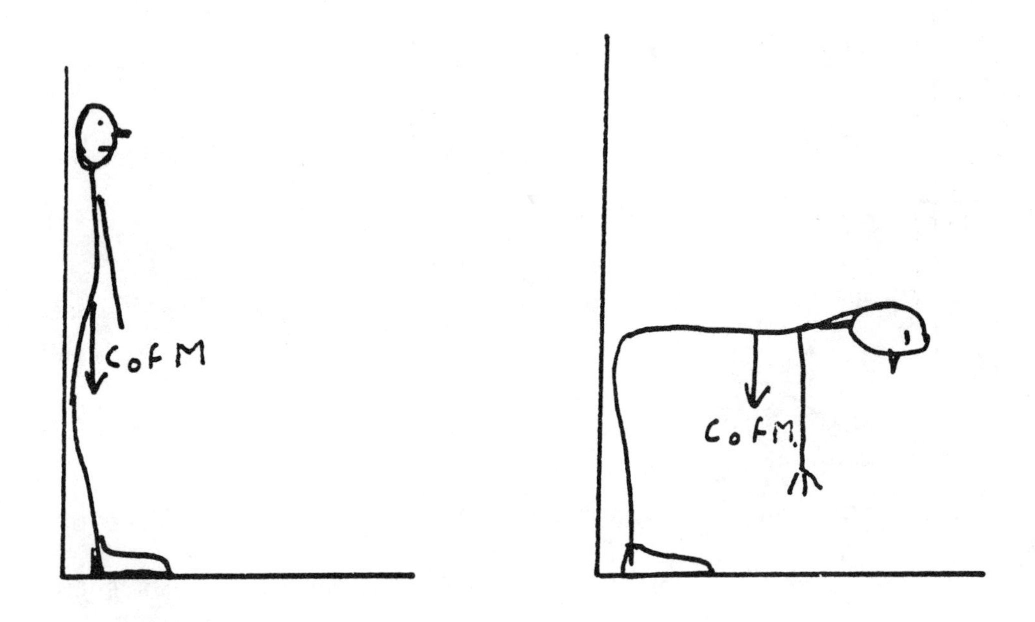

This is also why race cars have a low center of gravity so they won't turn over, where as a stagecoach will.

Or a tall lampstand is unstable.

An interesting experiment on balancing may be carried out using a marble and two straws. Fasten the marble half way up one straw, and try balancing it on one finger.

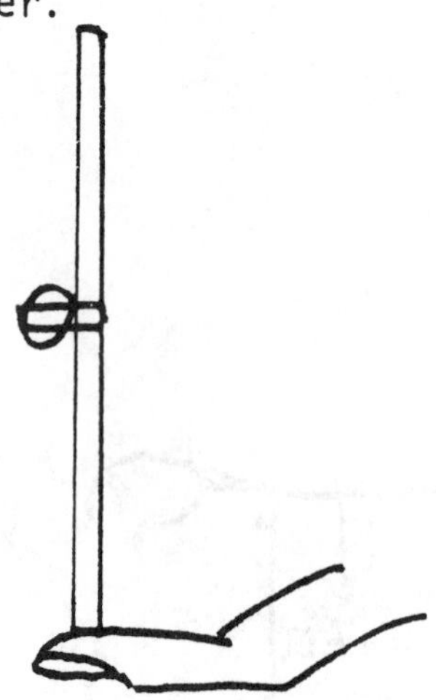

Now, push the end of one straw into the other (this is most easily done by removing the end of one straw by pushing it in with the thumbnail, as shown,) attach the marble at the end, and again try. It is much easier. Why?

The mass is essentially the same in both cases, but the second case is easier, because the moment of inertia about the finger is larger, so it takes longer to accelerate the marble, and you can move your fingers under the center of mass more easily.

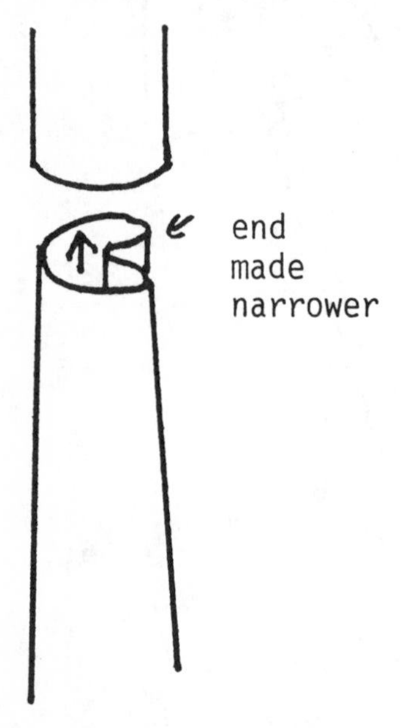

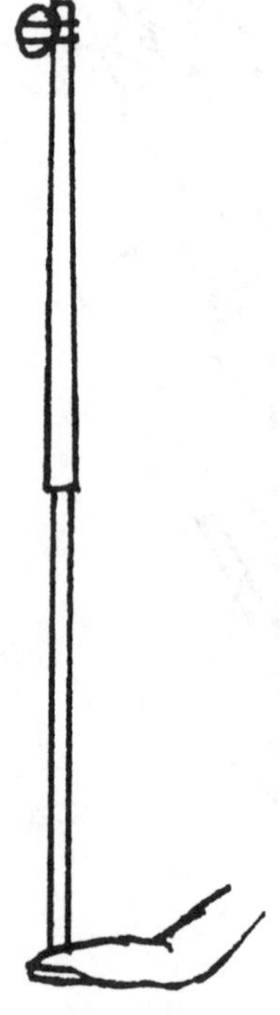

Some interesting center of mass experiments can be performed using the human body[1], hence lending themselves to class participation (see Chapter 14).

In the first demonstration (Fig. 1) a student is asked to stand on her toes normally, then with her toes against a wall. On her toes normally, the student automatically maintains the position of her center of mass directly above an axis of rotation through the toes. The torque acting on the student due to the force of gravity is then zero, and unstable rotational equilibrium results (the torque due to the force expected of the student by the floor is, of course, zero too). When the student stands against the wall, it is impossible to bring the center of mass far enough forward so that it lies above the toes. Hence, rotational equilibrium cannot be achieved, i.e. a person cannot stand on tiptoes with the toes against the wall.

A similar experiment is shown in the same figure. In each case the student can maintain equilibrium standing away from the wall, by positioning his center of mass above a rotation axis through a foot, (or feet) but cannot do so when standing adjacent to the wall.

Another demonstration (sometimes used as a party trick)[2] shown in figure 2, demonstrates the difference in position of the center of mass of a female and male student. A kneeling student first places her elbow, arms and hands together (as if "praying") with the elbows touching the knees and the forearms along the floor. A matchbox or other object is placed at the students fingertips. The student then clasps her hands behind her back, and is instructed to knock the matchbox over with her nose without her entire body falling over. In general, female students can perform the task whereas males cannot. Equilibrium in the kneeling position can be maintained as long as the center of mass does not move forward beyond a point above the knees, or backward beyond a point above the toes. Because the C of M of a male is closer to the head than a female, a typical male cannot knock over the matchbox without moving his center of mass forward of the knees, thereby tripping over.

1. Ernie McFarland, Physics Teacher 21, 42 (1983).
2. J. Watson and N.T. Watson Phys. Teach. 20, 235 (1982).

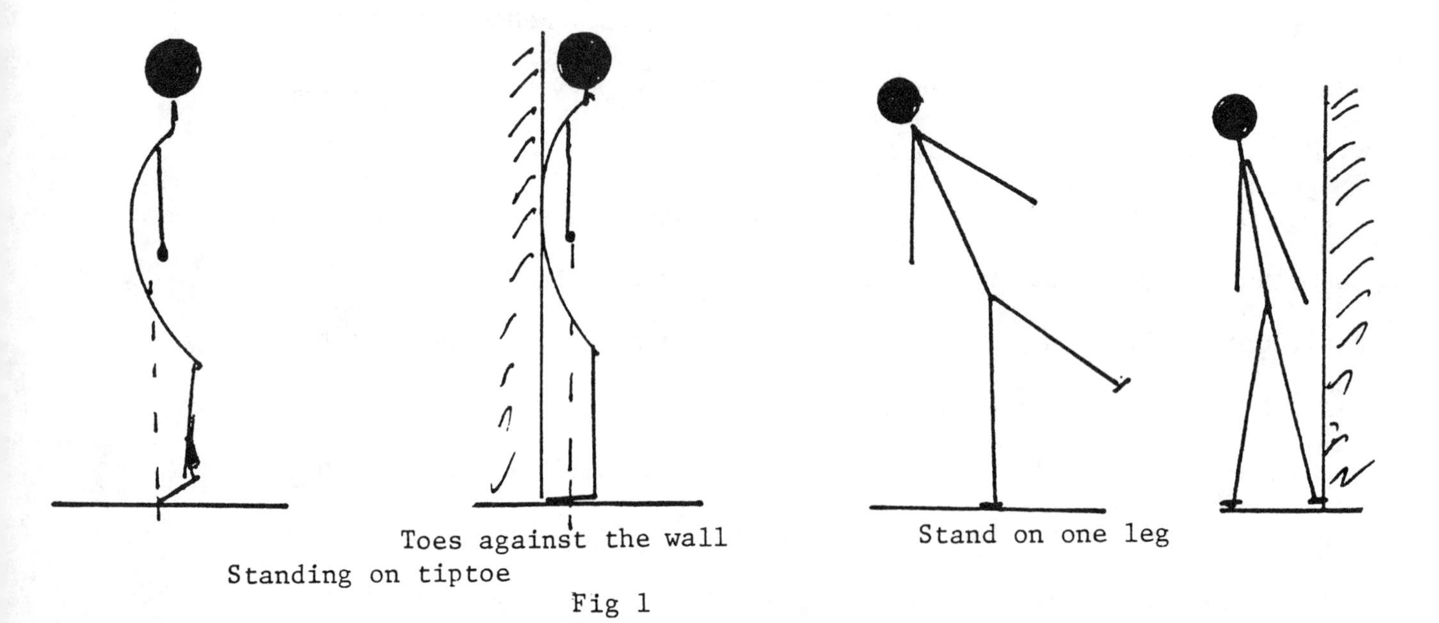

Fig 1

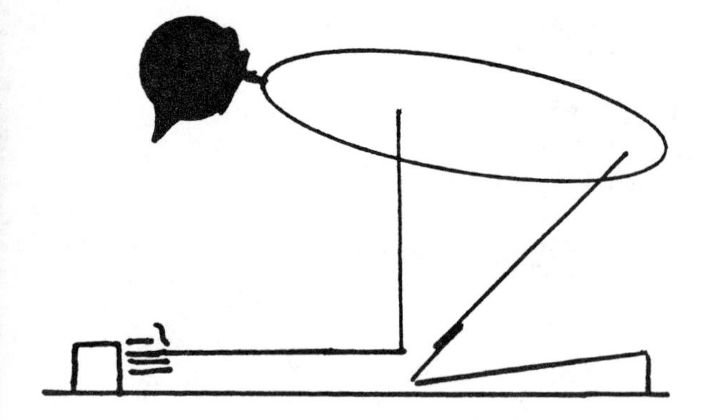
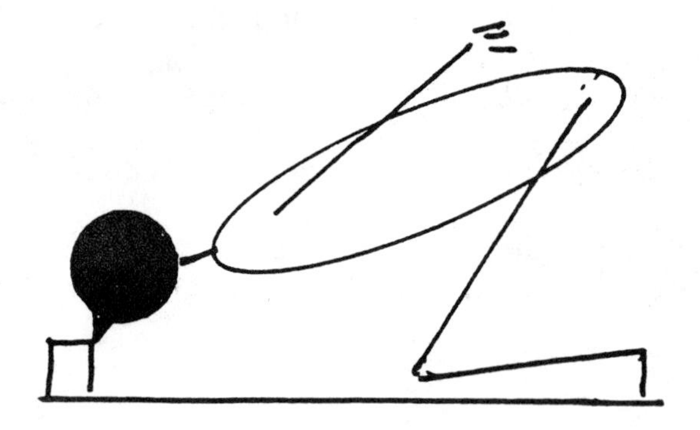

Fig 2

Experiment 1.16 Friction and the Climbing Monkey

Materials: Card, Tape, Straw, String

Akio Saitoh of Nara-Ken Japan provided this delightful experiment on friction which he calls "going up a tree." Figure 1 shows how to construct this apparatus.

1. Cut the sticky tape, attach it to the drinking straws, and fix them to the cardboard with an angle of about 20° between them.

2. Pass the string through the two straws, and hang the string on the nail as shown in Fig. 2.

3. Hold both ends of the string taut, pull on each end of the string alternately, and the cardboard will climb the string. You can draw a picture of a monkey as shown in Fig. 3 on the other side of the cardboard, and it looks quite amusing to see the monkey climbing.

When we pull on the left hand string, as shown in Fig. 4, the cardboard tilts to the left and the force normal to the straw on the right hand side, N_R, is larger than that on the left hand side N_L (Fig. 5). The left hand straw can slide up along the string, whereas that on the right cannot because the maximum force of static friction is proportional to the normal force.

What is the minimum tension in the string so that the monkey just begins to climb? If we assume that the string passes through the left hand straw without friction because the card is tilted, then the weight of the card is supported by friction with the right hand straw, so that $mg/2N_R = \mu$ where m is the mass of the card and μ is the coefficient of friction. But $N_R/T = \sin\theta$ where θ is the angle between the straws, if we assume the left hand straw in line with the string. So

$$T = \frac{mg}{2\mu \sin\theta}$$

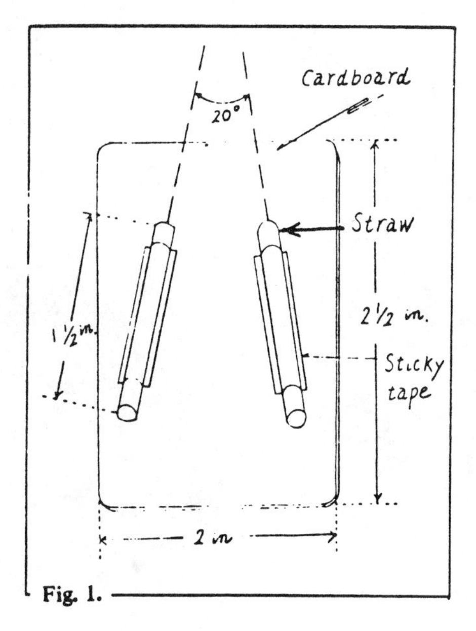

Fig. 1.

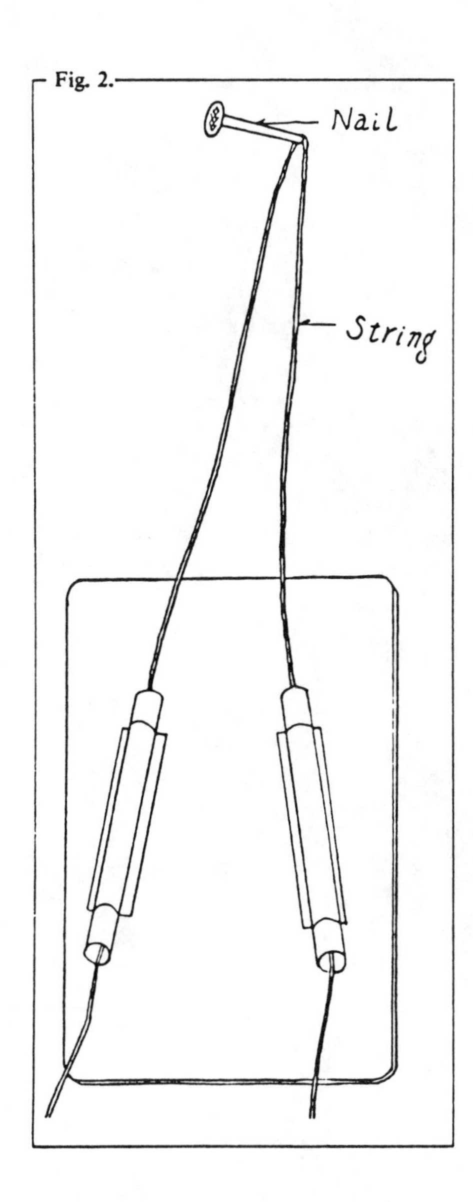

Fig. 2.

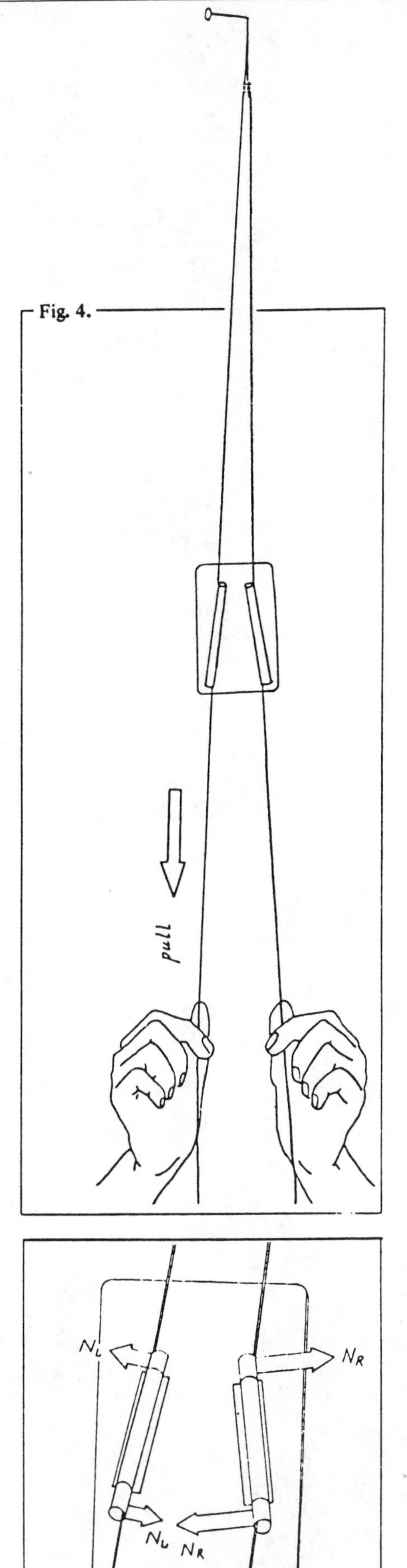

Fig. 4.

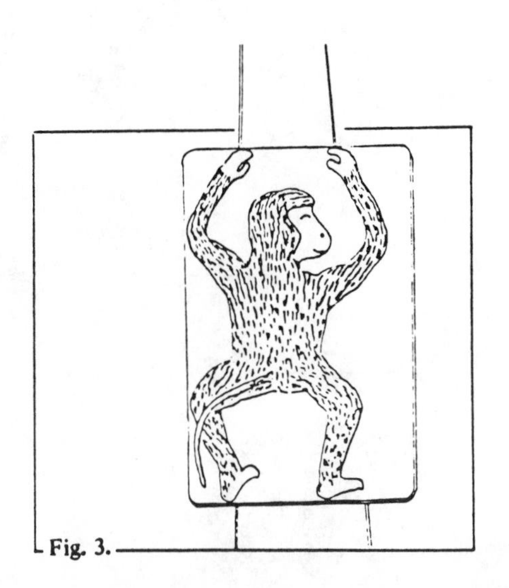

Fig. 3.

Fig. 5.

Experiment 1.17 Friction

Materials: cup, paper, marbles, books.

Procedure: Place four marbles in the styrofoam or Dixie cup. Put the cup on a book, and gradually tilt the book until the cup slides down it.

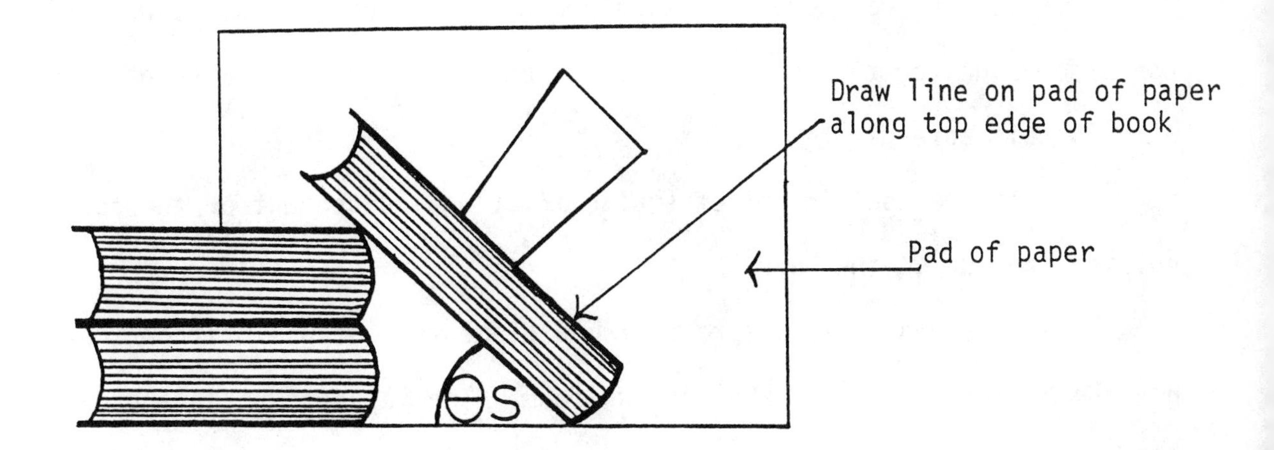

Repeat this several times, drawing a line along the top edge of the book placed against a pad of paper as shown each time. You can now measure the angle of tilt at which the cup starts to slide θ_s using the protracter shown at the beginning of this book. The tangent of this angle is the coefficient of static friction. This is the ratio of the force causing the cup to start sliding to the force holding it down on the surface as shown.

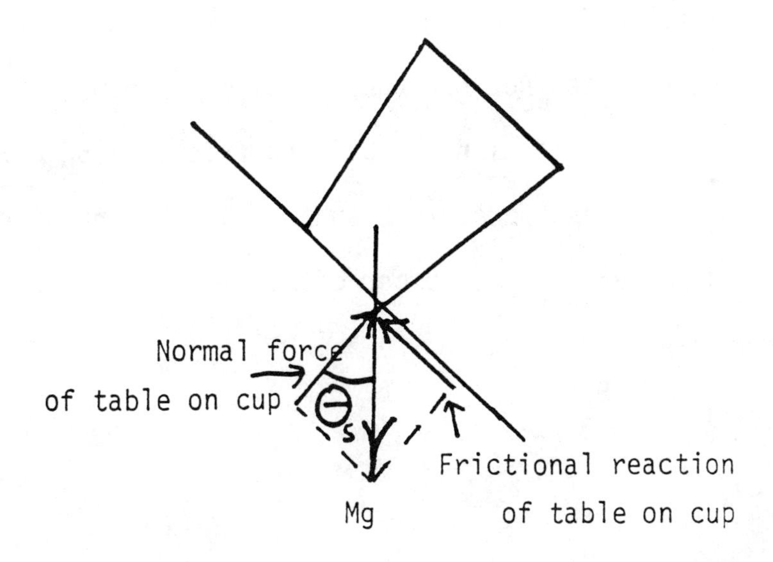

Now put a lot of marbles in the cup, and again measure the coefficient of static friction. Answer the following questions:

1. Do the measurements, taken under similar conditions, always give the same value for the coefficient of friction, or is there a spread. How wide is this spread?

2. Is the coefficient of static friction independent of the force holding the cup to the book?

There is generally a spread of measurements, because neither the book nor the bottom of the cup are very uniform. The value of μ_s is approximately independent of the number of marbles in the cup, at least over a fairly wide range, showing that it is the nature of the surfaces which counts. Now repeat the experiment using a different book - try to use one book with a rough cover, and one with a smooth cover. Does the value of μ_s change appreciably? It is generally large for rough surfaces, small for smooth ones.

Set the book at the angle for μ_s and try replacing the cup - it is very difficult to do, showing that, once the cup starts to move at all at this angle it will continue to slide. Reduce the angle of tilt of the book until, when given a slight push, the cup does not continue to slide on down the book. The tangent of this angle is the coefficient of sliding friction, μ_k. It is always less than μ_s, since it takes less force to keep the cup sliding than it does to start it. Is μ_k independent of the number of marbles in the cup? Again, this is approximately true.

With a pair of scissors, cut away half the bottom of the cup, as shown

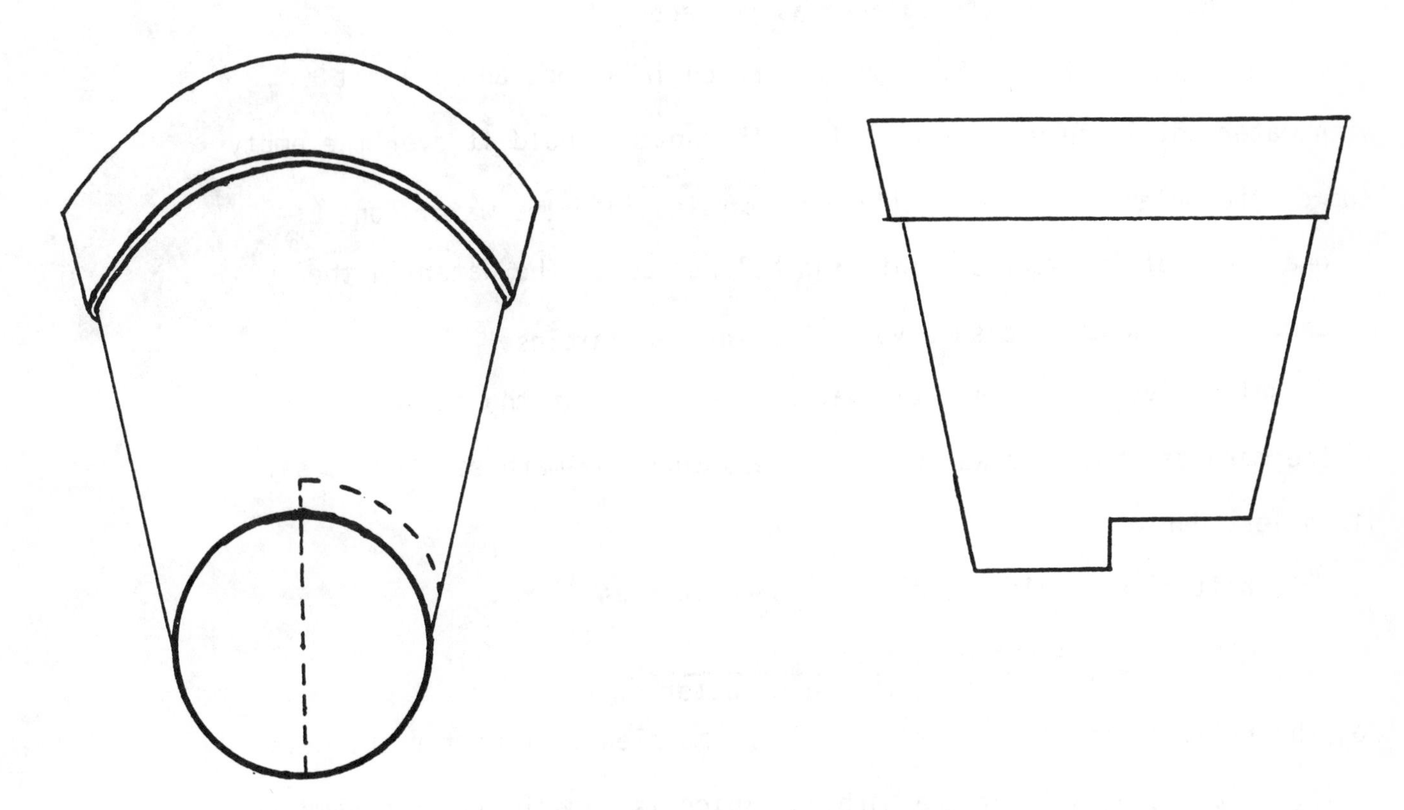

Place a second cup inside the first, put ten marbles in, and again measure the coefficient of friction.

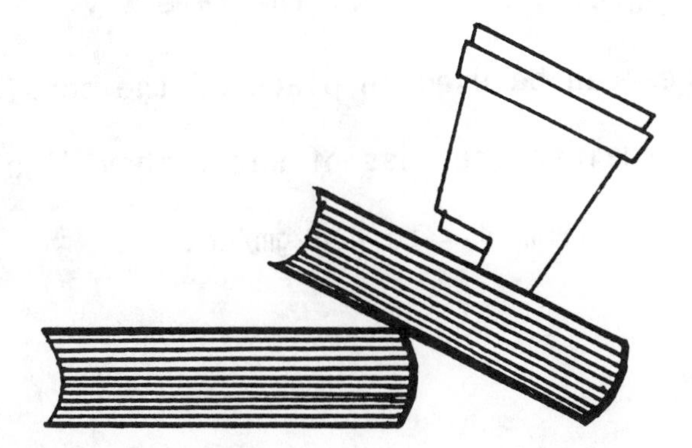

Is it the same as for the whole cup? Only half the area of the bottom of the cup is now in contact with the book. If the coefficient of friction is the same, it follows that it depends only on the nature of the surfaces in contact, not on their area.

Experiment 1.18- Relative density

Materials needes: - the torsion balance (or one of the other balances you have made), marbles, foam or paper cup.

Procedure: - Read the balance as you add from one to ten marbles. Now, cut a V notch in a cup, and fill it with water until the water runs out of the notch. Hold it over the empty cup on the balance, and carefully drop marbles into the water, one by one, so that it overflows into the balance cup. The water in the balance cup, now has the same volume as the ten marbles.

Qualitative - note that the water does not twist the straw as much as the marbles did. The water weighs less than the marbles, and we say it is less dense.

Quantitative - Relative density is defined as

$$\frac{\text{Weight of object}}{\text{Weight of a equal volume of water}}$$

So, the ratio of the scale reading for the marbles to that for the water is the relative density of the marbles, which is numerically the same as the density, (mass per unit volume) since water has a density 1. You can measure the density of other objects, coins etc. the same way.

Note: One of the other balances can be used in place of the torsion balance, provided you weigh the marbles against the mass of water they displace.

The density of common glass varies from 2.1 to 2.8 gm/cc.

The Hydrometer*

Seal one end of a straw with tape so that it is watertight. Attach a marble to the bottom of the straw with sticky tape so that it stands upright in the cup of water, floating about half submerged, as shown, Mark the straw at the water level with a pen. (If you have a tall cup or jar, it will let the straw float deeper).

Place the straw in a cup full of a second liquid of known density or specific gravity. The relative density or specific gravity (they are the same thing) is the ratio of the density of the material to that of water, and hence is a pure number. The density will have the same numerical value if the density of water is 1.0 which it approximates in the c.g.s. system, but it will be in units such as gm/cm^3, (the S.I. system uses Kg/m^3, for which the density of water is 1.0×10^3 Kg/m^3). Mark the level. Some examples of suitable liquids are glycerine (relative density 1.26) kerosine ($\sim$ 0.8), gasoline (care! - 0.68 - 0.72) or solutions of common salt (10% by weight, 1.07; 20%, 1.148) or sugar (10%, 1.038; 20%, 1.081; 30%, 1.270).

Knowing the density of water (1.00) and of the other liquid, you can now calibrate the straw to produce a hydrometer which can be used to measure liquids of unknown density.

The most common uses of a hydrometer are to measure the densities of battery acid or sugar solutions: knowing the density of an aqueous sugar solution, one can calculate the alcoholic content when the sugar is converted into alcohol by yeast. The alcoholic content is not measured directly by density because the specific gravity of the alcohol solution

---

* suggested by Michael Davis of North Georgia College, Dahlonege, Georgia.

(wine or beer) is too close 1. Thus at 20°C, the density of pure alcohol is 0.79 gm/cm^3. If the relative density of the sugar solution is 1.1, the potential alcohol is 12%, which would have a specific gravity of 0.9749. For 1.15 it is 17.2%, corresponding to a relative density of 0.964.

Batteries contain sulphuric acid. A 10% solution has a relative density of 1.0661, 20% 1.1394, and 50%, 1.21885.

If the volume of the hydrometer under the water (density 1 gm/cc) is V, and it is found that placing the straw in a vessel of liquid having a new density ρ causes the straw to rise a distance d (increase in density) the change in volume of the hydrometer is $-\pi r^2 d$, where r is the radius of the straw, and

$$\rho(V-\pi r^2 d) = m = V,$$

where m is the mass of the hydrometer. Hence, increasing V also increases the sensitivity of the instrument, for

$$d = \frac{V(\rho-1)}{\pi r^2}$$

Attaching something of density near 1, (e.g. a lump of wax) to the hydrometer will increase V, and the sensitivity.

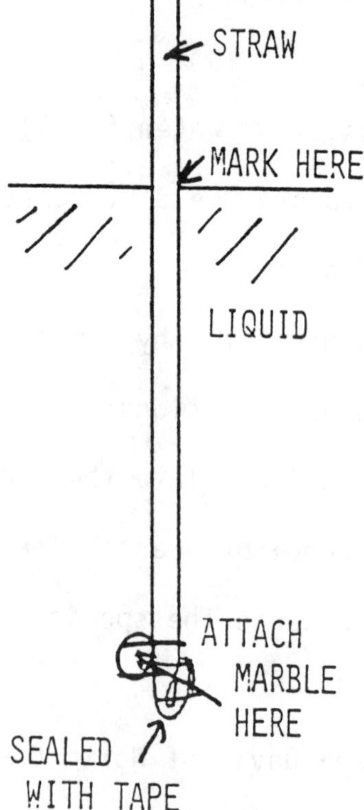

Experiment 1.19 . Buoyancy

Materials: Cup, metric ruler, 2 paper clips, 8 marbles, stickytape, 2 rubber bands, water.

Procedure: Attach a rubber band to a paper clip that has been opened out into an L shape. Secure 8 marbles to the paper clip base with tape, leaving the rubber band free. (Without the paper clip base the marbles will slip off when wet.)

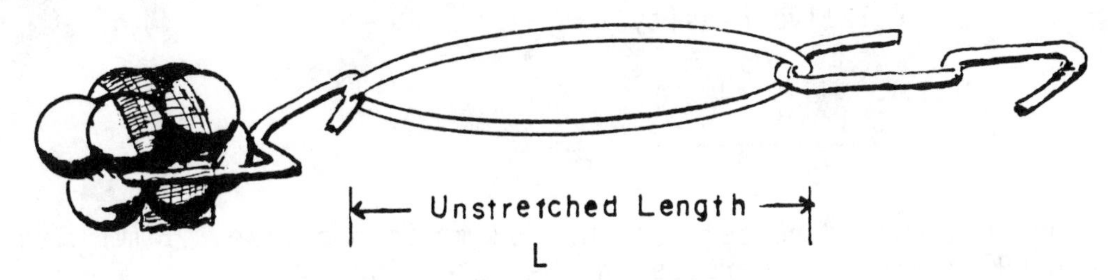

Open the second paper clip to form a hook for the top of the rubber band. Place the ruler in the cup straight up. Put about 7 cm of water in the cup. Measure and record (in mm) the unstretched length of the rubber band and the length of the band stretched while held out of the water, with the marbles on it. Lower the marbles into the water so that the bottom of the rubber band just reaches the surface of the water. Again measure the length of the stretched rubber band and record (in mm).

Qualitative: You will notice that the rubber band stretches less when the marbles are in the water. This is because the marbles are pushed upward by a force equal to the weight of the water they displace. This upward force is buoyancy that makes objects appear to lose weight in water.

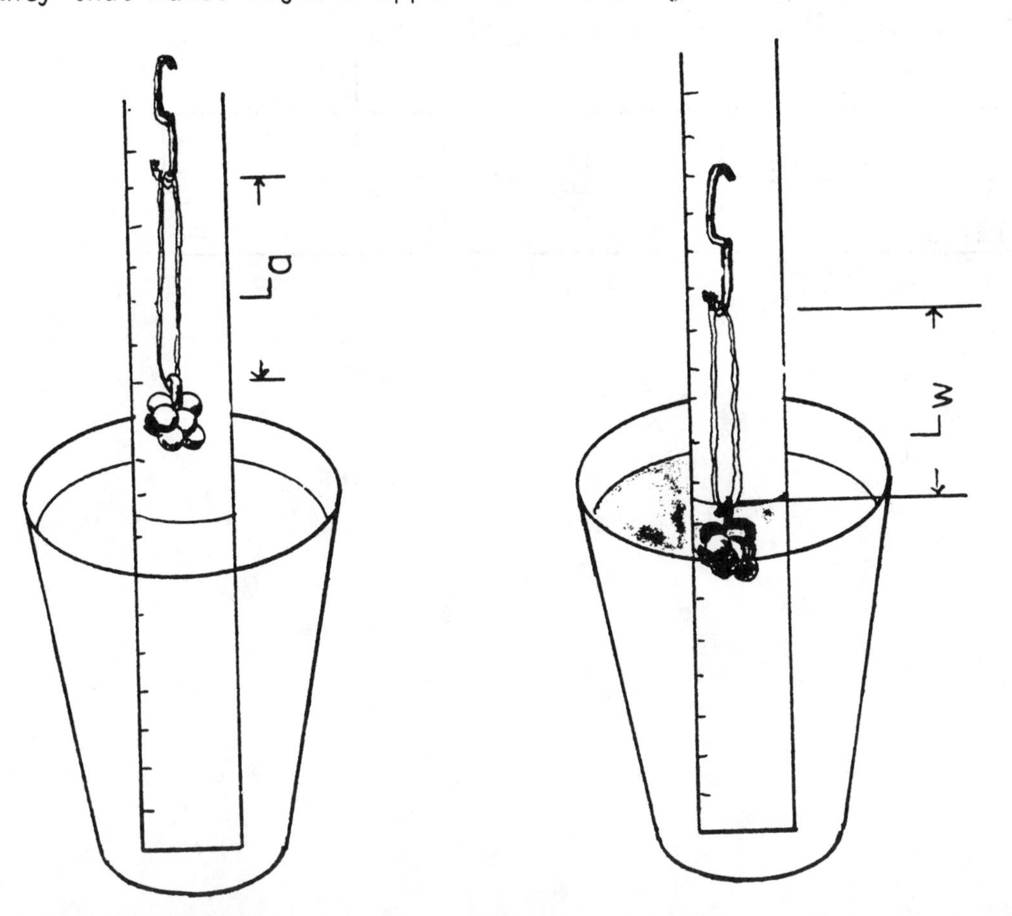

Thought question: If you snip off a small piece of the cup and put it on the water, why does it float?

Quantitive: The relative density of any object is $\frac{\text{Mass of object}}{\text{Mass of equal volume of water}}$

In our case, the mass of the object is proportional to the amount the band stretches in air, d_1, and the mass of the equal volume of water is proportional to the difference between this and the distance it stretches with the object underwater, d_2, so the relative density =

$$\frac{d_1}{d_2 - d_1}$$

For example, the unstretched rubber band was 62 mm in one case, 75 mm with the marbles on, and 69 mm with the marbles in the water, so

$$\text{relative density} = \frac{13}{6} = 2.17$$

Length in air L_a		$L_a - L = d_1$	
Length in water L_w		$L_w - L = d_2$	
Unstretched length L		$\frac{d_1}{d_2 - d_1} =$	

Z Eaves

Experiment 1.20. Velocity

Materials needed: - ruler, one marble, sticky tape.

Procedure: - Stick the half inch wide tape across the groove in the ruler at zero inches, 3 inches, 6 inches, and 9 inches. Tilt the ruler so that, when released from the top cellotape, the marble just makes it to the bottom. Now set it going, and listen to the clicks as the marble drops off each piece of tape and strikes the ruler - the clicks are always equally spaced in time, however fast you set the marble going with your finger. What does this show?

Try moving one of the pieces of tape. Are the times between clicks the same?

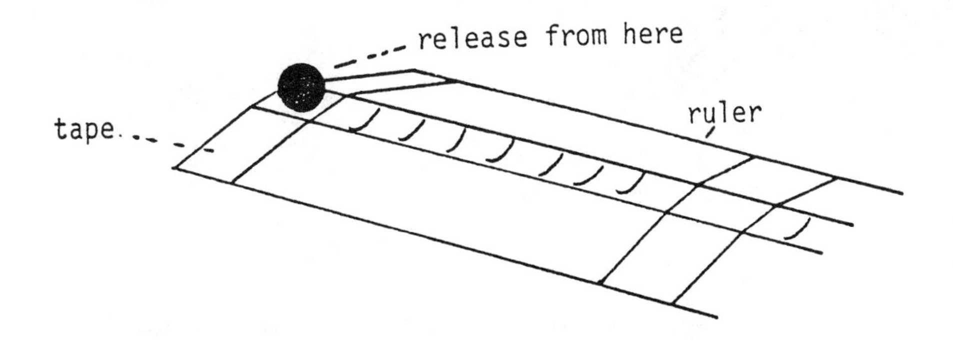

No horizontal forces act on the marble, which covers equal distance in equal times - this is called constant velocity. As we make the marble to go faster, the clicks are closer together, so higher speed, or velocity means the same distance in a shorter time, or a longer distance in the same time.

How fast does the marble travel? A crude estimate can be obtained by dropping a marble off the table at the same time you start rolling the one on the ruler. A normal table is 2 ft 6 in (76 cm) from the floor, so the marble takes 0.4 secs to drop. If the click on the ruler and that on the floor occur at the same time, the speed is the distance on the ruler divided by 0.4 in cm (or inches) per second.

Experiment 1.21 Acceleration

Materials needed: ruler, one marble, sticky tape

Procedure: Stick the tape across the groove in the ruler at zero inches, 3/4 inch, 3 inches, and 6 3/4 inches. Tilt the ruler by two or three books. Now release the marble from rest at the top, so that it rolls rapidly down, making an audible click as it passes over each piece of tape, and at the end as it rolls off the ruler. Listen to see if the clicks are evenly spaced. Does it make any difference whether the ruler has a steep slope or not?

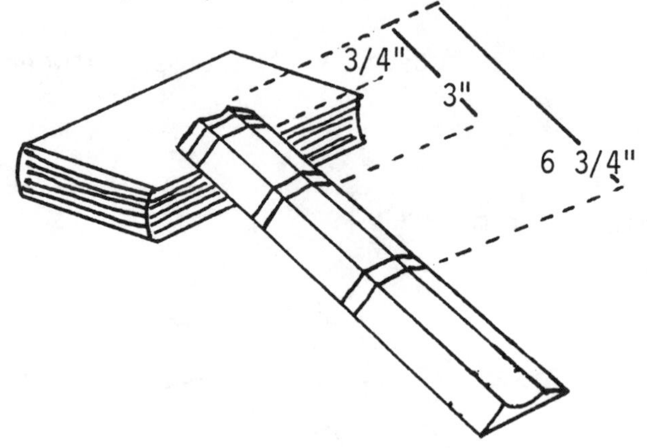

Qualitatively - as the marble travels down the ruler, it is subject to gravitational forces, making it cover larger distances (between pieces of tape) in the same time - so its speed (distance divided by time) is increasing - this is what we mean by acceleration under the action of a force.

Quantitatively - the pieces of tape are distant from the top 1, 4, 9 and 16 units. These are 1^2, 2^2, 3^2 and 4^2, so if the clicks are evenly spaced in time, the distance d travelled is proportional to the square of the time $d = kt^2$. The steeper the slope, the larger the acceleration, but the clicks are still even, so $d = k't^2$, with a new constant k'.

Experiment1.22 - Acceleration due to g.

Materials: - string, marbles, tape, ruler

Method: - The previous experiment, using marbles on a ruler, is not convincing to some students. The following experiment is a little simpler, and employs g directly.

Two strings are cut the height of the room, or as high as the student can reach standing on a chair. We will suppose this is eight feet. Marbles are attached to one string at two foot intervals. The other string has marbles attached at intervals proportional to the squares of the whole numbers, as follows

number	square	multiplied by 6	difference
0	0	0	
1	1	6"	6"
2	4	24"	18"
3	9	54"	30"
4	16	96"	42"

Now stand on a chair holding the two strings as shown in the following figure

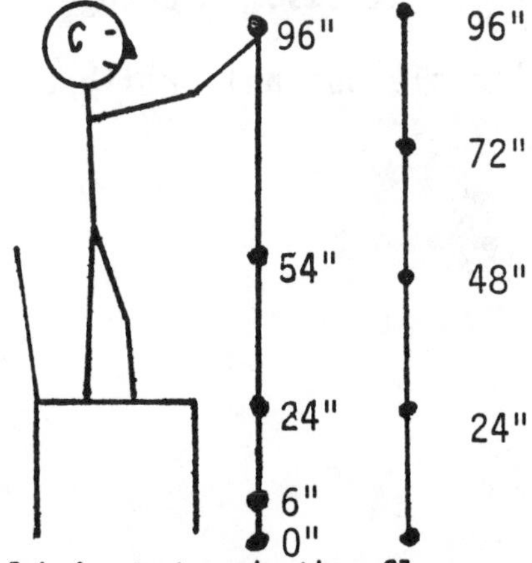

The bottom marble should just touch the floor. Drop the string having marbles at a uniform spacing. Now drop the other string. The clicks are easier to hear if you drop the string into a trash can.

Qualitative Question: - Do you hear the time between clicks get shorter as the higher marbles from the uniformly spaced string strike the floor? Does the time between clicks for the non-uniformly spaced string stay the same?

This is because the higher marbles are accelerated for a longer time and are traveling faster, covering the same distance in a shorter time as they approach the floor than the marbles starting near the floor.

Quantitative Question: - The formula for acceleration under g is

distance travelled $S = \frac{1}{2} gt^2$ (t = time taken)

By spacing the marbles at intervals such that the square root of successive distances is proportional to whole numbers, the time between successive clicks must be constant. In our case this time is, taking the lowest marble, $\frac{1}{2}$ foot.

$$\frac{1}{2} = \frac{1}{2} \times 32\, t^2$$

therefore $t = \frac{1}{\sqrt{32}} = \cdot 176$ seconds = time between clicks

Shift one of the marbles up or down the string, to tell how sensitive your ear is to changes in timing between clicks. A change in position of 20% in any one marble gives a quite noticeable non-uniformity between clicks.

Another way of measuring g is to poke a very small hole in the bottom of a cup. The drops fall onto a trash can, a sheet of paper, or anything that makes a noise. Move the cup up and down until the sound of the previous drop striking the can exactly corresponds to the time a drop leaves the cup. Then measure the height above the bottom of trash can, or sheet of paper, and the time for say 50 or 100 drops to fall. You can get the time it takes the drop to fall the fixed distance from this.

Time for 100 drops, T, = sec.

Height of cup, h, = m.

$g = \frac{2\,0000h}{T^2} =$ m./sec^2

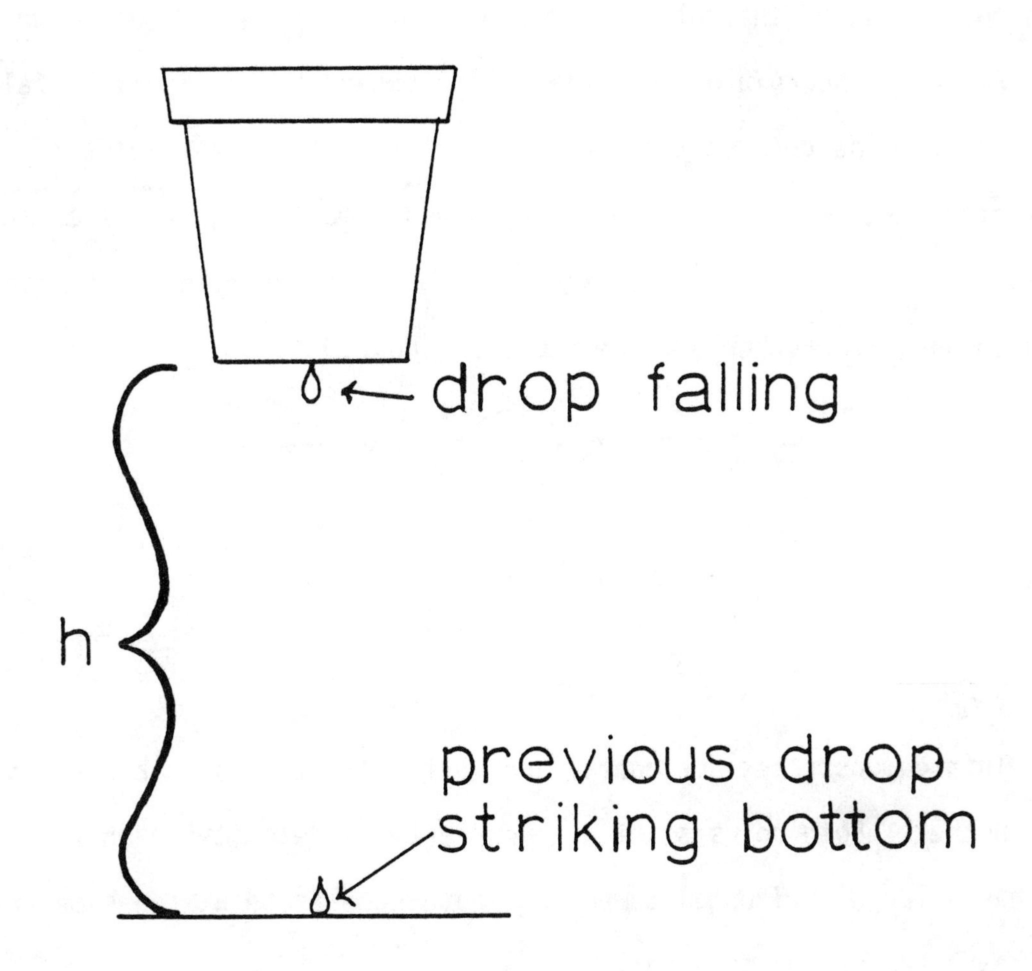

Experiment 1.23 Independence of vertical and horizontal motion

Materials required - Foot ruler, two coins or two pieces of chalk.

Procedure: - at a corner of the table, place a piece of chalk, or better, a coin. Hold a ruler behind it, as shown, and a piece of chalk, or another

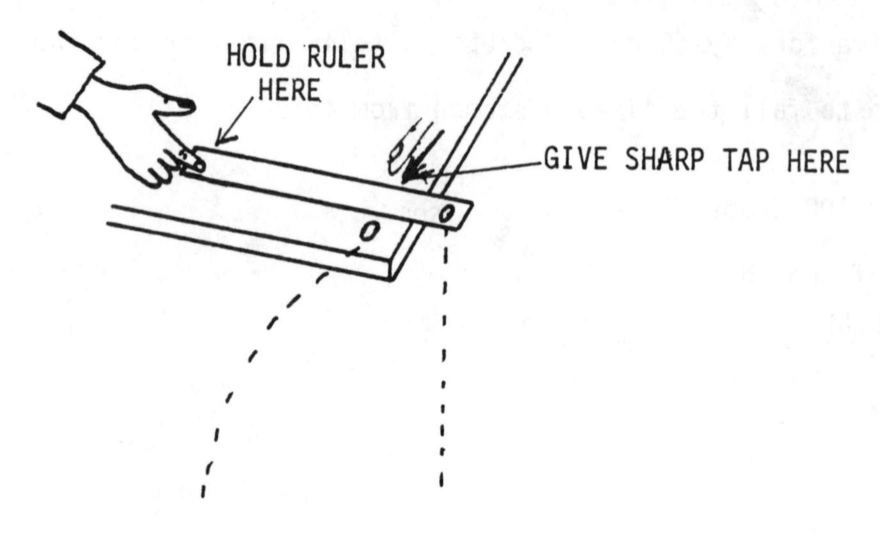

coin on the end of the ruler, over the edge of the table. Give the ruler a sharp tap. The coin on the ruler, left behind by its inertia, falls vertically. The coin on the table has a large horizontal velocity (but no vertical velocity). Both hit the ground simultaneously, as one can easily hear. You can do this experiment without the ruler by getting one coin to nudge the other off the table as shown.

What do we learn?

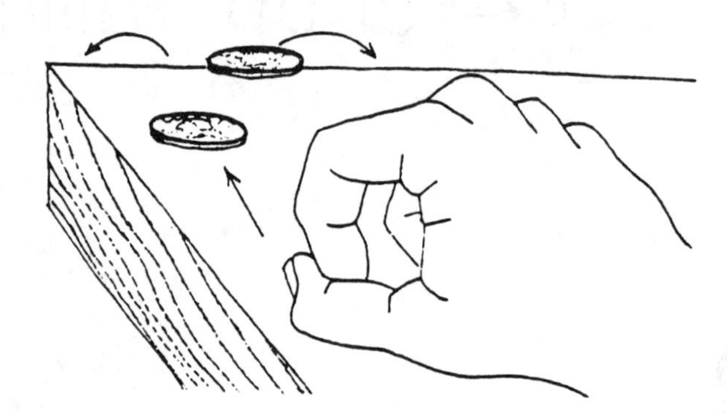

This demonstrates horizontal and vertical motion are independent of one another. Both coins start with no vertical velocity, even though one has a large horizontal component, both accelerate at the same rate and reach the ground together.

Experiment 1.24. Two dimensional kinematics using coins

Materials required: Two coins - two nickels (or quarters or pennies), and a nickel, a quarter and a penny. A piece of paper and pencil, or some other way of marking the coin's position on the floor.

Procedure: Collisions in two dimensions can be studied using coins on a sheet of paper. First, let us look at the angles in the collision process. Put one coin a few inches behind a similar coin (pennies or nickels are good) on a sheet of paper, and draw round them, marked 1 and 2 in figure 1. Flick one coin at the other, so that they travel a fair distance but do not leave the paper. You can propel the coin either by using a ruler rather like a pool cue, or by holding it at one end so that it lies perpendicular to the direction in which the coins should travel, then giving the other end, close behind the coin, a sharp tap. Draw around the coins in their final position. This is marked 3 and 4 in figure 1, which was an actual experiment using nickels. The rotational motion occuring during the collision may be studied by noting the orientation of the two coins before they start (e.g. the position of the "L" in "Liberty" on the head side of the coin) and when they finish. Draw lines 5 and 6, extending them backwards as shown, so that these lines are tangent to the original and final position of the struck coin. Place the initial coin between these lines so that it touches 2, and draw around it, as the dotted circle 7. This is the position of the striking coin when the collision occured. Now draw lines 8, 9, 10 and 11, extending 8 and 9 so that the angles θ and ϕ can be recorded by a protractor. The distances d_1 traveled after collision by the striking coin, and d_2, traveled by the struck coin should also be measured. V is taken as the velocity of the first coin just before the collision, and v_1 just after the collision, with v_2 the velocity of the second coin just after the collision. It is convenient to use a set of axes parallel (y) and perpendicular (x) to the direction of collision, which is that of v_2. This selection of axes has the advantage that to a first approximation no angular momentum is transferred between the coins from the effects of the collision in the y direction, and only angular momentum is transferred in the collision by motion in the x direction. Furthermore, since there is no linear transfer of momentum in the x direction, our consideration of any inelasticity in the collision can be restricted to the y axis, which is parallel to the line of impact. Conservation of momentum in the x direction then gives us

$$V \sin \theta = v_1 \sin (\theta + \phi) \tag{1}$$

and in the y direction

$$V \cos \theta = v_2 + v_1 \cos (\theta + \phi)$$

At this point we must consider the energy equation. The audible click as the coins collide shows an energy loss on collision, a measure of which is the coefficient of restitution e, defined as the ratio of the difference in velocity of the colliding objects along the line of impact (i.e. the y direction in our case) after the collision to that before the collision. Experimentally, this is found to be a constant. Considering now the motion in the y direction

$$e = (v_2 - v_1 \cos (\theta - \phi)) \; / \; V \cos \theta$$

$$= (V \cos \theta - 2\, v_1 \cos (\theta - \phi)) \; / \; V \cos \theta$$

(Since $v_2 = V \cos\theta - v_1 \cos(\theta + \phi)$)

$= 1 - 2v_1 \cos(\theta + \phi) / V \cos\theta$

$= 1 - 2 (V \sin\theta/\sin(\theta + \phi)) \cos(\theta + \phi)/V \cos\theta$

(Since $v_1 = V \sin\theta/\sin(\theta + \phi)$

$= 1 - 2 \tan\theta/\tan(\theta + \phi)$

If $e = 1$, the collision is perfectly elastic, $\tan\theta/\tan(\theta + \phi) = 0$, so $\theta + \phi = 90°$. If $e = 0$, the collision is perfectly inelastic, $2 \tan\theta = \tan(\theta + \phi)$ and $\theta = \phi = 0$.

For the example shown in figure 1, $\theta = 31°$, $\phi = 35°$, $e = .46$. We have found collisions between nickels to be rather more reproducible than with other coins.

Having obtained the coefficient of restitution from the angles, we can now examine the energy equation. If the coefficient of sliding friction is constant, independent of the velocity, the distance travelled by the coins should be proportional to their energy after the collision. If the distances are d_1 and d_2 as shown in figure 1, m is the mass of one coin and μ the coefficient of friction.

$$mg\mu d_1 = \frac{1}{2} m v_1^2$$

$$mg\mu d_2 = \frac{1}{2} m v_2^2$$

Since the relative velocity of the two coins in the y direction before the collision is

$$V \cos\theta = v_2 + v_1 \cos(\theta + \phi)$$

and after the collision is $v_2 - v_1 \cos(\theta + \phi)$

$$e(v_2 + v_1 \cos(\theta + \phi)) = v_2 - v_1 \cos(\theta + \phi)$$

$$\text{or } v_2 (1 - e) = v_1 (e \cos(\theta + \phi) + \cos(\theta + \phi))$$

$$v_1/v_2 = (1 - e) / (1 + e) \cos(\theta + \phi)$$

We can check this against the ratio $\sqrt{d_1/d_2}$. The results are generally not very reproducible because the frictional forces are not constant. In the case shown $\sqrt{d_1/d_2} = \sqrt{10 \text{ cm}/10.6 \text{ cm}} = .97$.

The angular momentum exchanged in the collision may justifiably be ignored, provided the collision is not too far from head on. As the collision becomes more glancing, the effects of angular momentum become more pronounced. If the coins are sufficiently smooth, they will slide on colliding, and again no angular momentum will be exchanged. However, if they stick, a tangential force

F will tend to rotate both coins in the same direction. This force may be replaced by a couple C plus a force F acting through the center of mass of the coin, such that C = Fr, where r is the radius of the coin, as shown in Fig.4 . The torque impulse leads to an angular velocity ω in the same sense for the two coins:

$$I\omega = Ct = Frt$$

where I, the moment of inertia of the coin is $\frac{1}{2}mr^2$ and t is the time for the collision. Since F is in the x direction, the change in momentum

$$mv_x = Ft, \text{ so } I\omega = mv_x r \text{ or } \omega r = wv_x \qquad (1)$$

where v_x is the change in velocity in the x direction during the collision.

To see how large this effect is in practice, we can employ the results of the experiment, shown in fig. 1. Let the angle of rotation of the struck coin be θ_2 before it stops, and of the incident coin θ_1. The frictional forces between the paper and the coins are independent of velocity, so there will be no net torque on the coins after the collision. Using the previous nomenclature.

$$d_1 = \frac{1}{2}at_1^2 = \frac{1}{2}\mu g\, t_1^2$$

$$d_2 = \frac{1}{2}\mu g\, t_2^2$$

where a is the deceleration and t_1, t_2 the time for the respective coins to stop after the collision

$$\theta_2 = \omega_2 t_2 = \omega_2 \sqrt{2d_2/\mu g}$$

$$\omega_2 = \theta_2 \sqrt{\mu g/2d_2}$$

Using the values from fig. 2, $\theta_2 = 130°$, $d_2 = 10.1$ cm and $\mu = .24$, we find $\omega_2 = .775$ rads/sec. From equation 1, $\omega_2 r = 2v_{2x} = .004$ m/sec.

This represents only five percent of v_{2y} and can be neglected to the degree of accuracy to which we are working.

It is possible to carry out the experiments using the transparent film on an overhead projector, drawing around the coins with a suitable marking pen. However, the coefficient of friction with the film is higher (about 0.4) and not as constant. The velocities at the time of collision may be estimated if the coefficient of sliding friction between the coin and the paper is known. This may easily be measured by tilting a pad of paper and dropping the coin at the top. If the coin slides barely all the way down without stopping, the tangent of the angle η the pad makes with the table is the value of the coefficient of friction. From figure 2, the reaction of the pad on the coin is mg cos η, and the force parallel to the pad is mg sin η, so $\mu = mg \sin\eta / mg \cos\eta = \tan\eta$. μ is generally about .24 between coins and paper, so traveling a distance of 10 cm after collision would require an initial velocity of

$\frac{1}{2} mv^2 = mg\mu d = m \times 9.81 \times .24 \times .1$

so $v = .7$ m/sec

or about 1.5 mph.

The angle can be measured using a ruler or a protractor, as shown in figure 2.

More accurate dynamics experiments can be obtained by placing the second coin right at the edge of the table, so that both it and the incident coin are knocked off the table onto the floor, as shown (fig. 3). A point directly beneath the coin at the table edge is marked on the floor, by dropping a coin there and seeing where it lands. A second coin is placed a few inches behind the first, and is propelled or flicked toward it in a direction normal to the edge of the table, as described earlier. The coins collide at some angle. Now, mark on the floor with pen, pencil or chalk where both struck the floor.

Make a diagram, as before, for the distances and angle. It is easier if two people mark on the floor where they see the coins fall, using chalk or pencil, and for permanence, use a large sheet of paper, such as a newspaper. Do this with two nickels, a nickel and a quarter, a nickel and a dime, etc. The masses of the various coins is given in experiment 1.01 p7

Qualitatively we learn:

1) When like coins collide, both fall off the table (except for a head-on collision) showing both travel forward.

If the collision were perfectly elastic, the angle between the trajectory of the coins would be 90°, and a head-on collision would leave one coin poised on the edge of the table. The mere fact that we hear a click as the coins collide shows that energy is being lost, and the coins travel off with an angle less than 90° between them.

2) When a heavy coin strikes a light one, both leave the table at considerable speed, even for a head-on collision.

3) When a light coin strikes a heavy one, the light coin stays on the table unless it delivers the heavy coin a glancing blow. Unlike (2) where the forward momentum is shared with the struck coin, here the struck coin carries the forward momentum with it, and the incident coin may recoil. Let us try to get a simple intuitive picture of this process. Imagine you are

sitting at the center of mass of the system, observing the two coins approaching you, the ratio of the speeds being inversely proportional to the masses, so the heavy coin approaches slowly, and the light coin rapidly. They collide where you are sitting, reverse, and go off with the same speed as before, apart from what is lost in the collision. The center of mass, where you are sitting, moves steadily forward during the whole of this process. Combining this center of mass motion with that of the coins about the center of mass, we see that after the collision, if the light coin is first, it has a fast forward motion superimposed on that of the center of mass, and will move rapidly forward, whereas the heavy coin moves forward sluggishly. If the heavy coin is first, the light coin will move backward, since its reverse speed is so much larger than that of the center of mass.

We can use our measurements to obtain a quantitative check on the theory. For example, if the table is h cm high (an average is about 74 cm) the coin takes a time $t = \sqrt{2\,h/g}$ to reach the floor, which is 0.39 seconds. If the horizontal velocity of the coin is v, it will travel a distance 0.39v horizontally before striking the floor. Since the horizontal component of momentum of the system parallel to the edge of the table is zero, the two coins, of masses m_1 and m_2 have equal and opposite momenta parallel to the edge. Hence

$m_1 u_1 + m_2 u_2 = 0$, where u_1 and u_2 are the respective velocities parallel to the edge of the table as shown in figure 4. Since $u_1 t = \ell_1$ and $u_2 t = \ell_2$, where ℓ_1 and ℓ_1 are the distances the coins travel in a direction parallel to the table edge, $m_1/m_2 = \ell_1/\ell_2$. Check whether this is true--it is independent of the elasticity of the collision since it depends on conservation of momentum. For an elastic collision between like coins, a rectangle can be drawn as shown, and $\ell_1 = \ell_2$. (Fig.3)

The coefficient of restitution e and the ratio of v_1/v_2 may be obtained from the angles, as for the coins sliding on the table. The values of v_1 and v_2 obtained directly from the impact points of the coins on the floor are more accurate than those obtained from sliding distances, because frictional forces vary somewhat.

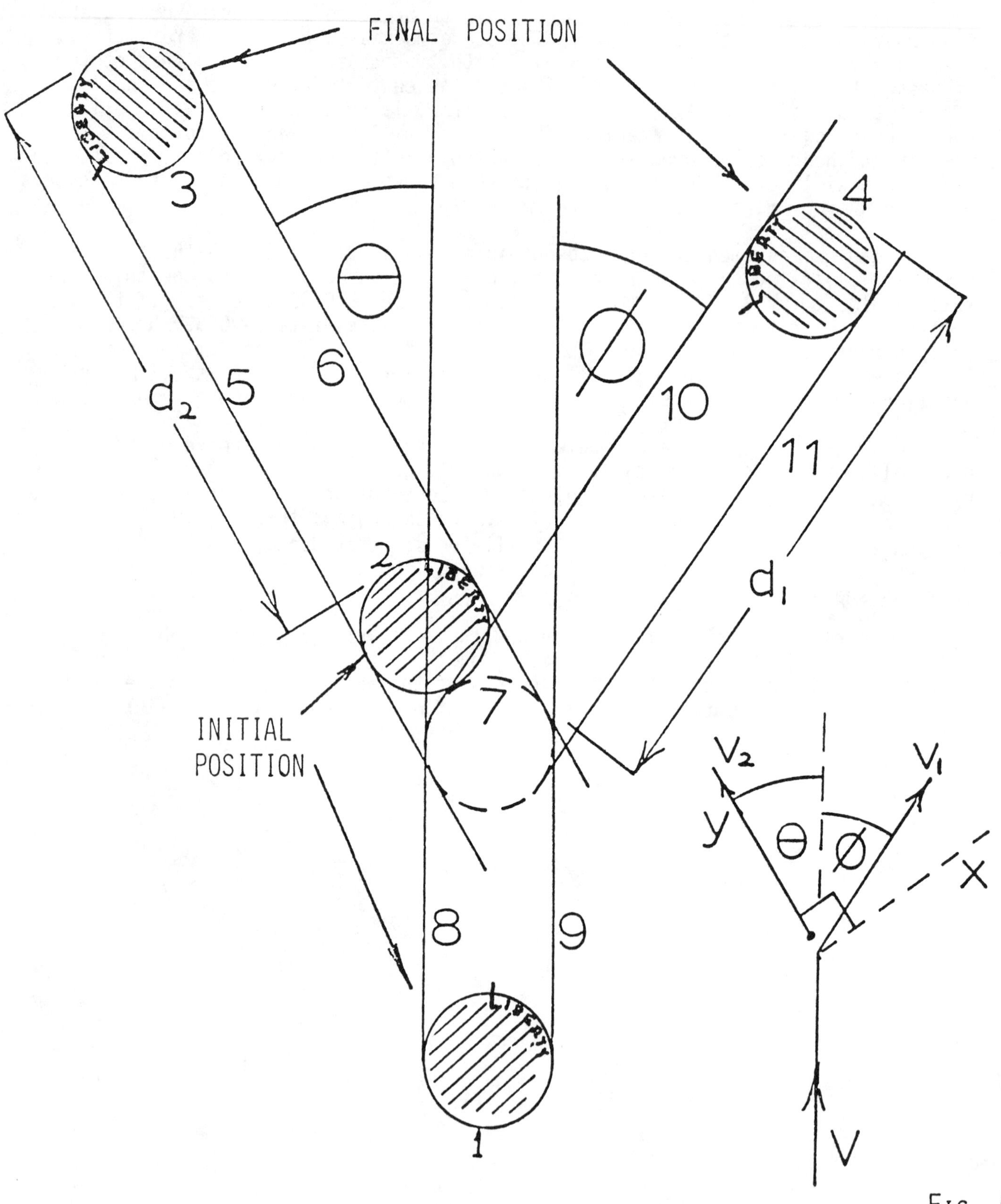
FINAL POSITION
INITIAL
POSITION
LIBERTY
1
2
3
4
5
6
7
8
9
10
11
d2
d1
Θ
Ø
V2
V1
y
x
V

FIG. 1

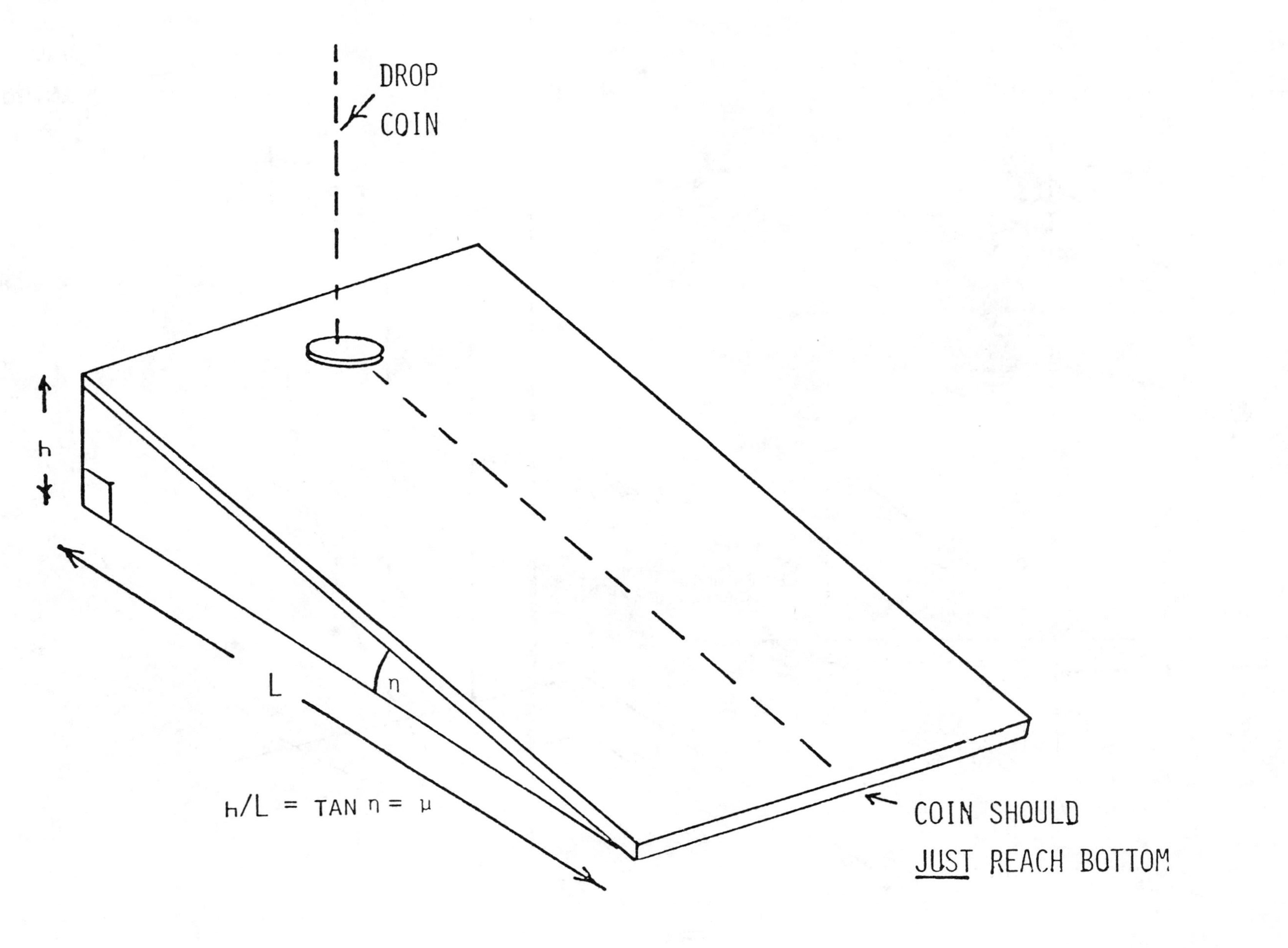

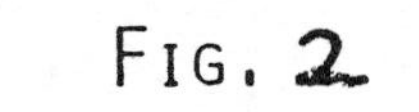

FIG. 2

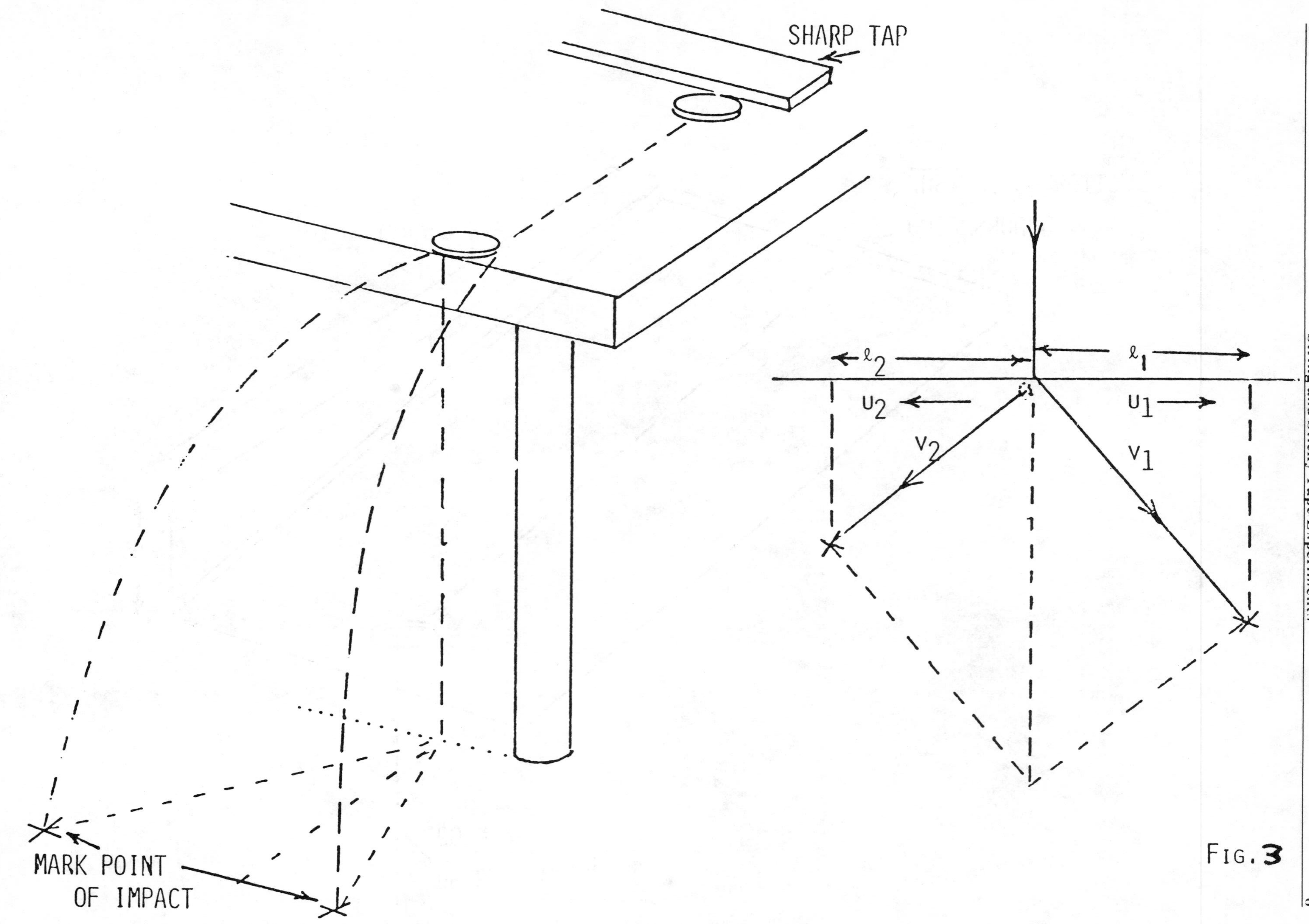

FIG. 3

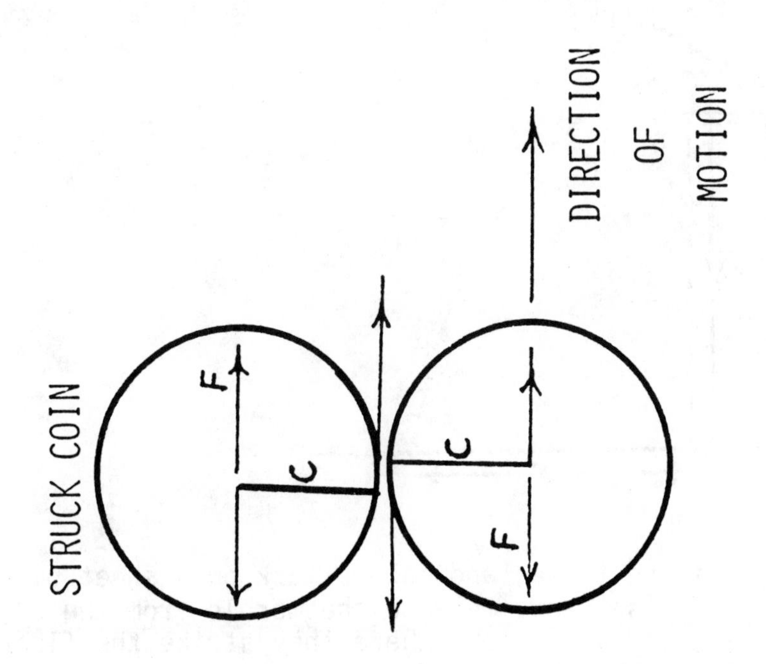

FIG 4

Experiment 1.25 Kinetic and Potential Energy

Materials: Ruler, two marbles, paper.

Procedure: We wish to study the relationship between potential and kinetic energy. Prop the ruler up against a book, as shown, so that the end is about two inches from the edge of the table, and the other end is about two inches above the table. Tape the lower end of the ruler to the table, to stop it from slipping.

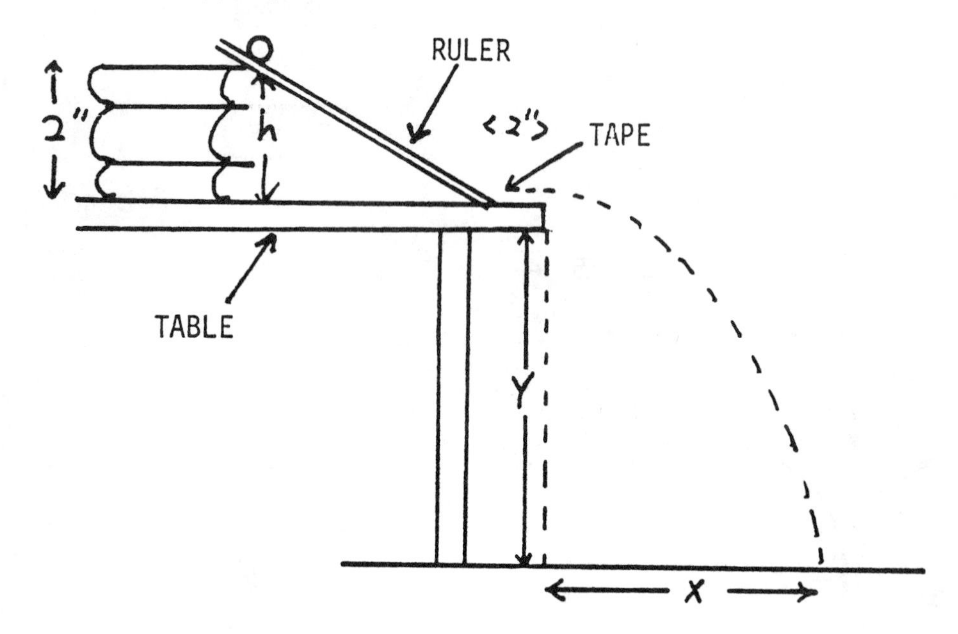

Drop a marble from the edge of the table, and make a mark on a sheet of paper taped to the floor, where it falls. Now, release the marble from the 3", 6", 9" and 12" marks on the ruler, and again note where they strike the floor, marking a sheet of paper taped to the floor. Repeat each measurement a few times, to check. You will need to know:

1. The height of the marble off the table.
2. The height of the table off the floor.
3. The distance the marble travels horizontally from the edge of the table to the point where it strikes the floor. i.e. the distance marked x on the diagram.

Qualitative: The marble has potential energy as it is held at the top. This potential energy is converted into kinetic energy as the marble rolls down the ruler and along the table. The marble leaves the table horizontally, so it has no vertical motion. Since we can separate vertical and horizontal motion, it always takes the same time to reach the floor from the table.

The potential energy increases as we move the marble up the ruler. This gives a larger kinetic energy, and the additional horizontal velocity resulting makes the marble travel farther horizontally before striking the floor.

Quantitative: The potential energy of the marble above the table is mgh. The kinetic energy of the marble as it leaves the table is equal to this

$$mgh = \frac{1}{2} m v^2$$

so the horizontal velocity is $v = \sqrt{2gh}$
where g is the acceleration due to gravity = 981 cm/sec^2 or 32 ft/sec^2.

The time taken to strike the floor from the table, since initially there is no vertical motion, is given by

$$y = \frac{1}{2} gt^2$$

now $x = v_x t$

so $v_x = \sqrt{2gh}$, $t = \sqrt{\frac{2y}{g}}$

$$x = 2 \sqrt{hy}$$

$$x^2 = 4 hy$$

Calculate x from h and y. Do the calculated and experimental values agree? Is the measured value for x larger or smaller than that calculated? Why?

Experiment 1.26 Inertia (Newton's 1st Law)

Materials needed - Dixie cup, card (or paper), piece of chalk or a coin.

Procedure: Place the coin or chalk on the card on the top of the cup, as shown. With the forefinger, tap the card sharply. It will slide off the top, and the coin will fall into the cup. This is a less drastic version of the table-cloth trick, where the cloth is pulled from under a set table (always pull the cloth downward).

Qualitative: "A body continues in its state of rest or uniform motion, unless acted upon by a force". In this case, negligible force acting on the coin when the card is removed, so its inertia keeps it where it is.

Experiment 1.27 Action and Reaction

Materials: Straws, marbles, rubber band, pencil, scissors, sticky tape

Procedure: Take two marbles, and attach each one to the end of a straw, using sticky tape. Cut a very small notch at the other end of the straw. Now place a

rubber band in the notch, stretch it, and place the notch of the other straw in the other end of the loop of the band, so it appears as shown.

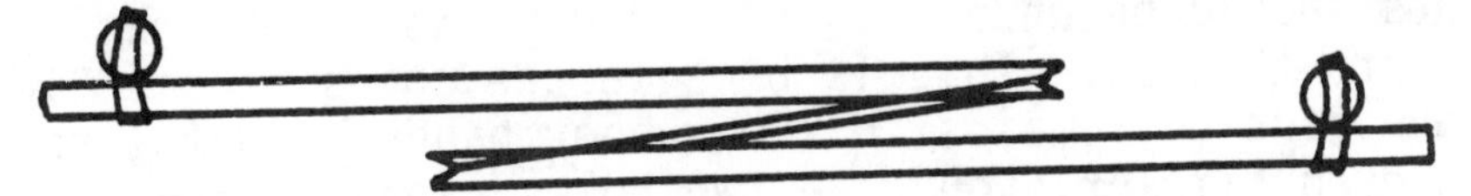

Place the system on a smooth table, or a smooth floor, and hold it in the loaded condition using a pencil. Suddenly release the pencil--the marbles will fly

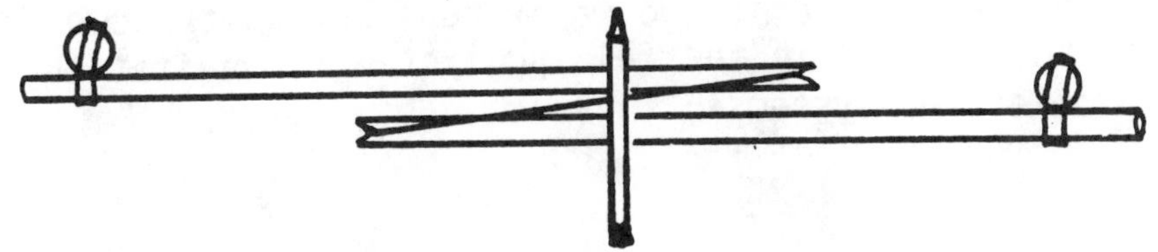

apart and travel across the floor. Measure the distance they travel. Do this for several tensions in the rubber band. Now attach a second marble to the end of one straw, and repeat the experiment.

Do not try to hold the straws down with the fingers--the fingers are too sticky and flexible to release straws at exactly the same time. It is possible to hold the straws in the tense position with a strip of cellotape, which is gradually pulled off by the rubber band until the system springs apart on its own. However, some effort is required to find the right piece of tape which is not too small that the band releases too early, nor too big so it never releases at all.

You can make a mark on the straw if you want to repeat the experiment with the same tension; draw the band to the same mark.

Qualitative

Questions: With just one marble on each straw, is the distance each travels the same? Why?

With two marbles, is the distance the same? Why?

Quantitative: The distance traveled is not exactly equal each time. Why? Is the ratio of distances more nearly one under high or low tension? Why?

With the two marbles on one straw, and one on the other, is one distance twice the other? Why? Is the ratio of the distance constant, independent of rubber band tension?

Experiment 1.28 Weightlessness

Materials: rubber bands, two coins, paper clip, tape, water

Method: It is difficult for students to understand the apparent absence of force in free fall. Here are two simple experiments

Two little weights are hung by rubber bands over the edge of a styrofoam cup. The weights used were small chemical balance weights, but instead, coins such as quarters or half dollars can be taped to rubber bands, leaving about an inch of free tape as shown in Fig. 1. The other ends of two rubber bands go through a hole in the bottom of the cup, and are held there by a paper clip. The bands should be under light tension. Now drop this system, and the weights hop into the cup. This is because the coins become "weightless" when dropped while the tension remains in the rubber bands. In the second experiment, which is even simpler, you poke two holes close to the bottom of a paper cup, as shown. The cup is filled with water, which pours out of the two holes. Now climb a step ladder, hold the cup high at arms length, and drop it, preferably into a trash can. No water runs out as the cup falls--again, the water is "weightless" in the falling system. Gravitational forces are absent in the accelerated system.

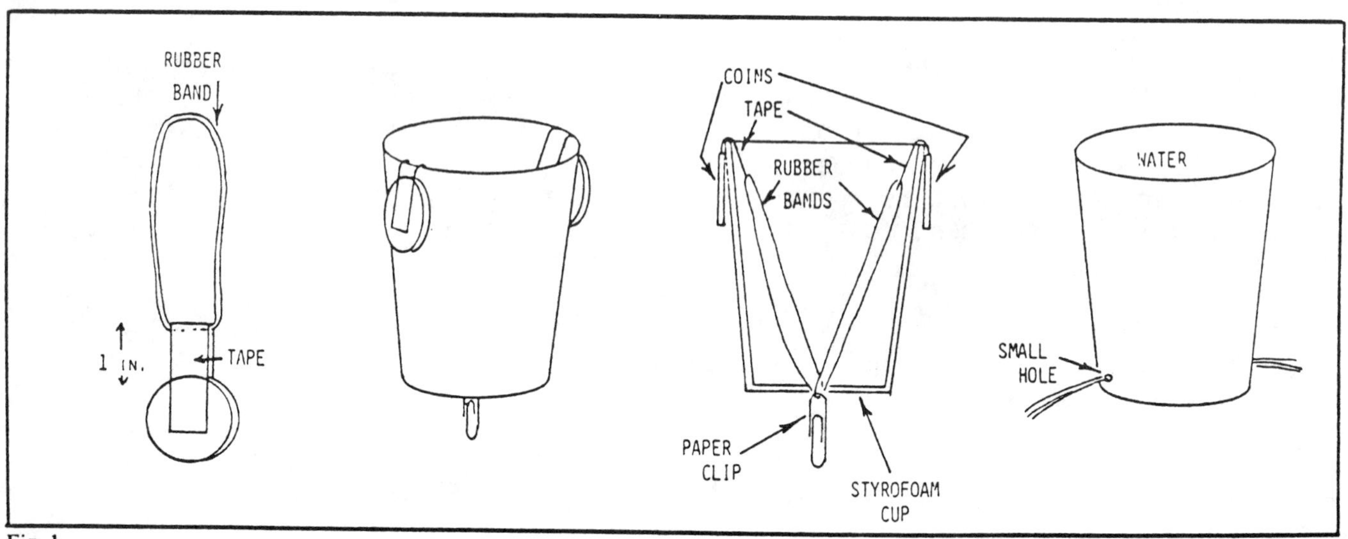

Fig. 1.

Experiment 1.29 Atwood's machine

Materials required: Two drinking straws, two foam or paper cups, marbles, string.

Procedure: Attach the cups to the ends of the string, as shown, and fasten the straws with tape to the backs of two chairs. Now, put 6 marbles

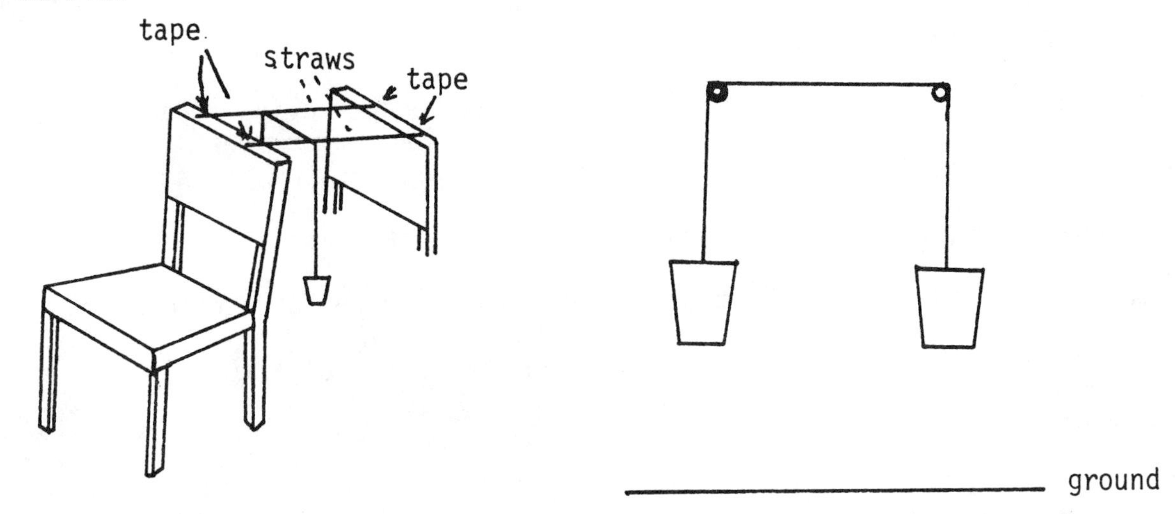

in one cup, and four in the other, and release the cups. Notice the heavier one moves faster and faster till it strikes the floor. Add marbles to the heavier cup, and again release.

Qualitative: The greater the difference in number of marbles between the two cups, the greater the acceleration - if the number of marbles is equal, there is no acceleration.

Quantitative: The mass being accelerated is the sum of the masses in the two cups, but the force accelerating them depends on the difference between the masses. Hence the acceleration is less than g, the acceleration due to gravity, by $g\frac{(n_1 - n_2)}{(n_1 + n_2)}$ where n_1 and n_2 are the number of marbles in the two cups. If you have a watch, time how long it takes the cup to reach the floor, with different numbers of marbles in the two cups. Measure the distance to the floor. Calculate the acceleration from the formula.

$$\text{distance to floor} = \frac{1}{2}\,\text{acceleration} \times (\text{time})^2$$

Use the previous formula to calculate g.

Note: There is considerable friction between the string and the straws. To reduce this, it is recommended that the pulley described on the next page be employed in place of the straws.

An even simpler version of this experiment uses paper clips. Hang two paper clips by a thread over a straw as shown, then add clips to the other side until, when set going, they just continue to move. Now add a few more clips to this side, and measure the time to reach the floor. The accelerating force is due to the added paper clips, or mg, and the mass accelerated is the total number of clips.

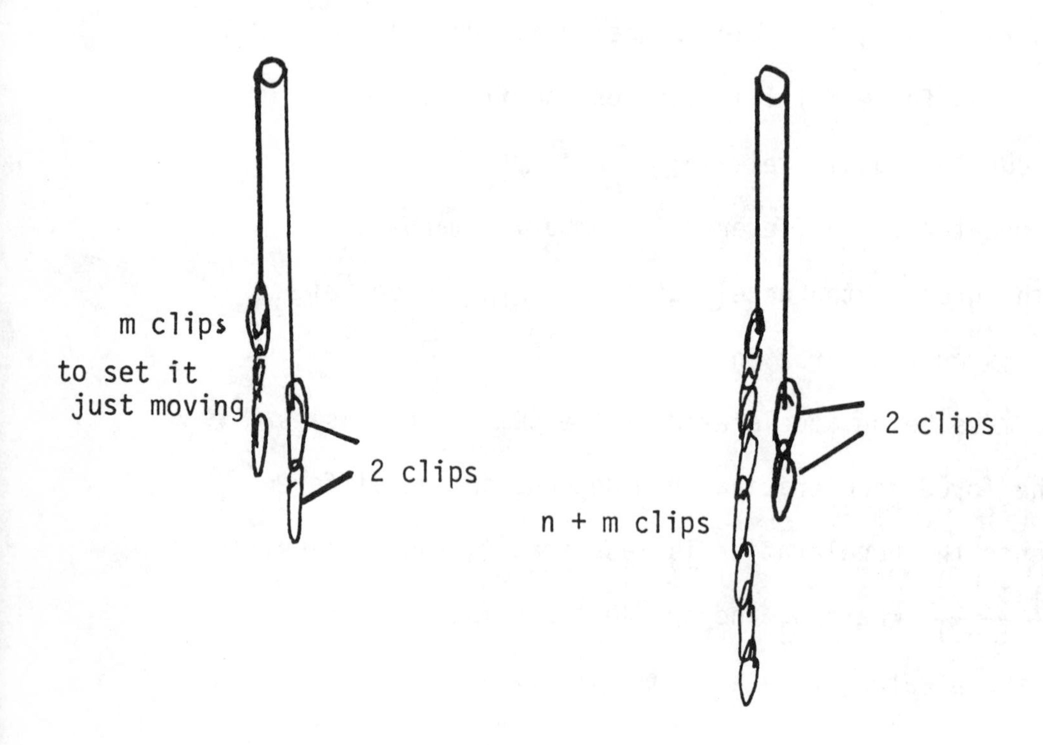

Construction of an (approximately) massless, friction free pulley.

Take two styrofoam cups, and poke a pencil point through the exact centers of the bottoms

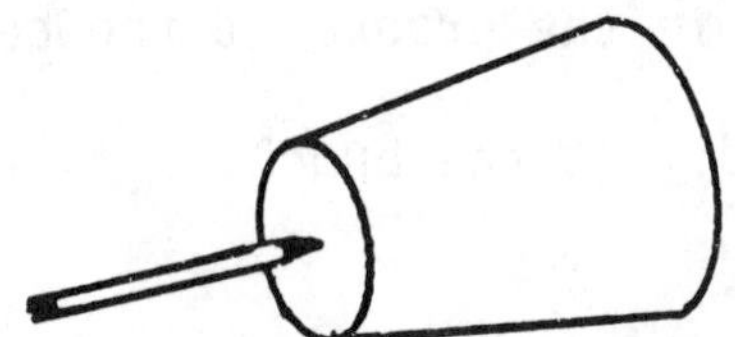

Now, push a straw through the bottom of one from the outside, and the bottom of the other through the inside, so the the two open ends face.

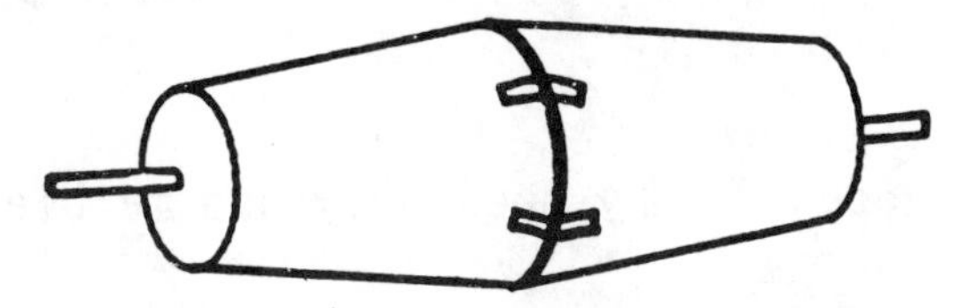

Tape the two faces tightly together with several pieces of sticky tape

Now, cut the bottom off another cup and slide it over one of the two.

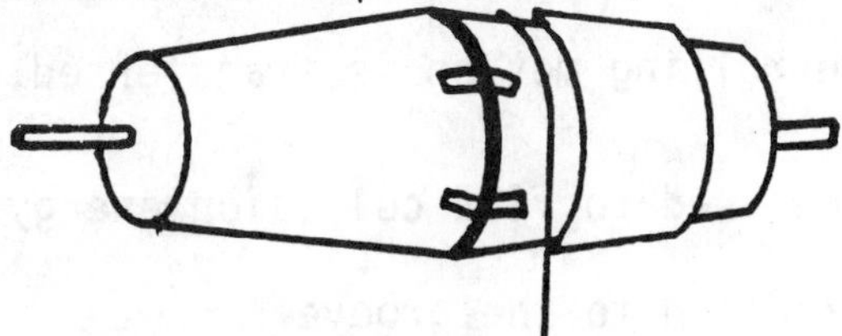

The pulley string runs between the two cups shown, and the straw is supported between the backs of two chairs, two tables, two stools, two desks, or anything convenient. The straw rolls, so that friction is very low, and the mass of the two and a half cups is very small indeed.

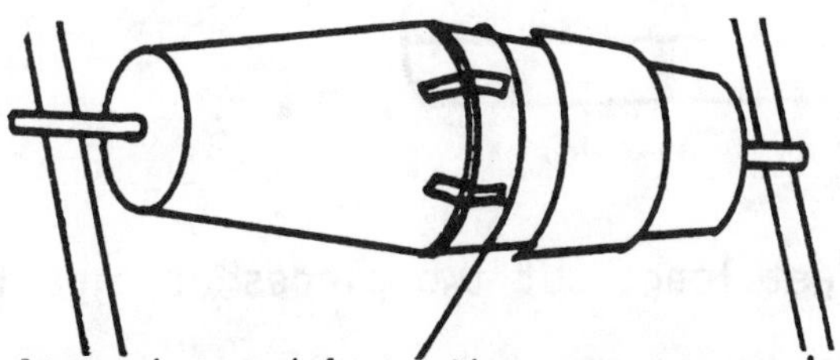

Using this for Atwoods machine, the cups carrying marbles must have their tops cut off, or they will bump.

Experiment 1.30 Collision of spheres -

Materials: marbles, string, tape, ruler

Procedure: Place five marbles in the groove at the center of the ruler and flick a sixth rapidly at the bunch.

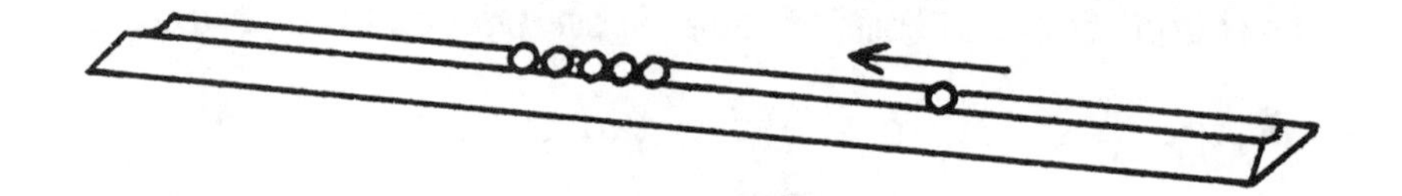

Qualitative question: What happens? Why does only one marble move?

If you roll the odd marble slowly down the groove, it may not stop dead on hitting the others, especially if there are only one or two and not five. Why? It is because the linear, but not the rolling motion is transferred.

This simple experiment cannot be used to find collision energy loss, since too much energy is dissipated in friction to the groove.

To study collisions between two marbles, a ballistic pendulum is used.

Take a straw, and cut slits at each end one inch deep

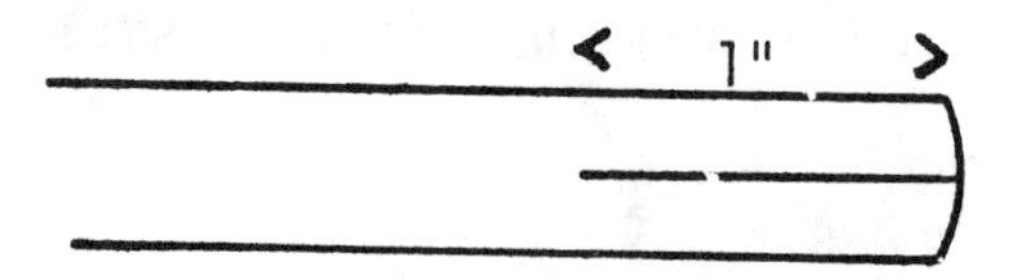

Cut two lengths of string four feet long, cut two pieces of tape 1 1/2" long. Stick the center of the string along the center of the tape.

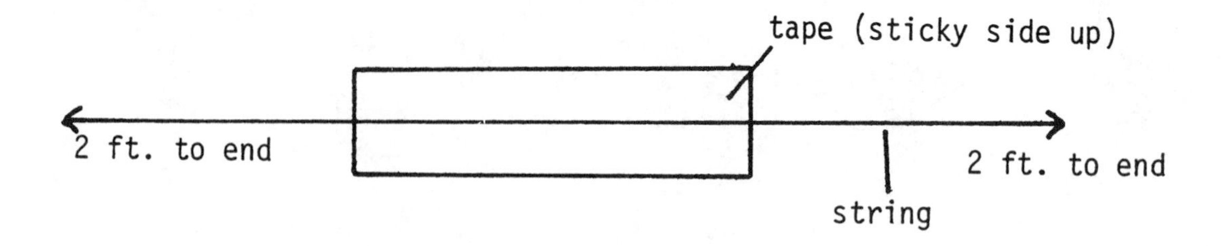

Now place a marble at the center of the tape, and press the tape to hold the marble like a stirrup

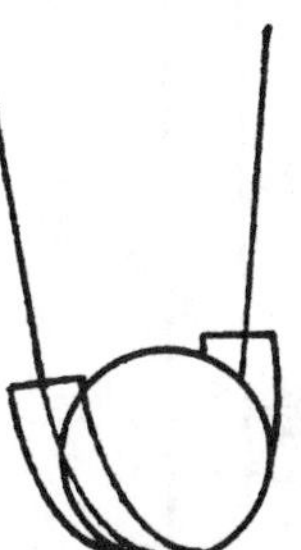

put the ends of the string in the ends of the straw, so there are two pendulums as shown below

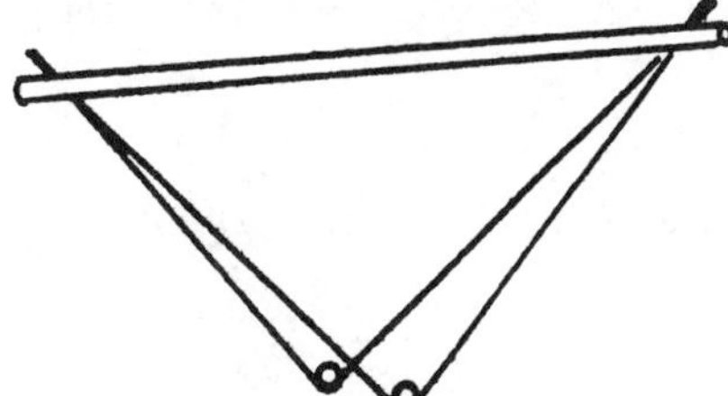

Now comes the tricky part. Carefully pull the strings through the slits until they are of approximately the same length. It is important that, when the two marbles collide, they must hit one another as close as possible to dead center. Attach the ends of the straws, using tape, to the backs of two chairs, or tables, at the same level. Hold the ruler so that

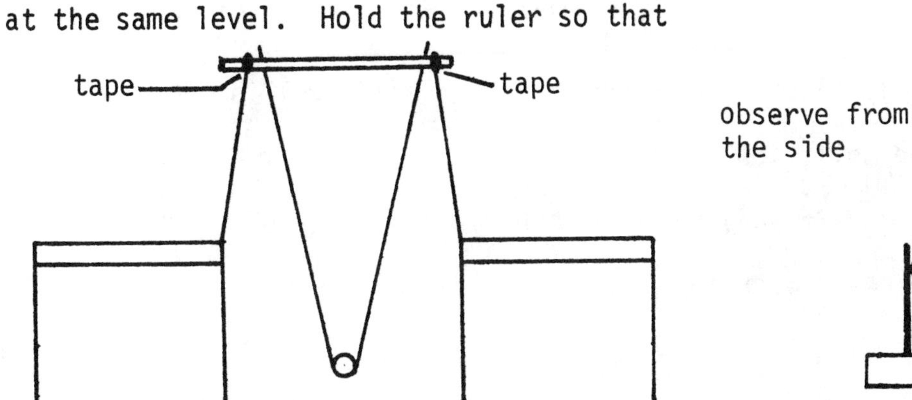

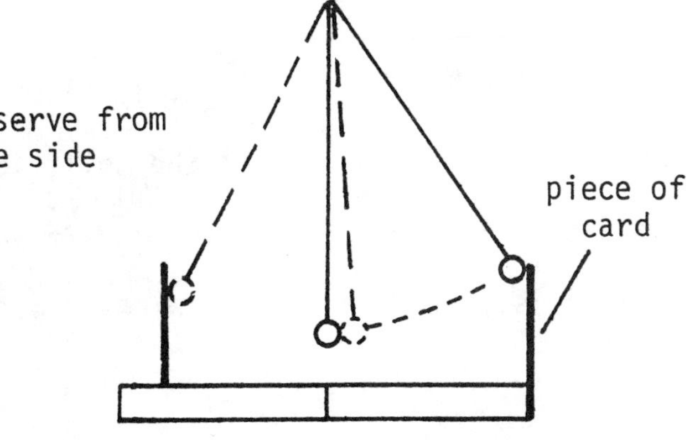

the 6" mark lies between the two marbles. Pull one back to 0 and release. How far does the other one travel? Make several experiments, and measure this as accurately as possible.

Qualitative: Why does the struck marble not travel as far, or rise as high as the incident marble? Where does the energy go? Are there frictional losses?

Quantitative:

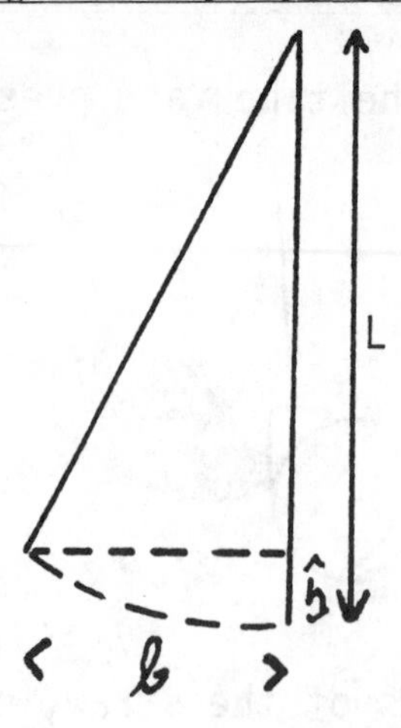

The incident marble drops a distance h before colliding. From geometry

$$(2L-h)h = \ell^2 \text{ or } 2Lh = \ell^2$$

Now, the kinetic energy just before collision = $1/2\ mv^2$ = mgh, the original potential energy $\frac{1}{2} mv^2 = mg\frac{\ell^2}{2L}$ or $v = \ell\sqrt{\frac{g}{L}}$

After collision, the ball travels a distance ℓ' where $v' = \ell'\sqrt{\frac{g}{\ell}}$

so $\frac{\text{velocity before collision}}{\text{velocity after collision}} = \frac{\ell \text{ before collision}}{\ell' \text{ second marble after collision}}$

if the first ball is reasonably stationary after collision. The ratio of the relative velocity of the spheres before and after impact is called the coefficient of restitution, R and measures how inelastic the collision is.

What would it be for a perfectly elastic, and perfectly inelactic collision?

How much energy is lost in the collision?

Note that if the string is 61cm (2ft)long,and the marble's mass is 5 gms.,the initial stored energy on drawing it back 15.24 cm. (i.e. 6") is

$$5 \times 981 \times \frac{(15.24)^2}{2 \times 61} = 9{,}344 \text{ dynes} \quad (0.09344 \text{ N})$$

The energy lost is then $9{,}344\ (1 - R)^2$ ergs. $(9.344\times10^{-4}(1-R)^2 \text{ J})$

Experiment 1.31 The Ballistic Pendulum

Materials: tape, marbles, string, rubber band, paper clip, cups

The ballistic pendulum is used to measure the speed of high speed projectiles by shooting the light projectile into a heavy hanging mass.

Procedure: Put about twenty five marbles in a cup, place another cup on top, and tape the two together, wrapping the tape all around, leaving a space at the top a little less than half to shoot in a marble. Tuck some toilet paper in the outer cup to stop the incoming marble. Hang this setup by two parallel strings from a table or chair back at an angle of about 45°.

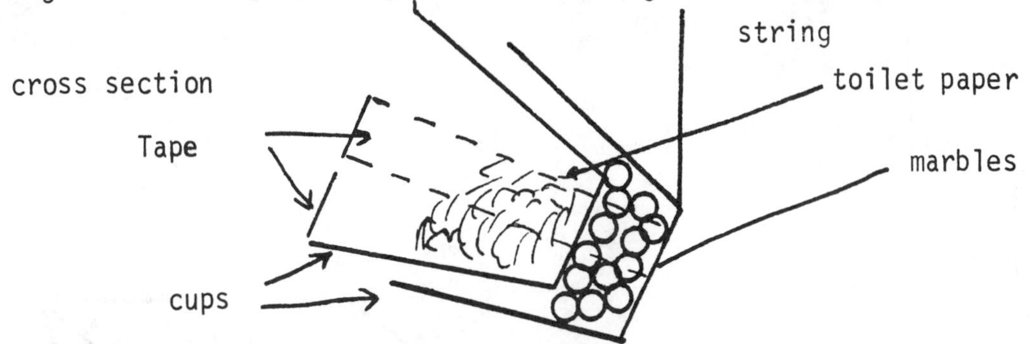

Now, bend a paper clip (as below) to hold a marble, and attach a rubber band so the marble can be shot.

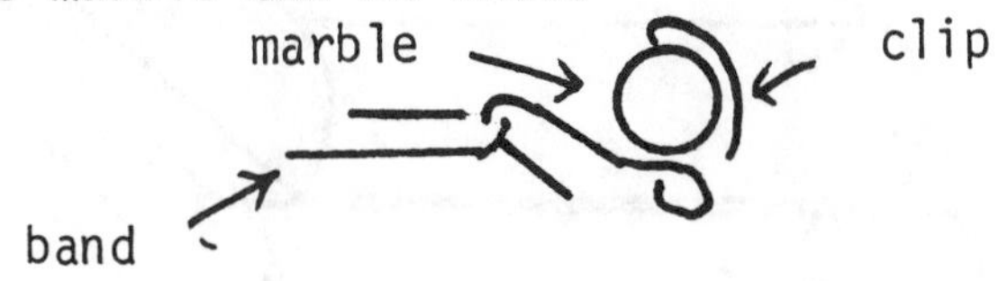

The other end of the band can be held by another clip, and the marble catapulted into the cup. It is easiest to shoot off a pile of books, as shown. The distance the cups travel is measured by a ruler underneath. You can fold a paper indicator to be pushed by the cups.
Measure the speed for different, measured stretches of the rubber band.

The analysis is given in experiment 1.27.

If L is the length of the string, ℓ the distance the cup travels, the initial velocity of the cup is

$$v_c = \ell\sqrt{\frac{g}{L}}$$

The initial velocity of the projectile, one marble, with twenty five in the cup, is

$$v_p = 26\ell\sqrt{\frac{g}{L}}$$

It is told the famous physicist, Lord Kelvin, used to perform the experiment by shooting an old, large-bore rifle into a suspended block of wood. One day he missed, and the bullet travelled through the wall, almost hitting the lecturer next door. Kelvin dashed round to find the petrified lecturer, and the class crying, "You missed, you missed. Try again."

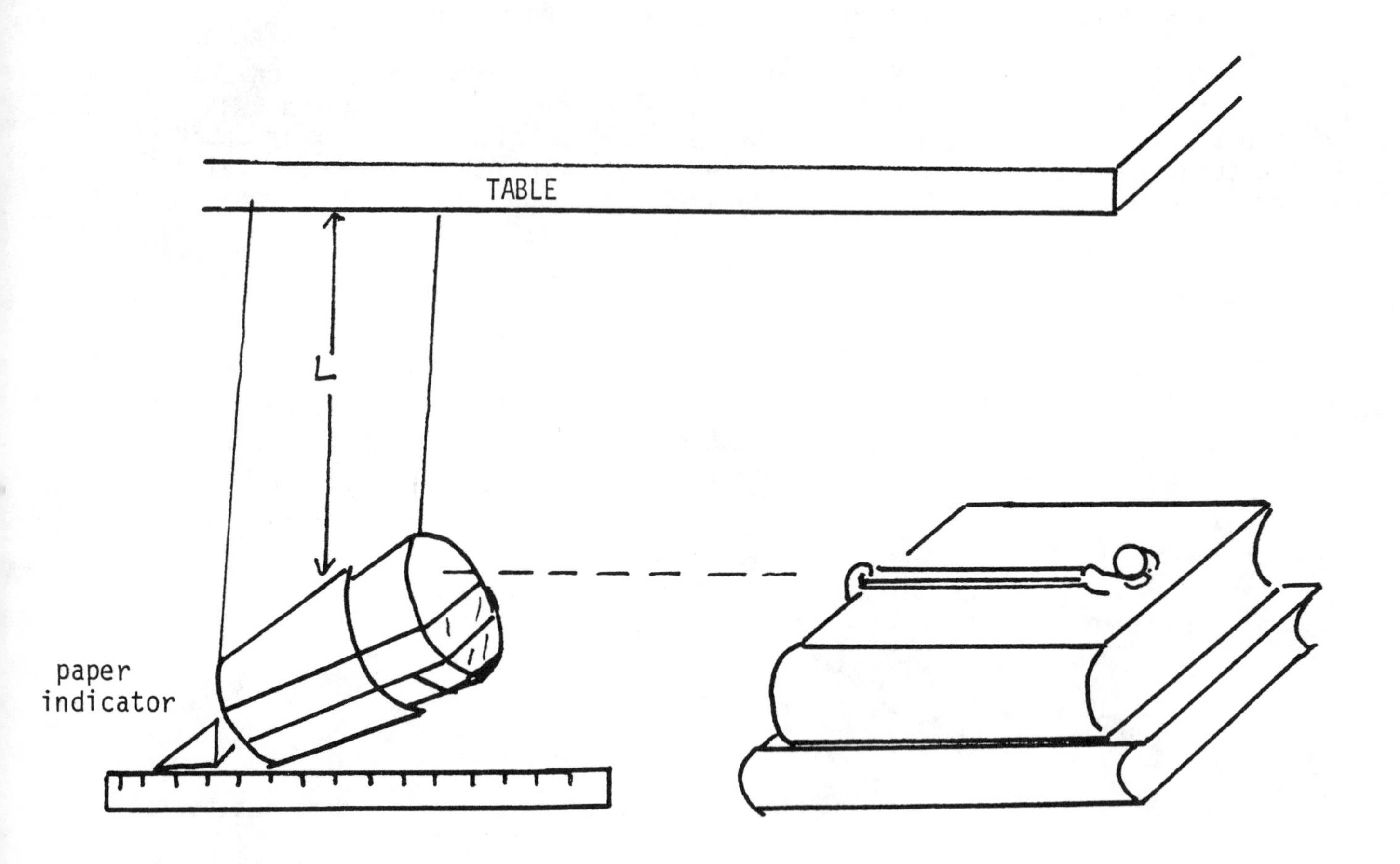

Experiment 1.32 Central Forces

Materials: - Soda straw, string, several marbles, sticky tape, cup

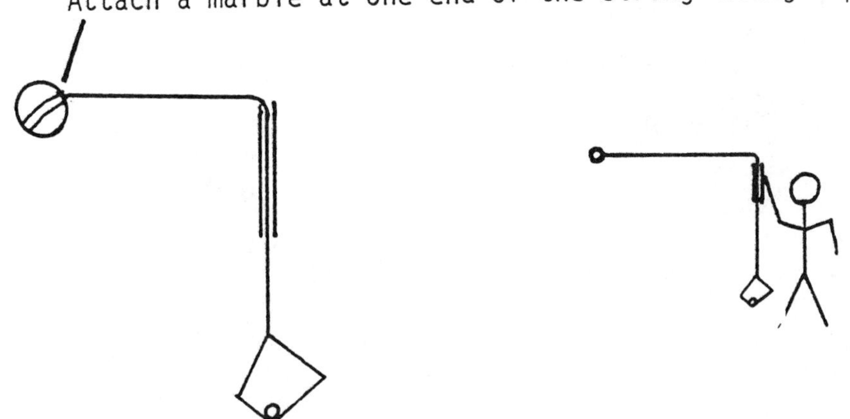

Attach a marble at one end of the string using tape.

Pass the string through a short section of soda straw, and attach a foam or paper cup at the other end, leaving about 4 feet of string between. Now whirl the marble around your head holding the straw. Place five marbles in the cup.

The force pulling the rotating marble inward is provided by gravity acting on the marbles in the cup. This is a constant force. By adjusting the speed of rotation, keep the marble at a fixed short distance from the center, and then allow it to move outwards, again holding it at constant speed. Put several marbles in the cup as a counterweight, and repeat the experiment. You will find it is easier to allow the marble to move out wards-friction raises problems in returning to the center.

What do you learn?

Qualitatively: -

The farther out the marble is, the slower you need to swing it around, to stop the counterweight pulling it inwards, so fewer rotations per second are needed to balance the couterweight. Putting more marbles in the cup makes you swing the marble round faster if you don't want it pulled inwards.

Quantitative: - Make a mark at a fixed distance, Say three feet, from the marble, using a marking pen, and count the number of rotations per second, with, say, five marbles in the paper cup, to hold the mark just above the straw. The acceleration for an object moving uniformly in a circle is given by acceleration = $\frac{v^2}{r}$ where v is the velocity of the object, and r is the radius of the circle in which it moves. The force providing this acceleration is supplied by the weight of the five marbles, so we have

$$5\ mg = \frac{mv^2}{r}$$

$$\text{or } r = \frac{v^2}{5g}$$

Suppose you count n rotations per second. (count for 20 seconds and divide by twenty). We may calculate n as follows $v = 2\pi rn$ then $r = 3 = \frac{(2\pi \times 3 \times n)^2}{5 \times 32}$ or n = 1.12 revs/sec.

You can easily check to find if this formula is correct by timing the rotations with difference radii of swing. You should find

$$g = \frac{(2\pi rn)^2}{Nr} \quad \text{ie. } n = \sqrt{Nrg}/2\pi r$$

when N is the number of marbles in the cup
g is the acceleration due to gravity (9.81 m/s^2)
n is the number of rotations per second
r is the radius of swing

Experiment 1.33 Moment of Inertia

Materials required: - pulley, straw, marbles

Procedure: - The experiment is to investigate the behavior of a rotational system set spinning by a weight. Make holes through the cup forming part of the "pulley" discussed in experiment 29 , and push a straw through as shown. Attach two marbles at each end

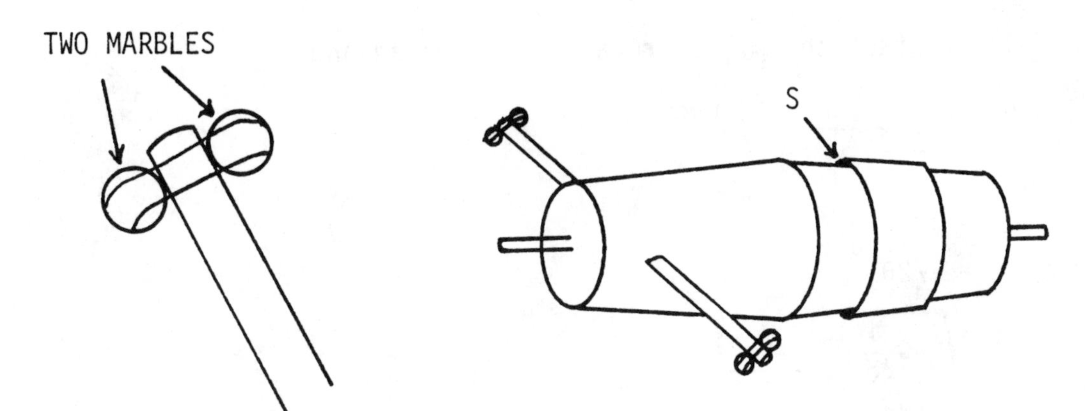

with tape, then adjust the straw back and forth until it balances.

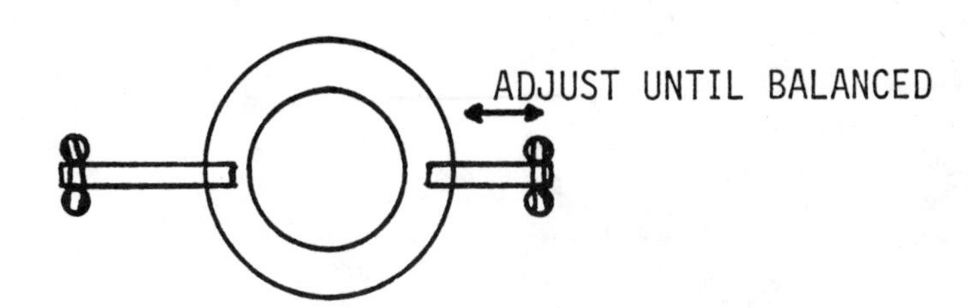

Fasten a string with tape to s, wrap around the groove, then hang a cup from the end. Time how long it takes the cup to reach to the floor with a marble in it.

Qualitative Question:

Explain why the dumbbell shaped device slows the acceleration of the cup, and the falling weights.

Quantitative Question:

If m is the mass of a marble, the moment of inertia of the dumbbell about the center is $4\,m\left(\frac{\ell}{2}\right)^2$ when ℓ is the length of the straw (generally 8 inches = 20.32 cm).

With one marble in the cup, the couple is

$$mgr = m \times 981 \times 3.8$$

So the total moment of inertia of the system is

$$I = 4\ m\ (10.16)^2 + m \times 3.82$$

$$\text{Couple} = I\ \ddot{\theta}\ (\ddot{\theta} = \text{angular acceleration})$$

$$\text{so } \ddot{\theta} = \frac{\text{couple}}{I} = \frac{981 \times 3.8}{4(10.16)^2 + 3.82} = 8.723 \text{ rads/sec}^2$$

To fall two feet the pulley rotates through an angle

$$\theta = \frac{24 \times 2.56}{3.8} = 16 \text{ rads}$$

in time t

$$\text{where } \theta = 1/2\ddot{\theta}t^2$$

$$t = \sqrt{\frac{2 \times 16}{8.723}} = \underline{\underline{1.91 \text{ sec}}}$$

What do you get?

Calculate the radius of gyration from the time it takes for the weight to fall.

Experiment 1.34. Moments of inertia of books and straws

Equipment: paper clips, straws, book tape

Method: The concept of inertial mass is often difficult to express. Generally I do it by talking about the difference between kicking a football and a rock of the same size. The resistance to motion of the rock is obviously greater. The moment of inertia is also difficult to conceptualize. A simple way to demonstrate moment of inertia uses two soda straws. Put a paper clip in each end of one, then straighten out the wide end of two other paper clips, and push the narrow ends into the middle of the second straw, using a third opened-up clip. Hold the straw between the thumb and forefinger, as shown in Fig. 1, and rotate it to and fro. There is very little resistance to rotation with the second one. Do the same with the first straw, and you will feel a strong resistance to rotation. The mass of the two straws is the same, but the one with the clips in the middle has a much smaller moment of inertia.

A good quantitative experiment measuring the moment of inertia of a book requires, in addition, only sticky tape and a watch. Tape the book closed (Fig. 2a). Now, support it by a length of sticky tape about 9 in. long from the edge of the table as shown in Fig. 2b. Start it into a twisting oscillation about the vertical axis of the tape and measure the period by timing ten oscillations. Do the same from the second edge (Fig. 2c). To measure it about an axis perpendicular to the others through the center, pass a piece of tape around the book as shown in Fig. 3. Measure the period about the third axis. If the same length of tape is used for the support, and the mass of the book is fixed, the restoring torque for a given angle of rotation is constant. This means the square of the period is proportional to the moment of inertia. If the length of the sides of the book are a and b, the ratio of the periods about the two edges and the center should be as

$$a : b : \sqrt{a^2 + b^2}$$

since the book may be regarded as a rectangular sheet. Results accurate to about 5% generally ensue from this experiment.

Using 3/4 in. tape, we can assume approximately that the book is supported by the two edges of the tape as shown in Fig. 4. The restoring torque as the book is twisted is provided by the fact that as the tape turns, the book is lifted. If the book is rotated by Θ, the restoring torque is Fd. F is the horizontal component of the force acting along the edge of the twisted tape. Since the total force acting along each edge is approximately $Mg/2$, this component is $(Mg/2) \sin \phi \approx Mg\phi/2$. Geometrically we have $\theta d/2 = \phi \ell$.

Thus: $F = (Mg/2)\,\phi = (Mg/2)\,(d/2\ell)\theta = Mg\theta d/4\ell$

The equation for the rotational simple harmonic motion is

$$I d^2\theta/dt^2 = Mg\theta d^2/4\ell$$

leading to a period $T = 2\pi\sqrt{4\ell I/Mgd^2}$

Now, $I = M\rho^2$, where ρ is the radius of gyration,

so $T = 2\pi\sqrt{4\ell\rho^2/gd^2} = 2\pi\rho\sqrt{4\ell/gd^2}$

By measuring T, you can find ρ and hence calculate I. After you have measured the moment of inertia of the book, you can try an interesting experiment on stability. This works best with a book whose long and short sides differ considerably. With the book taped shut toss it in the air so that it rotates about the long axis as shown in Fig. 5a. It will spin quite stably. Now hold it between two palms and spin it into the air (Fig. 5b). Again, it will spin stably. Lastly, toss it to spin parallel to the short side of the book (Fig. 5c). This proves impossible. Instead, the book twists and turns about no one axis. Whereas the first two axes are ones of minimum and maximum moment of inertia, and hence, stable, the third is not. It seems so intuitively obvious that it will spin stably about the third axis that it is very easy to win bets with this trick.

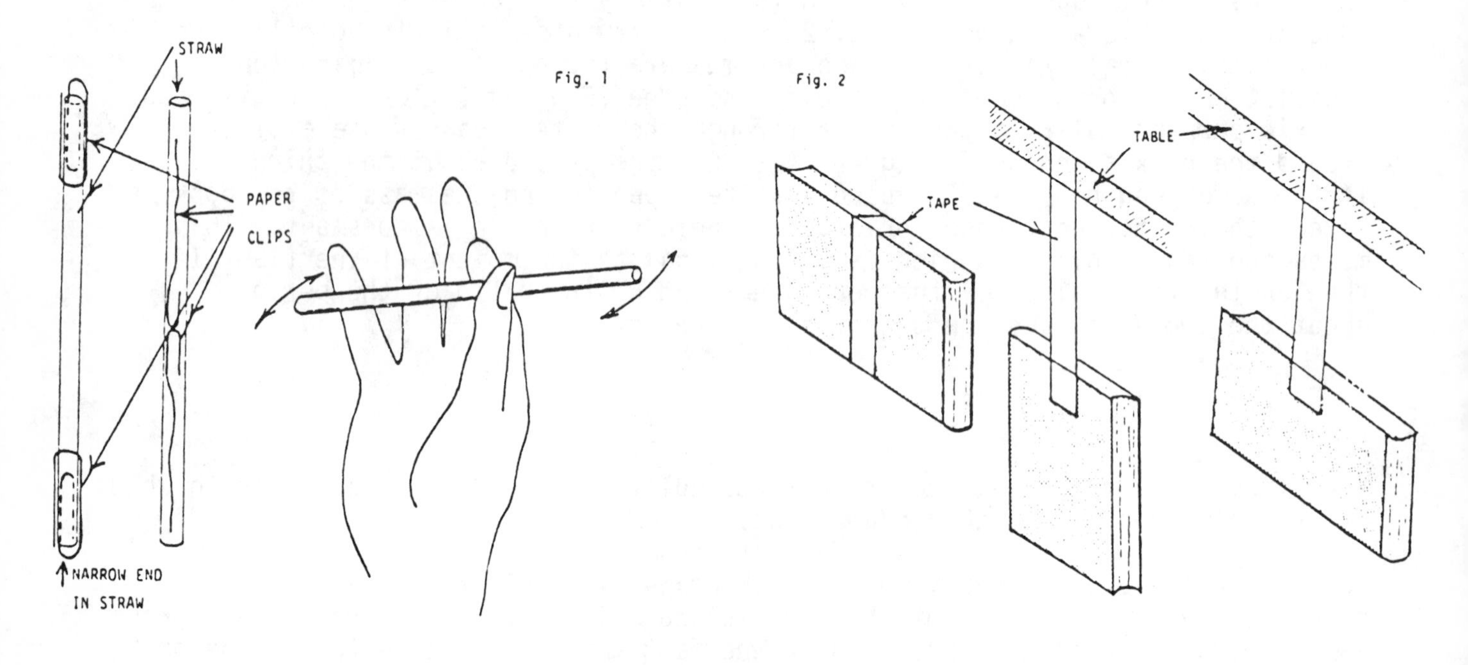

Fig. 1
Fig. 2

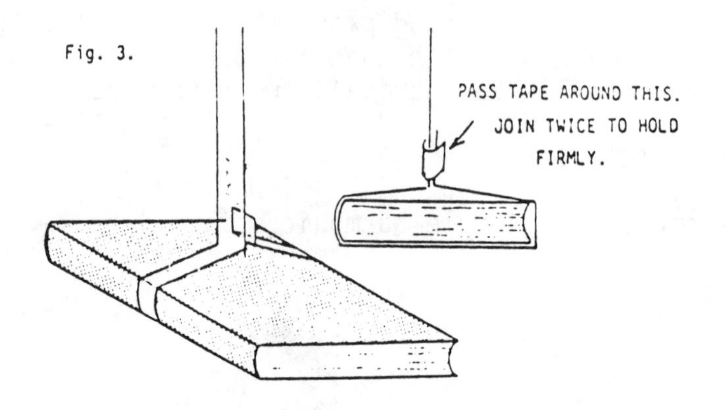

Fig. 3.

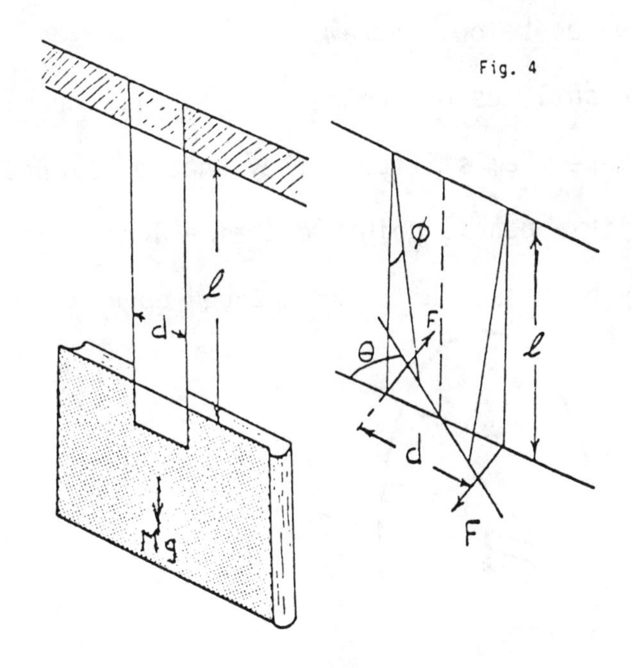
Fig. 4
ℓ
d
Mg
φ
θ
F
ℓ
d
F

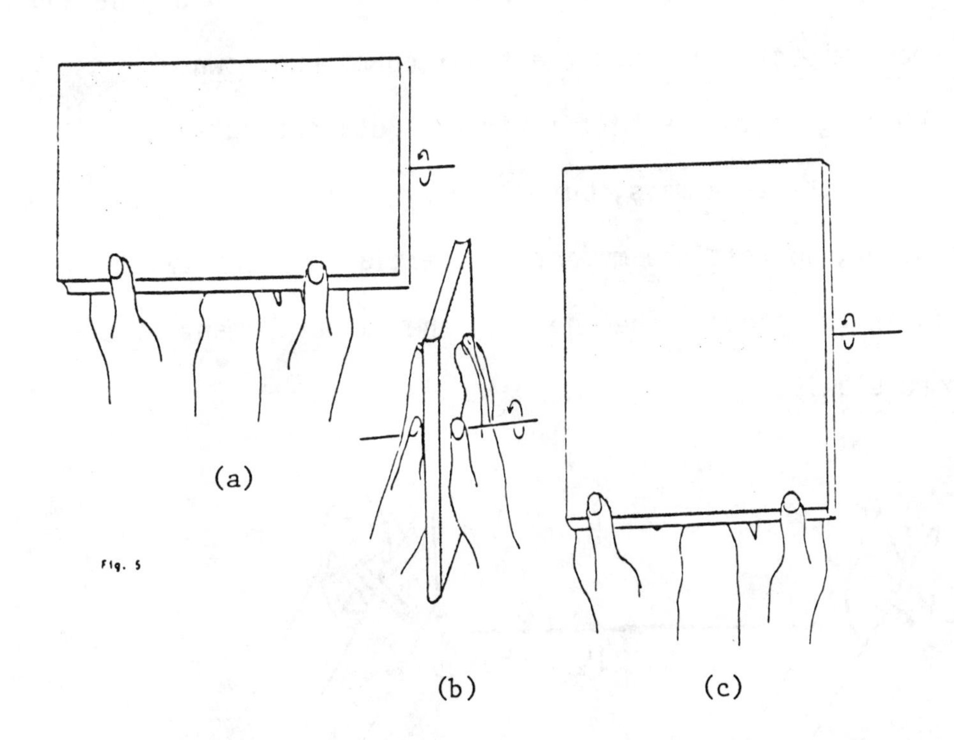
(a)
Fig. 5
(b)
(c)

Experiment 1.35 Moments of Inertia

Materials: - Cut outs provided below. Straw.

Procedure - Cut out the two outlines on the next page. If you wish, you can paste them on card to make them stiffer. They should balance on the 0 which can be punched out with a pencil point so that a length of straw about six inches long may be passed half way through as shown below

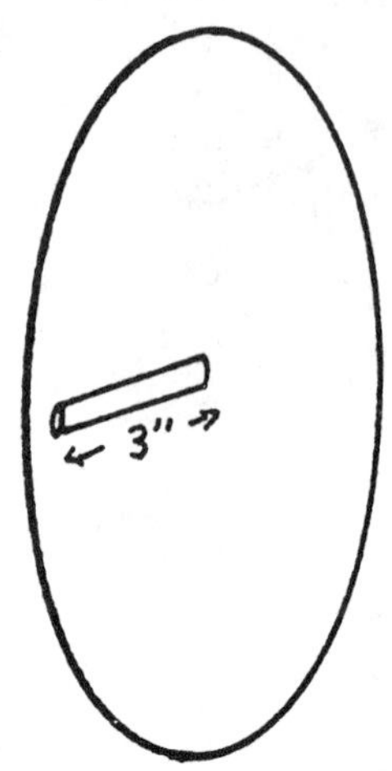

Now, fold two 8½ by 11 sheets and tilt them at the same angle (one end should be about 1½" above the other). Let the two figures run down.

Qualitative: Which gets to the bottom first? Both cut outs have the same area, and therefore the same mass, but the mass is more spread out in one case than the other, so that the moment of inertia is about twice in one case what it is in the other. The one with the larger moment of inertia accelerates more slowly.

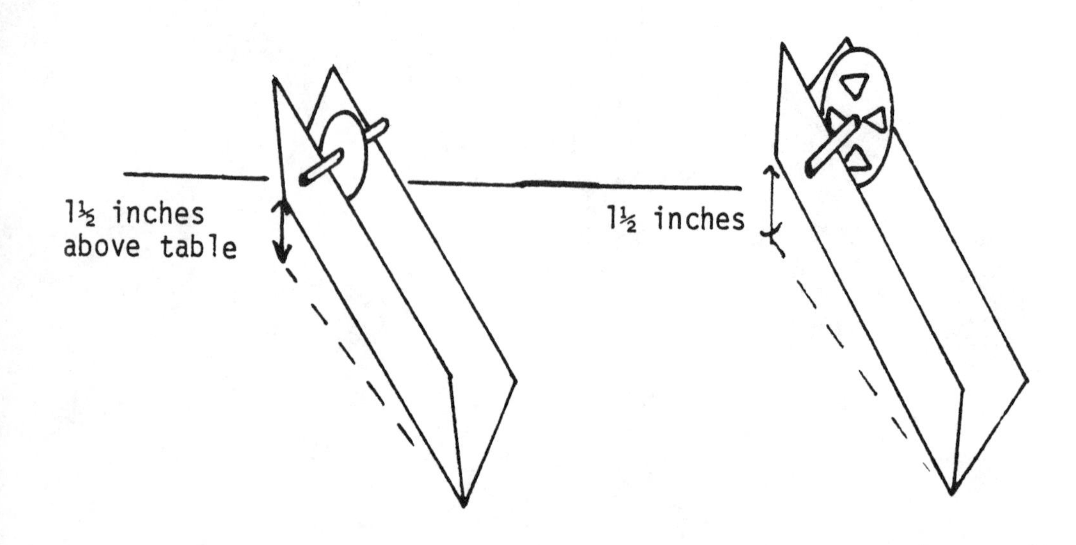

Quantitative Questions: - The moment of inertia of one disc is twice the other, and the masses are the same. Hence, the force is the same, and the couple spinning them also

$$C = I\ddot{\theta}$$

Hence $$\theta = \frac{1}{2}\frac{C}{I}t^2$$

C = couple
$\underset{..}{I}$ = moment of inertia
θ = angular

If one disc reaches the bottom in time t, the other disc, having rotated through half the angle, should only be half way down. How far down was it?

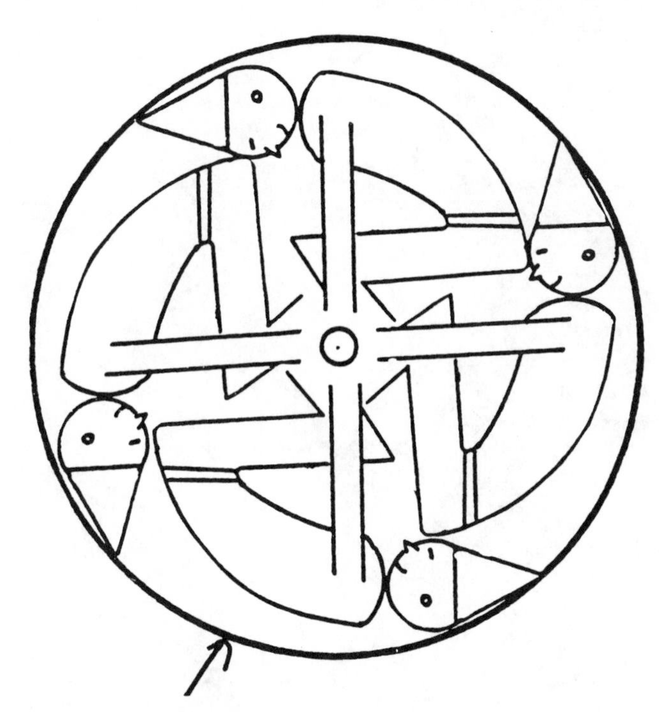

CUT OUT THESE TWO FIGURES

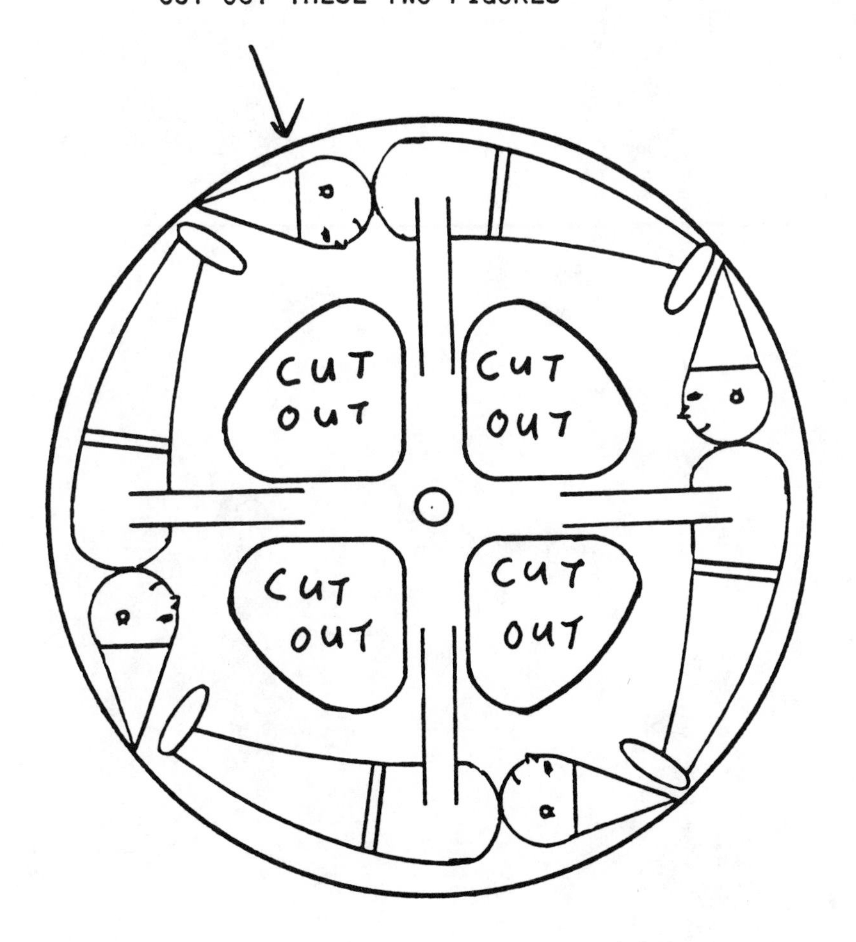

EXPERIMENT 1.36 Rotational Energy and Momentum

Materials required: cups, straw, string, ruler, about twenty marbles, books to prop up ruler, sticky tape.

Procedure: Attach a piece of tape about 4" long to each side of two cups.

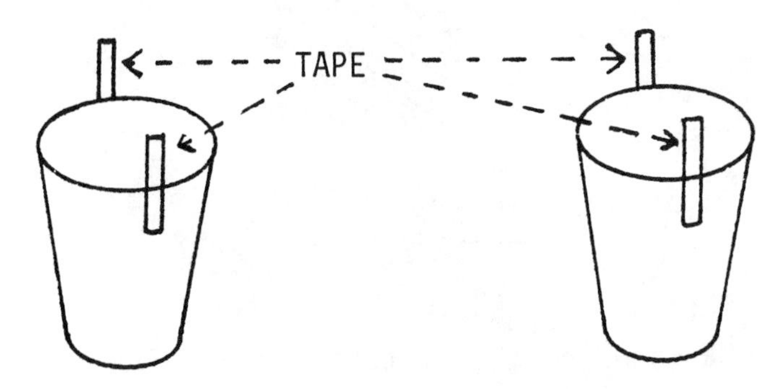

Wrap the tape over a straw, so that the cups are suspended at the ends as shown.

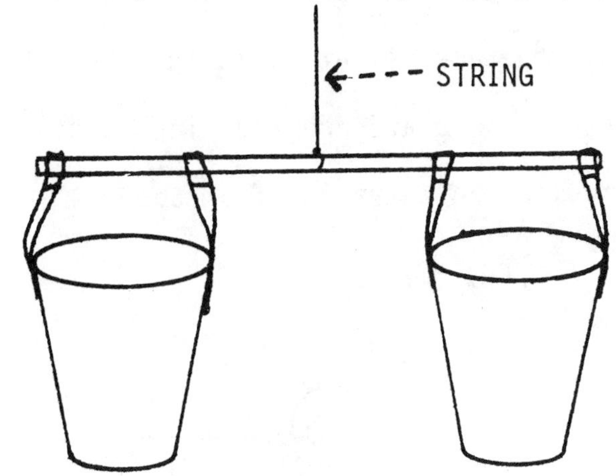

Now, tie a piece of string to the center of the straw, place ten marbles in each cup, and suspend just above the floor, from a table or chair back.

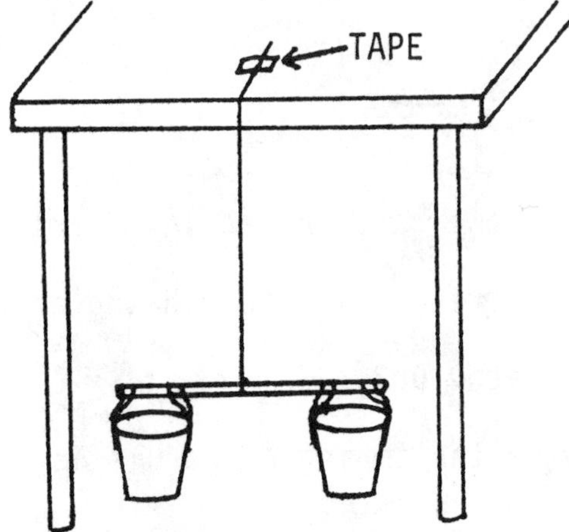

Allow the system to come to rest. Support a ruler on a pile of books so a marble can run down into a cup.

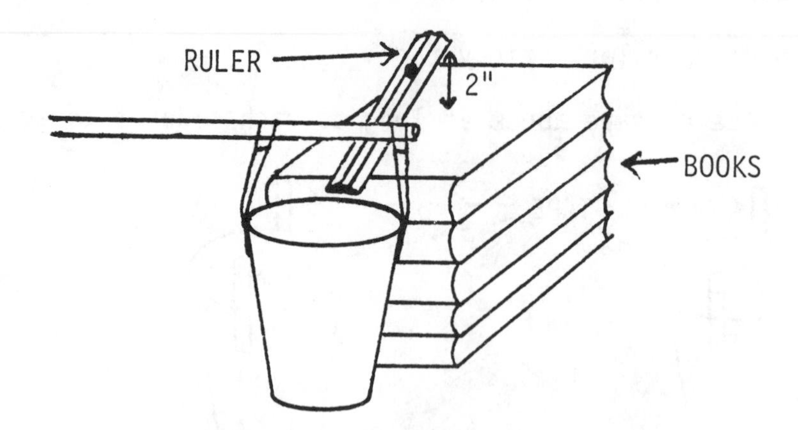

Let the marble roll down a known height, into the cup. Time half a rotation, and also measure the angle of rotation (number of turns) until the cups stop spinning. Do this twice from the same height, allowing the marble to enter the cup at the two edges, as shown. These are about 2½ inches apart, so the arm of the rotational impulse can be varied from about 4" down to 1 3/4".

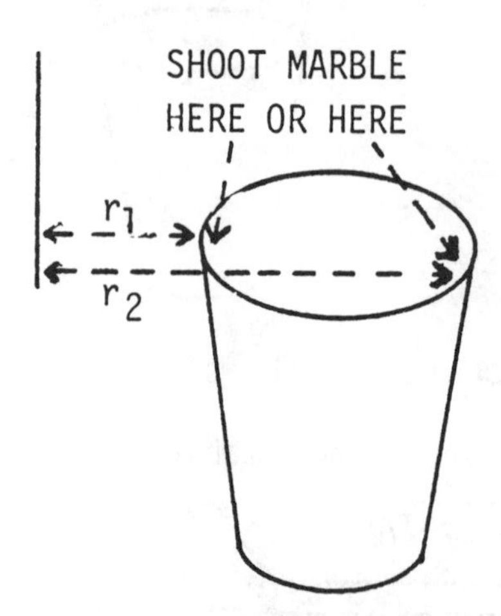

Qualitative: Why is the rotation slower and the angle of rotation smaller when the marble enters near the support string, even though the energy delivered is the same? Where does the excess energy go?

Quantitative: The kinetic energy of the impinging marble is $\frac{1}{2}mv^2 = mg\,h$ where v = velocity leaving the chute.

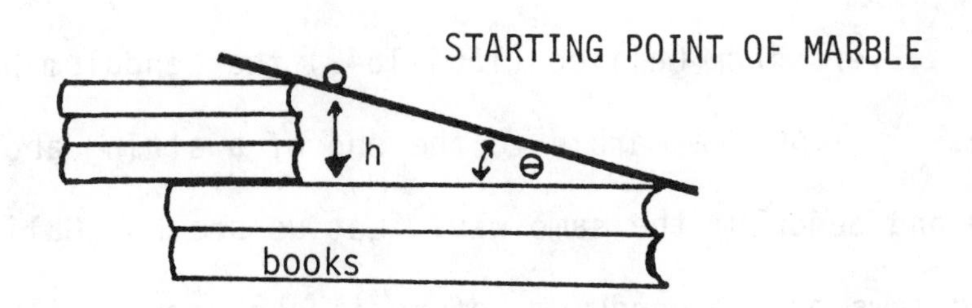

(more accurately $\sqrt{2gh}\cos\theta$ is the horizontal component of velocity). So the momentum

$$mv = m\sqrt{2gh}\cos\theta$$

and the moment of momentum = $mr\sqrt{2gh}$

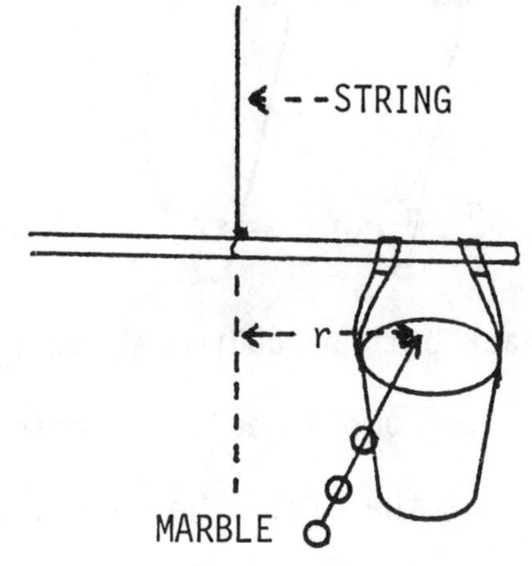

so $I\dot{\theta} = mr\sqrt{2gh}$ where $\dot{\theta}$ is the angular velocity, $d\theta/dT$

$\dot{\theta}$ is obtained from the time T for half a revolution.

$$\dot{\theta} = \frac{\pi}{T}$$

Hence I is obtained in terms of m. Is this roughly equal to (mass. of marbles) x (distance from center)^2? The kinetic energy of rotation is converted to torsional energy of the string. If ϕ = angle of rotation of the string,

$$\frac{1}{2} I \dot{\theta}^2 = \frac{1}{2} k \phi^2$$

Where k is the torsional constant for the string i.e. couple = $k\phi$

Hence, ϕ should be proportional to $\dot{\theta}$, and also proportional to the value of r (shown in the figure) if we release the marble from the same height each time. Is this true for the two cases where the marble enters at opposite sides of the cup?

Experiment 1.37 Time and the pendulum

Materials: Marbles, string.

Procedure: Ever since Galileo (1564-1642) the pendulum has been associated with time. Tape a marble to the end of a string about four feet long. Make a second pendulum the same way. Let us see how Galileo was able to deduce the laws of the pendulum before clocks were available. Make both pendulums the same length, and hang from a suitable support - anything will do, as a last resort, use your hand (it is, however, not very steady!)

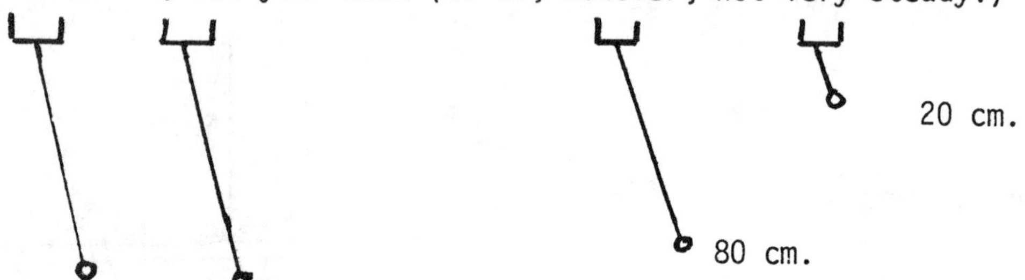

Now pull back both pendulums so that one swings through a larger arc than the other (a larger amplitude). Release them at the same time. Do they keep time, independent of the arc? They should, since the period (time of one oscillation, to and fro) of a pendulum is independent of the amplitude (if it is small!)

Now attach two marbles to one pendulum bob, and repeat the experiment. Is the period dependent on the masses of the bob?

Lastly, make one string four times the length of the other, say, 9 inches and 36 inches. How many swings of one bob equal that of the other? Make one length nine times the other--say 15 cm. and 135 cm. Again, how many swings of one bob occur for one swing of the other?

The time of one oscillation, T, the period, depends on the length of the pendulum l. Is the period proportioned to l i.e. double the length, double the period? Is $T^2 \alpha\ l$ i.e., double the period, four times the length or $T^3 \alpha\ l$ double the period, eight times the length?

Is the period constant for very large arcs, say 70° each way? It is not, and, in fact the period is more accurately given by

$T = T_o(1 + \frac{\theta^2}{16})$ where θ is the amplitude in radians.

Experiment 1.38 Oscillations of a spring -- Inertial and Gravitational Mass

Materials: ruler, stickytape, marbles, string and a paper or foam cup

Procedure: Pass a string through the top of the cup and attach it to the end of the ruler, as shown. Hold two inches of the other end of the ruler as

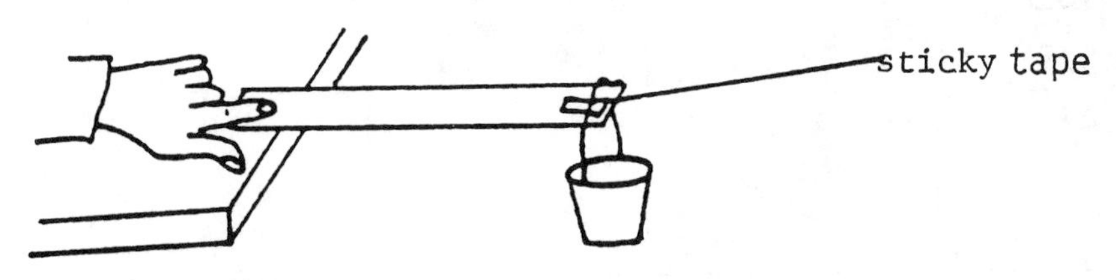

tightly as possible to the table top. Place ten marbles in the cup. Flick the far end of the ruler, and note how the vibrations produced gradually die away. One oscillation is one complete up-and-down motion at the ruler end. Count the number of oscillations in ten seconds, or five seconds if the oscillations die away too quickly. The oscillations are very rapid with few marbles, and the end of the ruler must be examined very closely to count them. If difficulty is experienced, use two rulers taped tightly together (making it about two feet long), or a yardstick or meter ruler, if available. You may use a watch to time the oscillations, but if it lacks a second hand, tape a marble to a piece of string, and have one student hold it 39 inches (99.4 cm) above the marble. Each swing (two swings to one oscillation) takes one second, so one observer calls out at the beginning and end, after ten seconds, and the other counts the number of oscillations of the ruler.

Put fifteen, twenty, and then, if possible, twenty-five marbles in the cup and again count the number of vibrations in ten seconds.

Now count the number of oscillations in five seconds for very large vibrations when you pull the ruler way down before releasing (large amplitude) and for tiny vibrations, where you barely touch the ruler to set it going. Lastly, shorten by two inches the length of ruler vibrating, (i.e. hold four inches of ruler against the table) and again count the number of oscillations in ten seconds.

Qualitative Questions: Note how the oscillations become slower, the more mass is at the end of the ruler, and become faster, the shorter the ruler, (the stronger the springiness).

Does increasing the amplitude of the vibrations make them faster or slower?

Quantitative Questions: The frequency f is the number of oscillations per second - i.e. the number you counted divided by five or ten. The period T (the time of one oscillation) is $\frac{1}{f}$

Theory shows that $T = 2\pi \sqrt{\frac{m}{k}}$

where m is the mass on the spring, and k is the spring constant. Plot on the graph paper on the following page the period T against the square root of the mass on the spring (the number of marbles). The dots on the graph are typical examples. In this experiment, the ruler is the spring.

no. of marbles	square root of number of marbles	number of oscillations in five seconds n	frequency f (n/5) oscillations per second	period 1/f seconds
(5)	(2.2561)			
10	3.1622			
15	3.873			
20	4.472			
25	5			

Is the result a straight line?
What does this show?
Draw a line through the points.
Does it go through the origin (no mass, T = 0)?

As a further exercise, you can check whether $T^2 \alpha 1/k$. You measured T for two different values of k (lengths of ruler). Use a lot of marbles in the cup. Now you need to find k, which comes from the equation

force = k x extension of spring

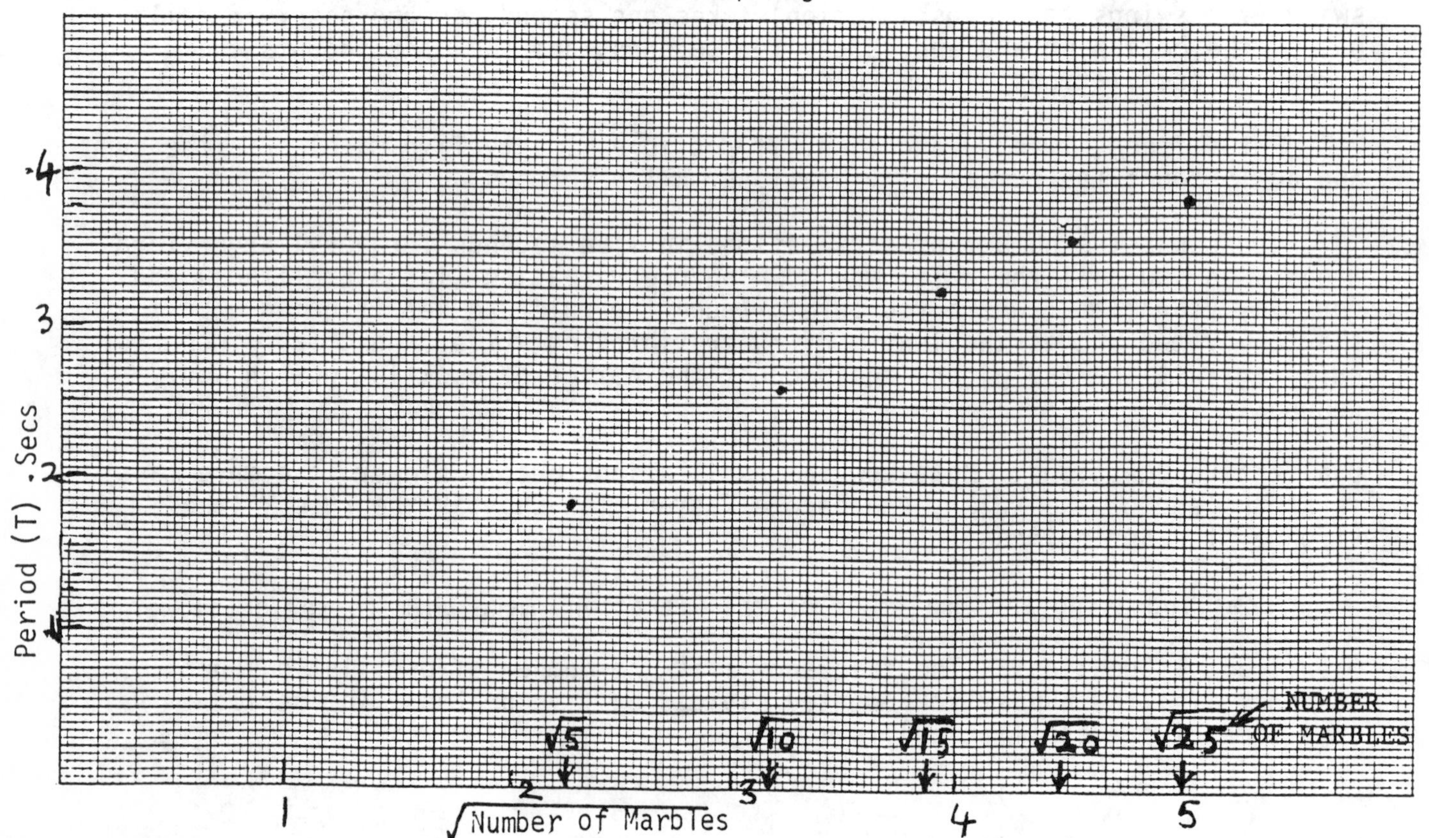

To find the ratios of the two values of k, hang a weight on the end, and notice how much it deflects the end of the ruler with first 10cm. and then 20 cm.of the ruler extending over the edge of the table to which it is held.

$$\frac{k_2}{k_1} = \frac{\text{deflection with 10 cm.}}{\text{deflection with 20 cm}} \qquad (k \propto 1/\text{deflection})$$

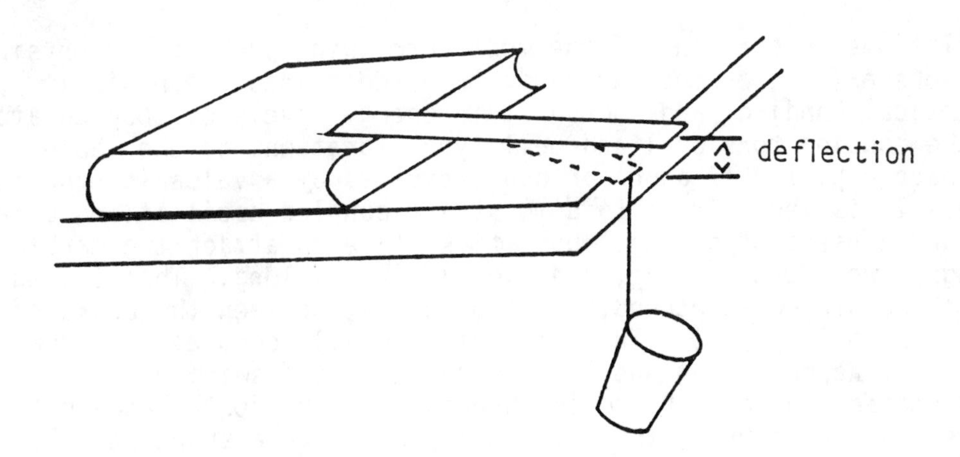

The ratio of the deflections is the ratio of the values of k. See if this ratio is the same as the ratio of the two values for T^2 obtained with two and four inches of the ruler pressed to the support.

Notice that we can obtain the ratio of inertial and gravitational masses, since $k = \frac{mg}{\text{deflection}}$ where here m is the gravitational mass and

$$T = 2\pi\sqrt{\frac{m}{k}}$$ where in this case m is the inertial mass.

This ratio is invariably one.

An interesting extension of this experiment shows students the relationship of musical pitch to frequency.

Take the cup away, and examine the vibration as that portion of the ruler vibrating over the book (or the edge of the table to which it is held) is shortened. With the 25 cm of the ruler extending over the table edge, it is easy to observe the ruler vibrating, and it may even be possible to count the oscillations. As the extension is shortened, the vibrations clearly increase in frequency, until a note can be heard. The frequency can be determined by adjusting until it is in unison with a soda straw 18.3 cm long which vibrates if blown* at 900 Hz. This shows high frequency, high pitch and a shortend ruler go together. The restoring force increases a sthe ruler is shortened, as does the vibrating mass, so the frequency is a strong function of the length of the ruler - for the wooden rul er I was using, I find, if the period in seconds is T and the length of the ruler L cm

$$T = 0.0000417\ L^{1.8}$$

* Experiment 7.03

EXPERIMENT 1.39: Coupled and Forced Oscillations

Materials Required: Marbles, string, tape, heavy weight (such as a drink can)

Coupled oscillations present one of the most intriguing problems in physics. The ramifications extend, on the one hand, from the quantum mechanical aspects of chemical bonding, and the shift in energy levels of coupled atomic systems, to the coupled torsional and bell type vibrations of the whole earth, which have a period of about an hour, excited by a volcanic eruption or earthquake. It is very simple to demonstrate coupled oscillations using four marbles and a piece of string. Use sticky tape to attach the marbles to four strings, two 30 cm. long, and two 15 cm. long. Then tie the strings to a third string stretched, not too tautly, between the backs of two chairs, as shown in fig. 1. The pendulums are all "coupled" together by this string. What happens when one long pendulum is set swinging? The small pendulums are unaffected, and the motion handed back and forth between the long pendulums - and of course, the reverse happens when a short pendulum is disturbed. The normal modes of the coupled system can then be demonstrated by holding the two long pendulums at the same distance transverse to the string between the chairs, and releasing them together as shown in fig. 2. They continue to swing with a constant period until the motion damps out. For the other normal mode, also shown in the figure, the two pendulums are held out an equal distance on opposite sides and released simultaneously. Again, there is no transferring the motion back and forth. Measure the period of the two normal modes. The mode when only one pendulum is released is a combination of equal parts of the two normal modes, which beat against one another, so the frequency difference of the two normal modes is the frequency at which the motion is transferred too and fro between the two pendulums. By moving the pendulums closer together, or apart on the string between the chairs, you can show that tight coupling (closer together) gives a shorter period for transfer of the motion, and weak coupling (farther apart) gives a longer period.

Forced oscillations can be demonstrated using the same system. One pendulum has its marble replaced by a heavy weight - a soft drink can (full) is convenient. The other pendulum can have three paper clips in place of the marble. These are more heavily damped, and the addition of a small piece of paper to the clip leads to even heavier damping. The arrangement shown in fig. 3 works well. The string to the clips passes through another clip taped to the cross string. The length of the light pendulum may then be easily changed by raising or lowering the clips, without affecting the heavy pendulum. Quantitatively then, the observations to be made are the amplitude and phase of the paper clips as the pendulum supporting them is lengthened. Fig. 4 shows these as a function of the length of the pendulum, for three paper clips, and three paper clips plus a piece of paper 5 cm square. The resonance is more highly damped in the latter case. The heavy pendulum was 60 cm long. At resonance, the phase lag of the clips is $\pi/2$, 90°, with the result that when the heavy weight is in the middle, the clips are at the end, and when the weight is at the end, the clips are in the middle. This device is a modification of Barton's pendulums, when not one, but many pendulums of different length are supported on the cross string, so that the amplitude and phase can be compared simultaneously. Resonance occurs when the force has the same frequency as the undamped free oscillations, not the damped oscillations. The highly damped pendulum has a much smaller amplitude at resonance, but the resonance is broader.

An even simpler demonstration can be made with a piece of string and two weights, one heavy and one light. The heavy weight could be a rock, book, or soft drink can. The light weight can be three or four paper clips. The soft drink can can be supported by passing the string through the ring on top. Tie the string as close to the center of the top as possible and fasten the thread to the paper clips to the center of the bottom with sticky tape. This way, when the can spins it will not affect the oscillations. Suspend the heavy weight from a suitable support--a coat rack, table, etc. by a thread, or fine string about two feet long. I found it easiest to bend a paper clip into a hook, attach the string to it and hang the can from the top of a door way, as shown in fig. 5.

First start the paper clips swinging. The free oscillations on their own decay rapidly. This is a "transient". Now start the heavy weight swinging. This produces the periodic force, which causes the clips to oscillate, their amplitude slowly building up. When the string to the clips is short, they stay in phase with the heavy weight, and their amplitude of oscillation is small, relative to the can. As the second string is lengthened, the amplitude of the oscillations builds up, until at resonance, when the two strings are of approximately equal length, starting from rest the oscillations of the clip will build up to a swing about ten times wider than those of the can. This is a very effective demonstration. The clips are also obviously 90° out of phase with the can, being at the end when the can is in the middle, and vice versa, as shown in fig. 6. Lengthening the thread to the clips still further, causes the oscillations to die down, until when the thread to the clips is twice as long as the string to the can, the oscillations will appear as in fig. 7 with the clips approximately 180° out of phase to the can.

Quantitative Solution

The Quantitative part of this problem can get quite involved. It is first necessary to examine the solution to the equation for forced oscillations. The equation is

$$F_o \cos \omega t = d^2\theta/dt^2 + \gamma d\theta/dt + \omega_o{}^2\theta$$

where $\omega_o/2\pi$ is the free oscillation frequency of the paper clip, ($\omega_o = \sqrt{g/\ell}$ where ℓ is the length of thependulum), γ is the damping constant per unit mass, F_o is the amplitude and $\omega/2\pi$ the frequency of the applied force per unit mass (i.e. force/mass of the paperclip) and θ the angular displacement of the string to the paper clips. When $F_o = 0$, the transient solution to this equation is

$$\theta = A \exp(-\gamma t/2) \cos(\sqrt{(\omega_o{}^2-\gamma^2/4)}\ t - \psi)$$

where ψ is an arbitrary phase angle. The solution with the applied force present, ultimately tends to

$$\theta = A\cos(\omega t - \phi)$$

where $$A = F_o/\sqrt{((\omega_o^2 - \omega^2)^2 + \gamma^2\omega^2)}$$

and $$\tan\phi = \gamma\omega/(\omega_o^2 - \omega^2)$$

The significance of these two solutions is clear. The first is simply a decaying sinusoidal oscillation, and the second a steady oscillation to which the clips will ultimately settle down. However, if we start the can swinging with the clips initially at rest, we get a combination of two equations of the form

$$\theta = A\{\cos(\omega t - \phi) - \exp(-\gamma t/2 \cos((\sqrt{\omega_o^2 - \gamma^2})\ t - \phi)\}$$

To find γ, set the clip swinging alone, and note the time in seconds required for the amplitude to decay by a half, $T_{1/2}$. Since exp $(-\gamma T_{1/2}/2) = 1/2$, $\gamma = 2\log 2/T_{1/2} = 1.386/T_{1/2}$.

For three paper clips swinging from a 60 cm thread, $T_{1/2}$ turns out to be about 10 seconds and γ equals .139. If the can is also swinging from a 60 cm thread $\omega = 2\pi/T = \sqrt{(g/\ell)}$.T, the period is 1.55 seconds, giving $\omega = 4$ rad/sec. This also equals ω at resonance. We now have all the required parameters, and the curve derived from these is the one shown in fig. 8c. 8a and 8b are the corresponding transient and continuous solutions. Now, try the experiment yourself--it only requires a few paper clips, string, and a soft drink can (actually, a can of beer was employed for these experiments.) Fig. 9 shows beats occurring when $\omega = 3.5$, rather than 4.0, ω_o remaining the same. This is very noticeable experimentally. Damping reduces the effect of the beats.

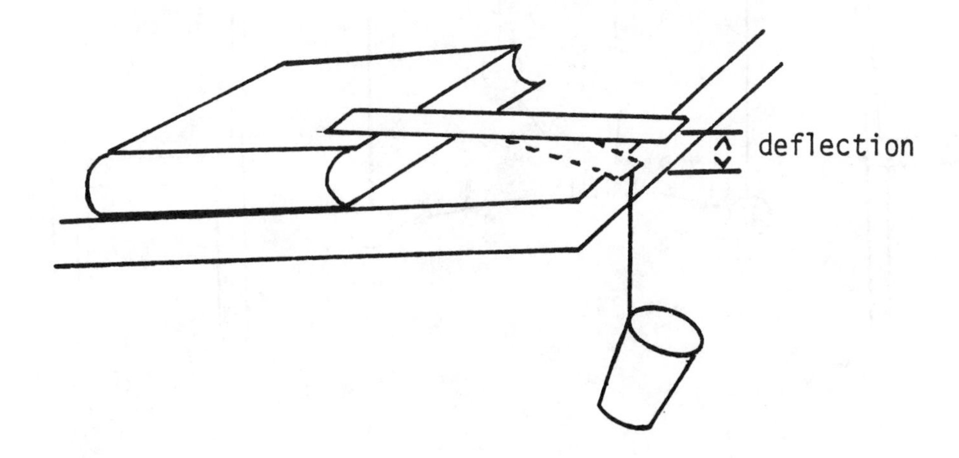

The ratio of the deflections is the ratio of the values of k. See if this ratio is the same as the ratio of the two values for T^2 obtained with two and four inches of the ruler pressed to the support.

Notice that we can obtain the ratio of inertial and gravitational masses, since $k = \frac{mg}{\text{deflection}}$ where here m is the gravitational mass and

$$T = 2\pi\sqrt{\frac{m}{k}}$$ where in this case m is the inertial mass.

This ratio is invariably one.

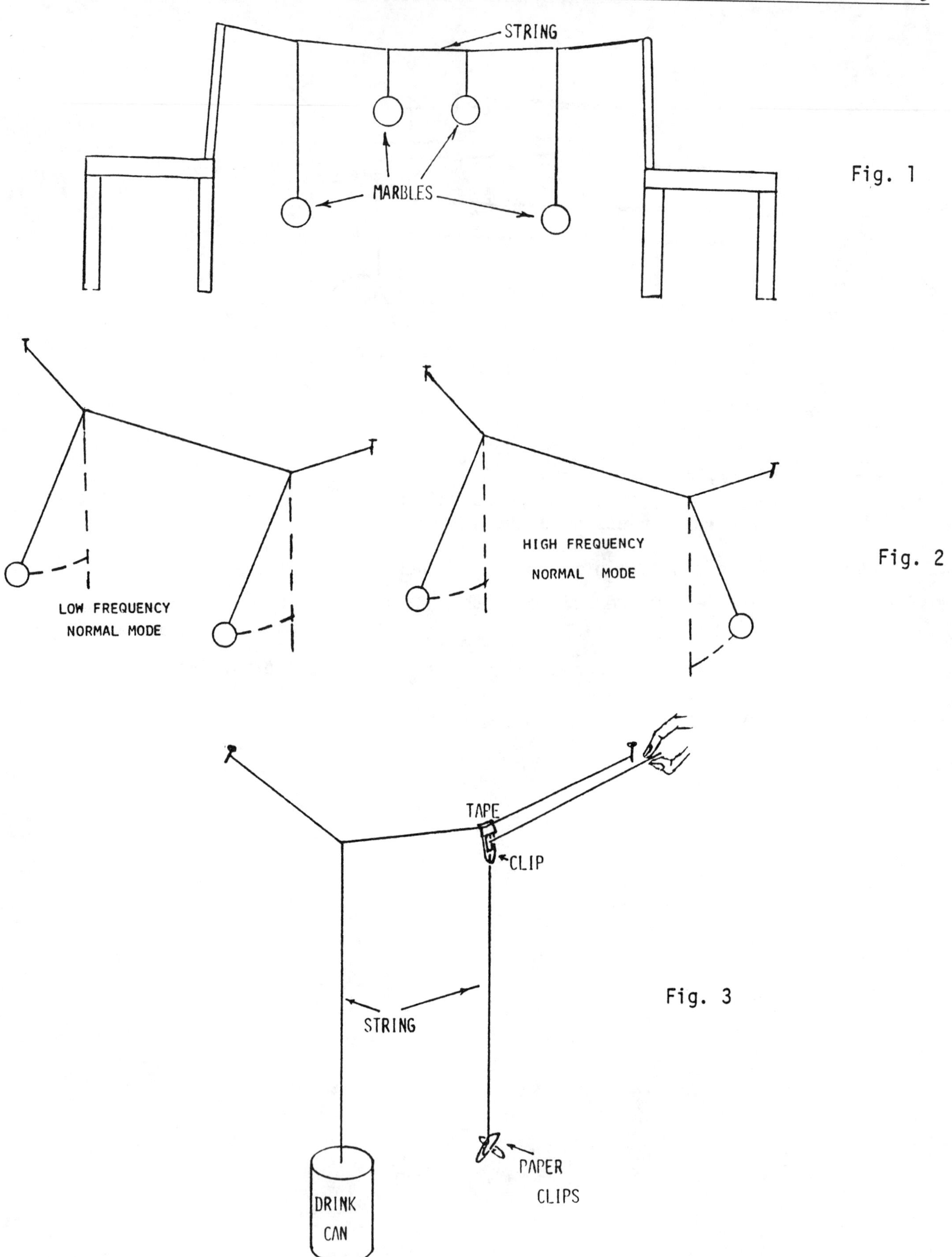
STRING
MARBLES
Fig. 1
LOW FREQUENCY
NORMAL MODE
HIGH FREQUENCY
NORMAL MODE
Fig. 2
TAPE
CLIP
STRING
Fig. 3
DRINK
CAN
PAPER
CLIPS

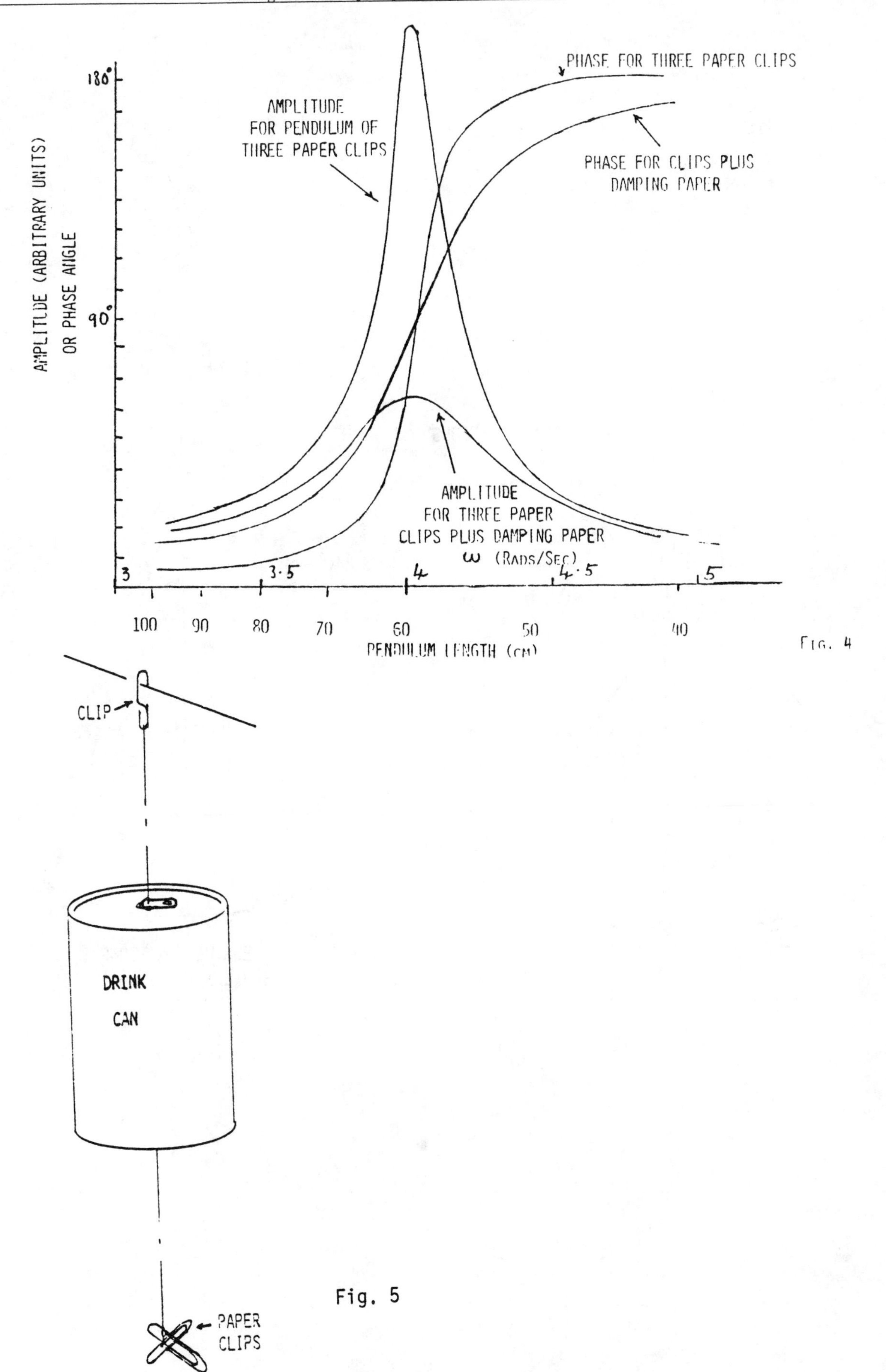
180°
90°
AMPLITUDE (ARBITRARY UNITS)
OR PHASE ANGLE
AMPLITUDE
FOR PENDULUM OF
THREE PAPER CLIPS
PHASE FOR THREE PAPER CLIPS
PHASE FOR CLIPS PLUS
DAMPING PAPER
AMPLITUDE
FOR THREE PAPER
CLIPS PLUS DAMPING PAPER
ω (RADS/SEC)
3
3.5
4
4.5
5
100
90
80
70
60
50
40
PENDULUM LENGTH (CM)
CLIP
DRINK
CAN
PAPER
CLIPS

FIG. 4

Fig. 5

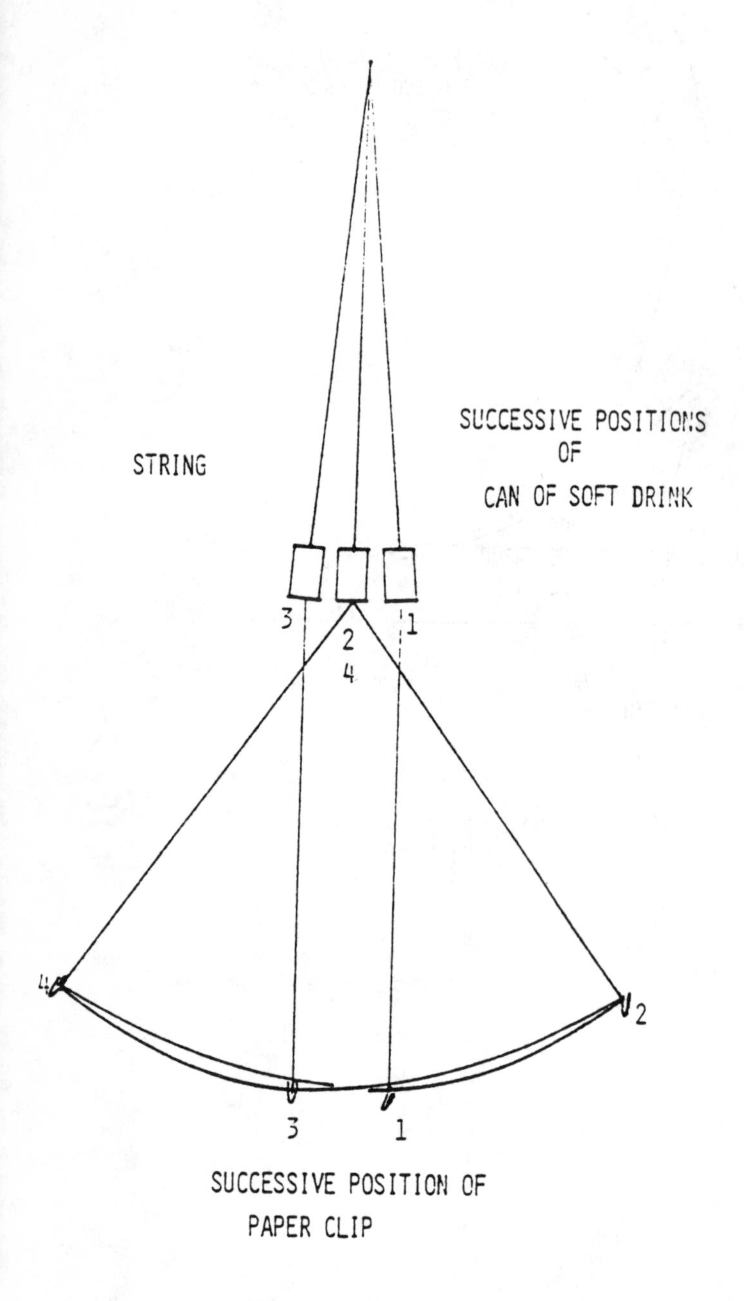
STRING
SUCCESSIVE POSITIONS
OF
CAN OF SOFT DRINK
3
2
4
1
4
2
3
1
SUCCESSIVE POSITION OF
PAPER CLIP

Fig. 6

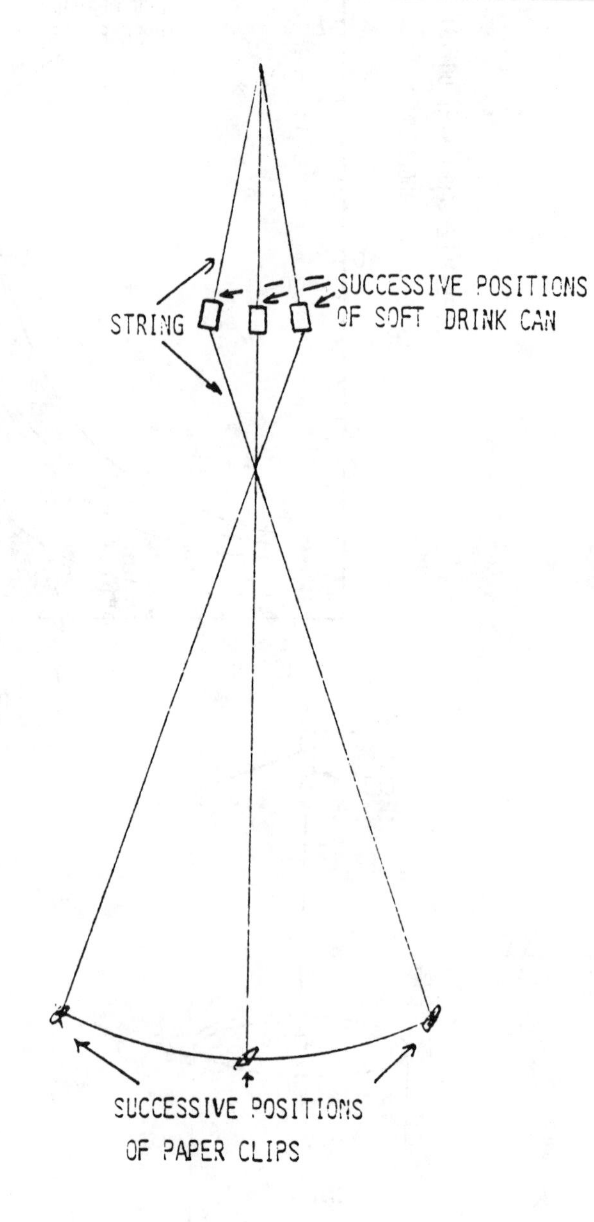
STRING
SUCCESSIVE POSITIONS
OF SOFT DRINK CAN
SUCCESSIVE POSITIONS
OF PAPER CLIPS

Fig. 7

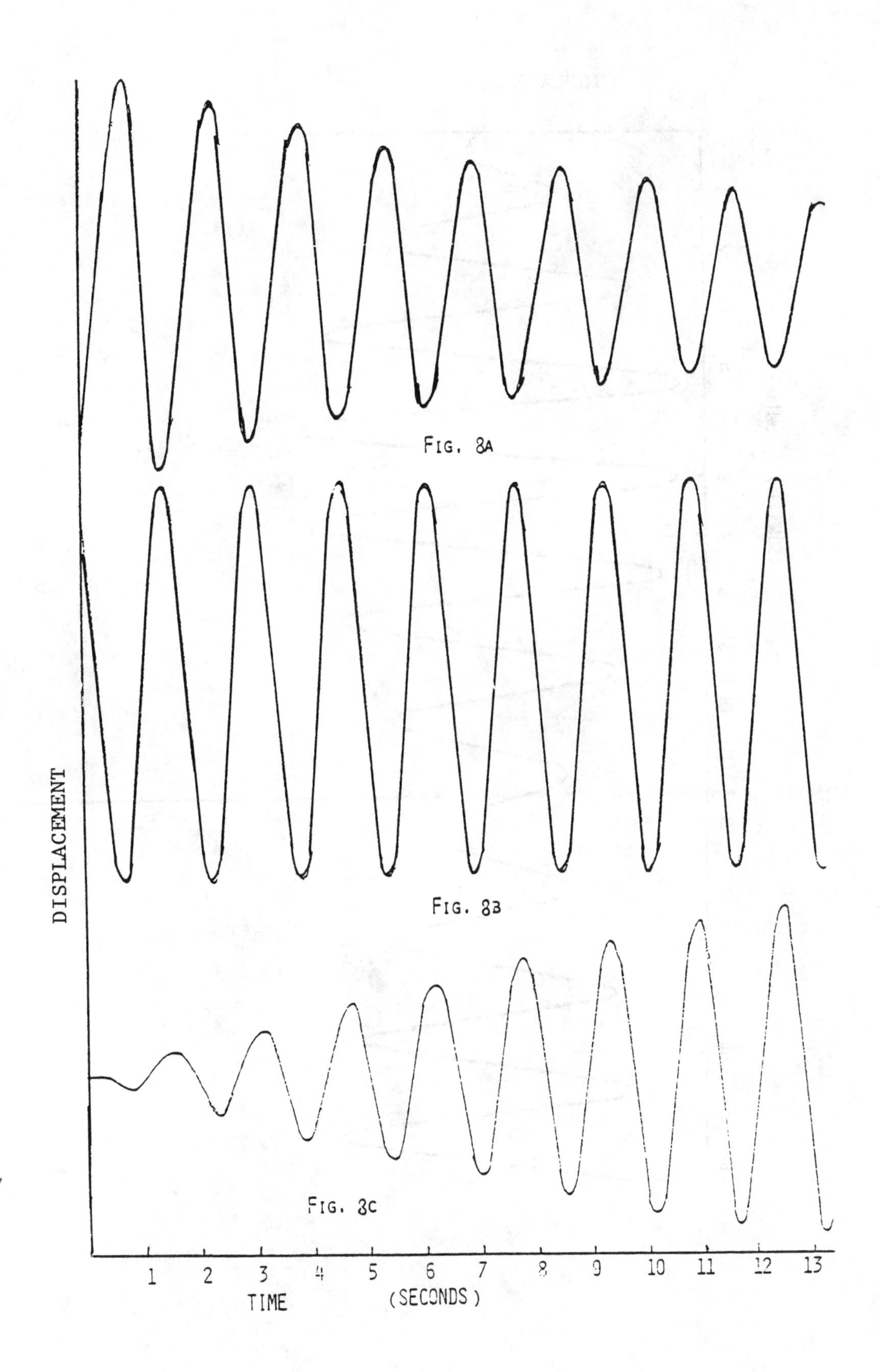
DISPLACEMENT
FIG. 8A
FIG. 8B
FIG. 8C
1
2
3
4
5
6
7
8
9
10
11
12
13
TIME
(SECONDS)

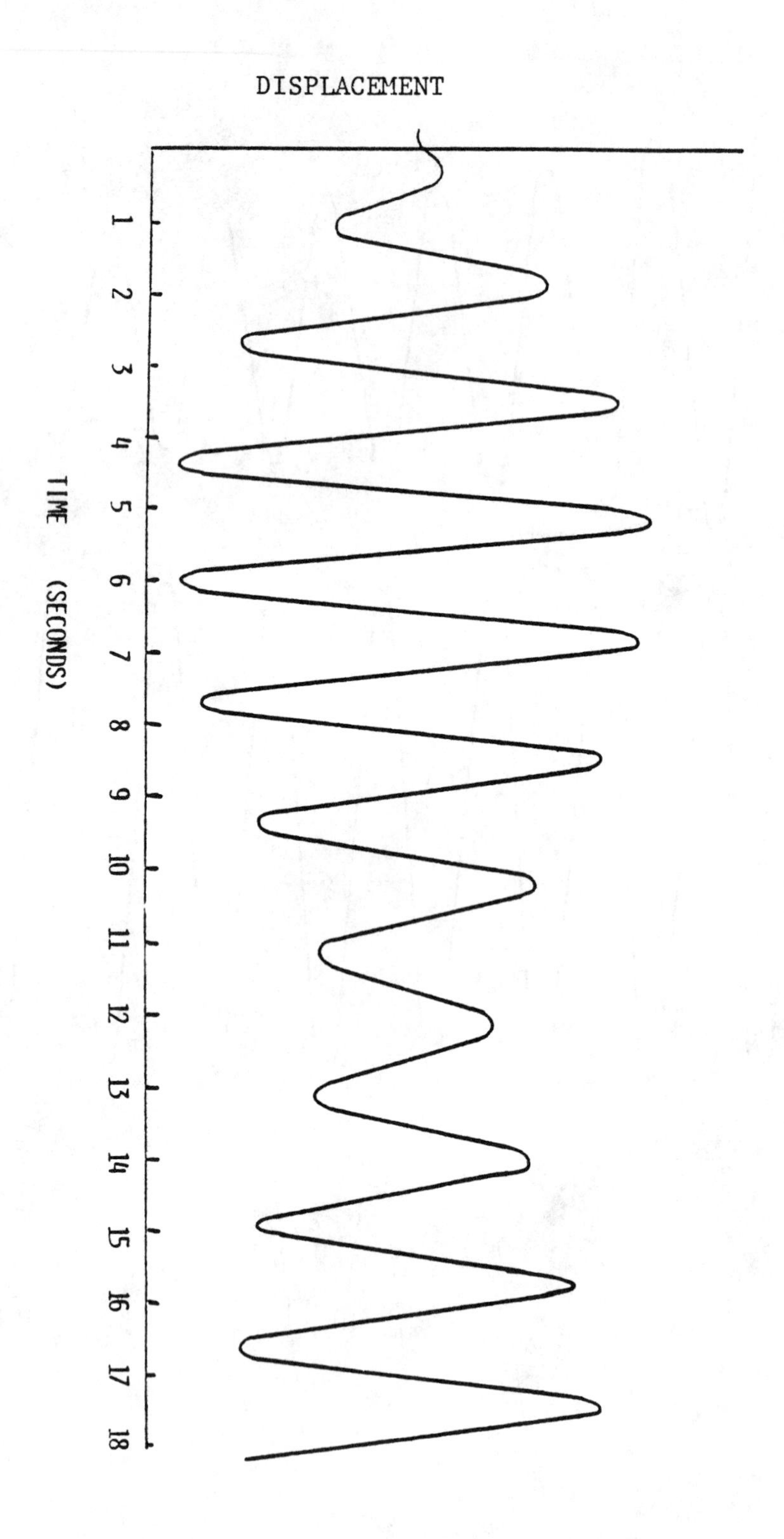
DISPLACEMENT
TIME (SECONDS)
1
2
3
4
5
6
7
8
9
10
11
12
13
14
15
16
17
18

Fig. 9

Another interesting coupled system is shown in the figure. A cup of marbles is supported from a table by a rubber band, and a second cup of marbles supported from this also by a rubber band, as shown. By applying a periodic force using a third rubber band, as shown, it is possible to set the system in oscillation. The modes are very similar to the pendulums, two normal modes, with the cup in or out of phase, in addition to which are complex modes composed of these two.

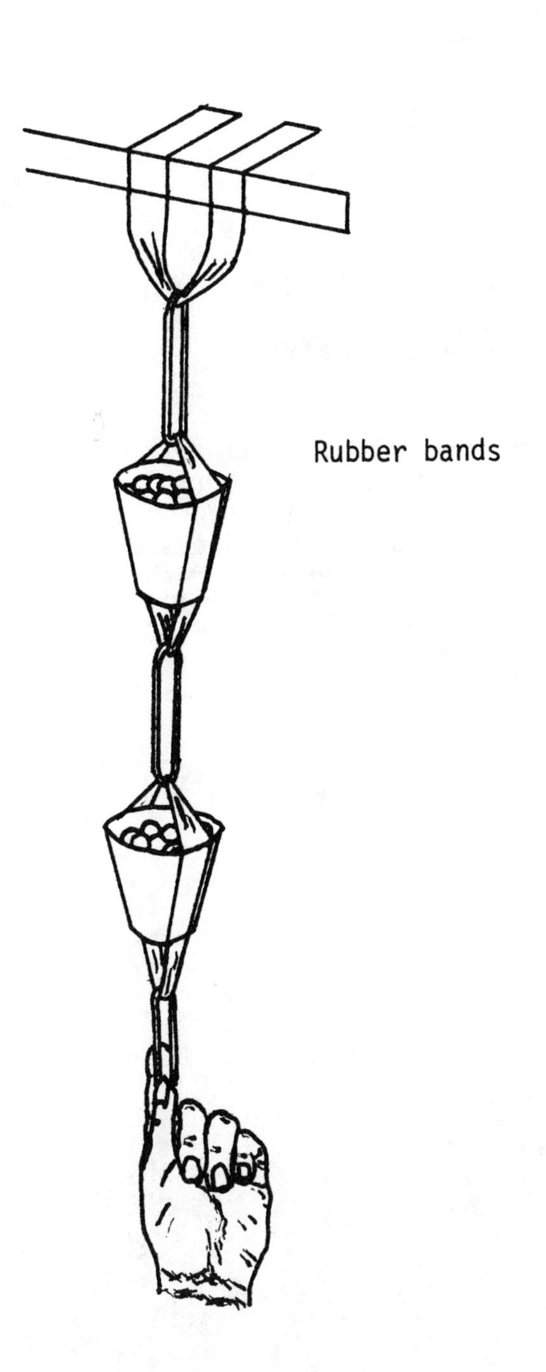

EXPERIMENT 1.40 Foucault Pendulum

Materials: Four straws, a marble, sticky tape and thread, or fine string.

Procedure: Fasten the straws into a rigid tetrahedron, as shown, using tape. Attach a marble to a piece of thread using cellotape, then fasten it to the apex of the tetrohedron so it can swing freely. Better, the ends of the straws may be attached together using paper clips as shown below, and the thread held by a clip.

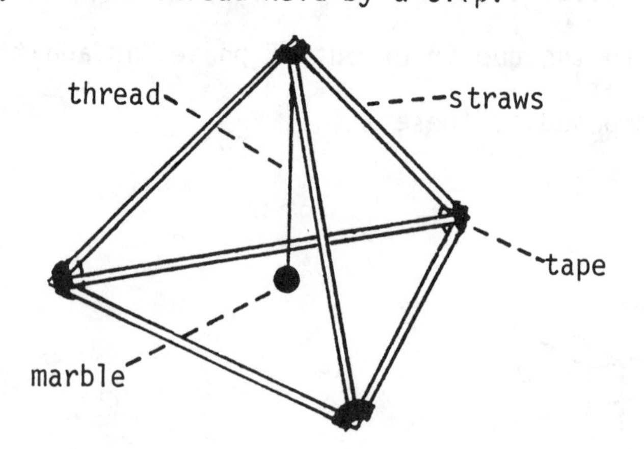

Set the pendulum swinging and rotate the bottom of the tetrahedron.

Qualitative Question: Is the swinging of the pendulum affected by rotating the base?

The equivalent pendulum on earth would be at the north or south poles. Would the swing of the pendulum rotate with the earth? To demonstrate this, place the pendulum over the center of the polar map of the world on the next page. The fictitious force which appears to rotate the phase of the pendulum is called the Coriolis force.

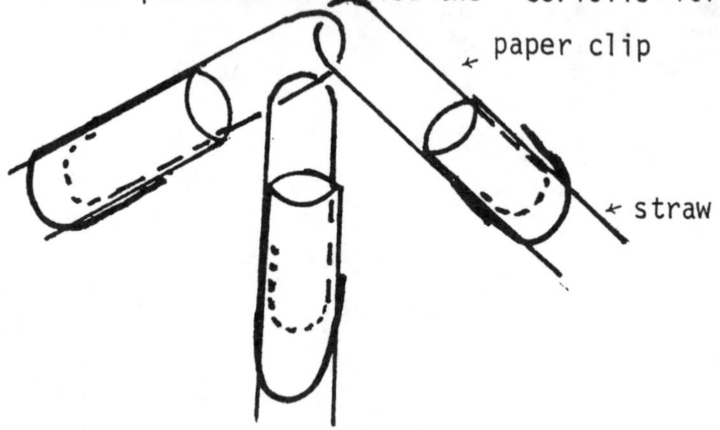

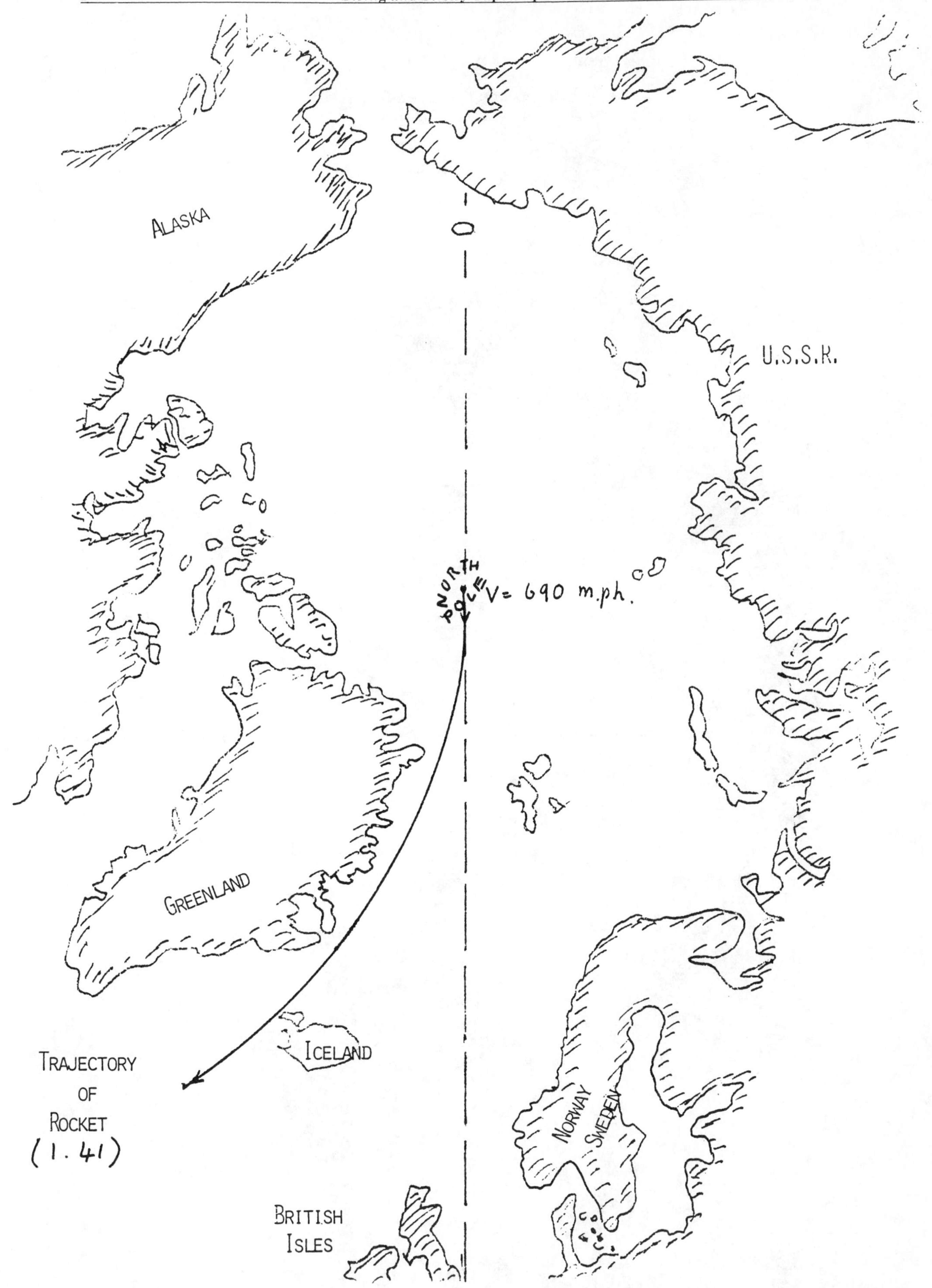

ALASKA
U.S.S.R.
NORTH POLE
V= 690 m.ph.
GREENLAND
ICELAND
TRAJECTORY OF ROCKET (1.41)
NORWAY
SWEDEN
BRITISH ISLES

Experiment: 1.41 The Coriolis Effect

Materials: String, book, tape, marble, paper, cup.

Method: Trying to explain why an object which you might think should travel in a straight line appears to move in a curved path on earth's surface is frequently a real problem. Here is about the simplest way you can demonstrate it--and showing how it occurs really helps. Tape a fairly heavy book closed and suspend it level, as shown in figure 1, by four strings fastened underneath which tie together to one string attached to a support such as a table top. Tape a sheet of plain paper to the top of the book. Dip a marble in a cup of water, start the book spinning, and roll the marble onto the paper on the book. The water on the marble will leave a quite obvious track on the paper. Replacing the plain paper with a copy of the map of the north pole, given in the Foucault's pendulum experiment is a help when referring specifically to the earth. Try rolling the marble more or less rapidly, and spin the book fast and slow. It is also interesting simply to drop the marble on the paper, and see how it travels outward. The experiment is qualitative--the general mathematics of rotating systems is quite involved and probably is not worth tackling at an elementary level, except perhaps for special cases, such as a marble with uniform velocity v starting from the axis of rotation of a book spinning with angular velocity ω. Then, its position after time t in rotating polar coordinates on the paper stuck to the book is $\theta = \omega t$, $r = vt$. An example is a cannon fired from the north pole.

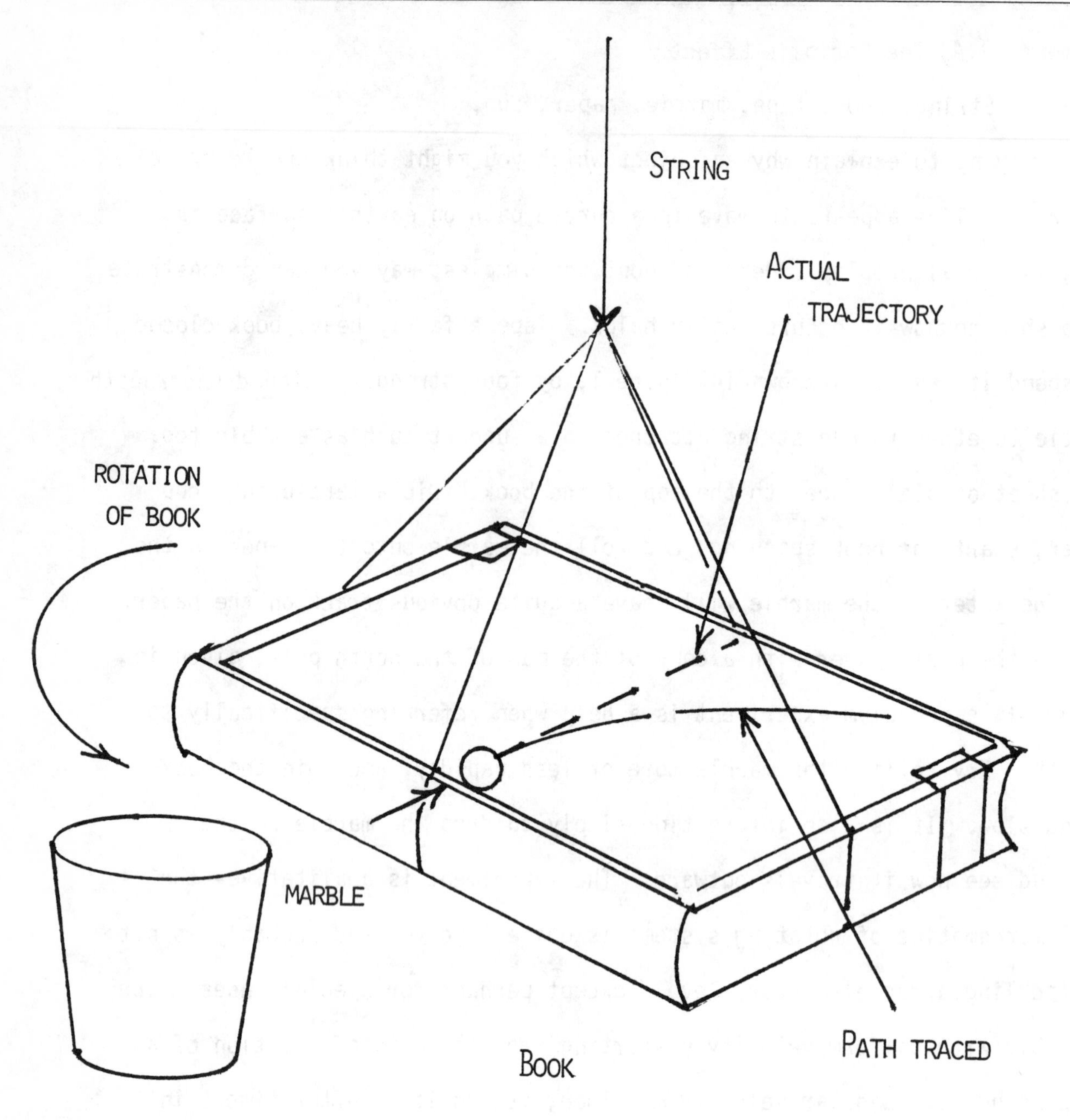
STRING
ACTUAL
TRAJECTORY
ROTATION
OF BOOK
MARBLE
BOOK
PATH TRACED

Fig. 1

Experiment 1.42 Law of Equal Areas

Materials: cups, string, straws, paper clip

Procedure: Conservation of angular momentum under a central force leads to Kepler's law of equal areas - that a line from a planet to the sun sweeps out equal areas in equal time. This law applies to all orbits under central forces. For example, if we observe an elliptical pendulum from above illuminated by a strobe light, we get the picture shown below. The two triangles shown are equal in area.

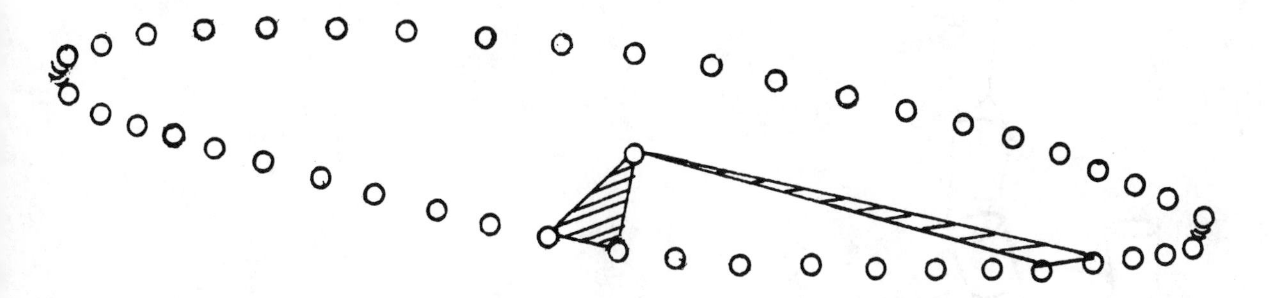

The following experiment confirms the same law in the absence of a strobe light. This is rather a messy experiment, and should be done where spilt water will do no harm. With a needle, poke a hole in the bottom of a cup. Put about fifteen marbles in the cup which is attached by a string to a suitable support, a chair back, the edge of a table, etc. Set the system going to ensure you can swing the cup in a suitable ellipse. Spread a large sheet of newspaper on the floor. Fill the cup with a diluted mixture of ink and water. The hole should allow a few drops a second to emerge. The elliptical pendulum traces its path by the ink drops falling on the newspaper, and since there are equal time intervals between the separation of the drops, the velocity of the pendulum will be proportional to the reciprocal of this distance.

Viscosity causes the amplitude of the pendulum to decay, so several orbits may be mapped.

Another method of performing the same experiment is to replace the ink in the cup with water, and enlarge the hole somewhat. After the pendulum is set swinging in an ellipse, two cups are placed under it, one at the end, and one about the middle of the swing, so that water goes into both cups as shown.

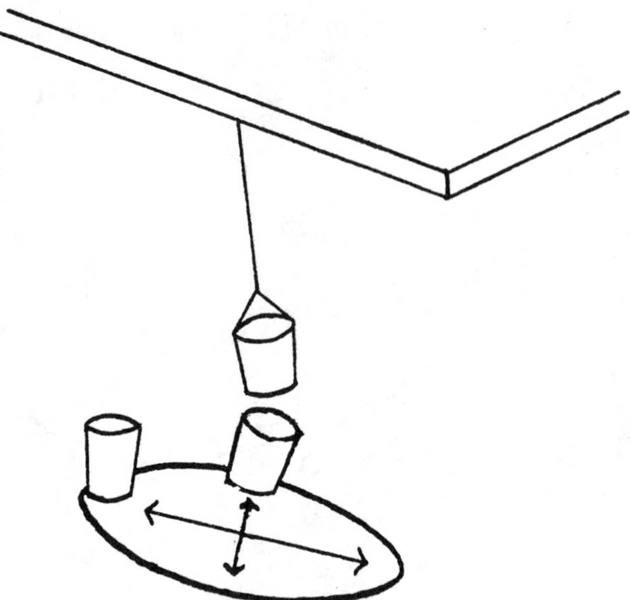

The trajectory will be mapped out by water drops on the floor, and you need to know the distance from the center, and the angle, for the two cups.

The volume, or mass, of water in each cup needs to be measured. If you let a lot run out, you can just measure the height in each cup, but then you have a real mess on the floor. If you let a little run into each cup, you can suck it up from the cup in a straw, and measure the length,

(taking care not to suck it into the mouth!) The volume measured is proportional to the time the pendulum spends over each cup, and hence the reciprocal of the velocity. The area of the triangle representing the angular momentum is therefore, proportional to $\frac{\text{moment to the center } (r \sin \theta)}{\text{mass of water in the cup}}$
Note: the trajectory must cross the center of the cup, or less will go in.

Note that the central force in this experiment is elastic- that is,

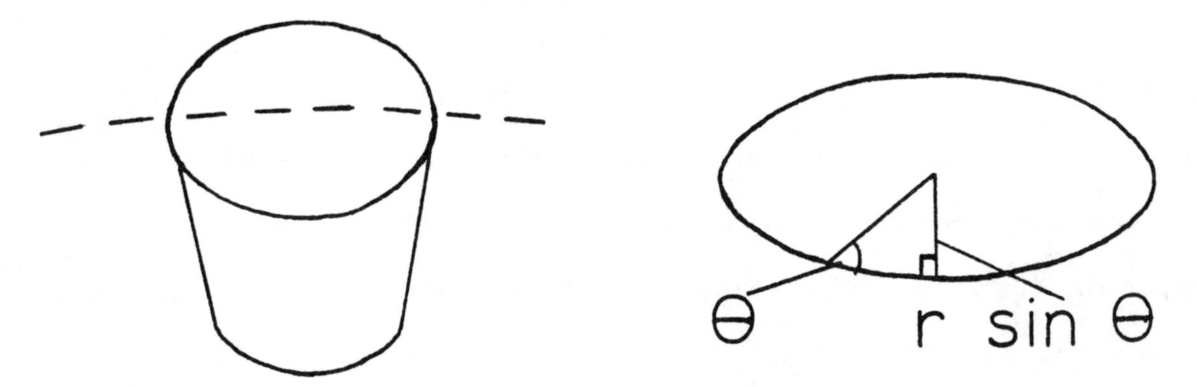

the force is proportional to the displacement, and directed towards the center. This gives an elliptic orbit with the force directed to a point at the center of the ellipse. With an inverse square law, such as provides planetary orbits, the force is directed towards the focus of the ellipse, which is close to one end for an eccentric orbit. Since angular momentum is conserved in both these cases, the law of equal areas applies. For planetary motion, the sun is the focus.

Experiment 1.43 Inertia

Materials: Tape, scissors, marbles, cup.

Procedure: Hang a cup, full of marbles, from the table, as shown, using sticky tape. Now cut two thirds the way through the tape at A and B making sure the cuts are identical. Give a very sharp pull at B. Which tape breaks? Repeat, pulling very slowly. Does the same tape break? Why not?

You make a similar experiment using rotational inertia every day. When you pull the toilet roll slowly, it unrolls without tearing, because the force on the paper, and the torque on the roll is small. Then you jerk the paper and it tears - the moment of inertia of the roll opposes the sudden torque on it, and the increased tension in the paper tears it. Why is this much more difficult when the roll is almost empty?

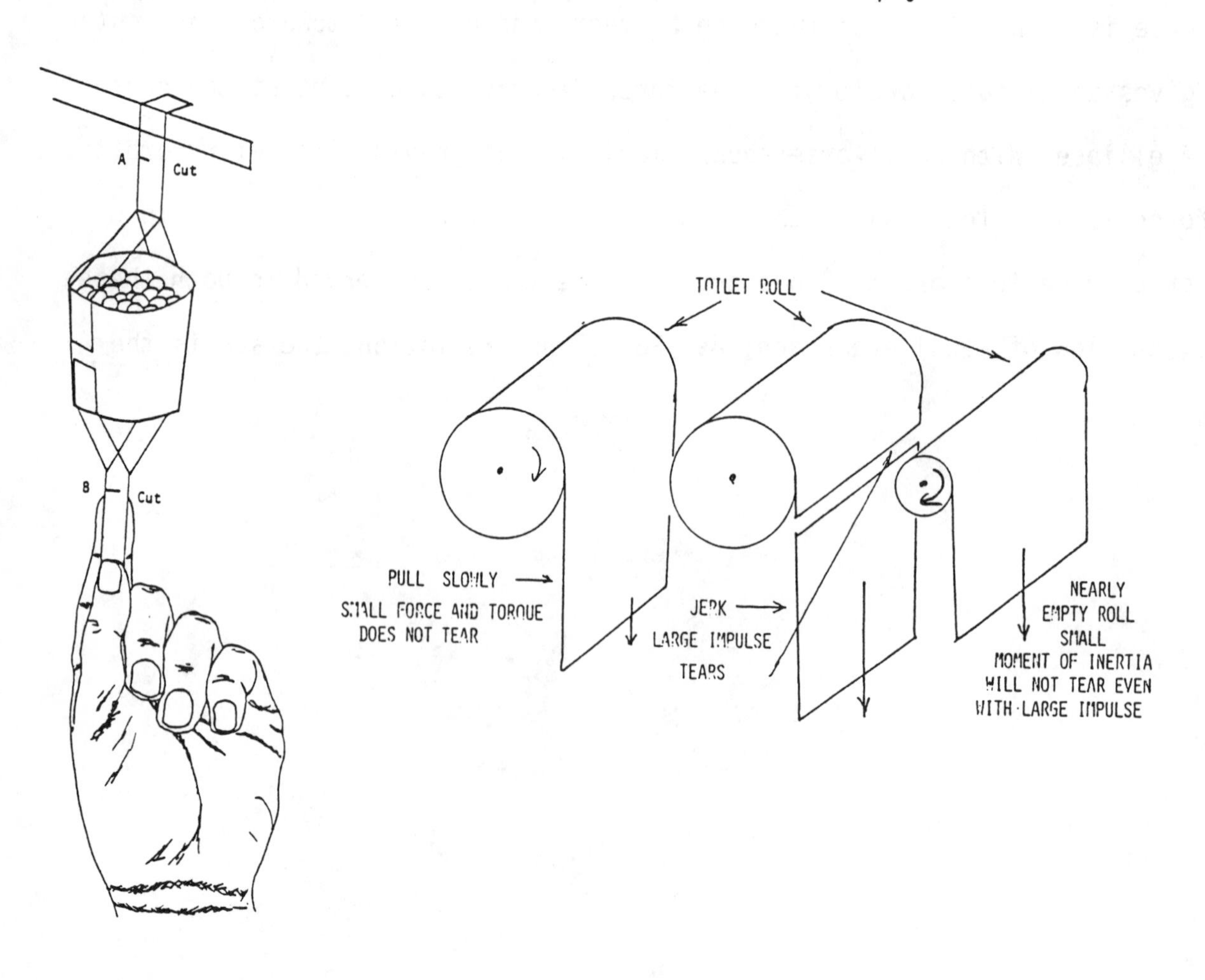

Experiment 1.44 Car accelerometer

Materials: two paper clips, soda straw.

Procedure: Straighten out the paper clips and insert them in the ends of the straw. Then bend the straw to the shape shown below. Tape the device to the side window of a car, filling it with water level with the zero. Note how high the water rises when accelerating along a level road. The device can be used on a bicycle, but is very difficult to read without falling off.

If the acceleration of the car tilts the water to angle θ,

$$\tan \theta = \frac{\text{car acceleration}}{\text{acceleration due to gravity}}$$

What is the maximum acceleration of the car in gs (acceleration due to gravity)? How long would it take to reach 100 km./h (63 mph) from rest with this acceleration?

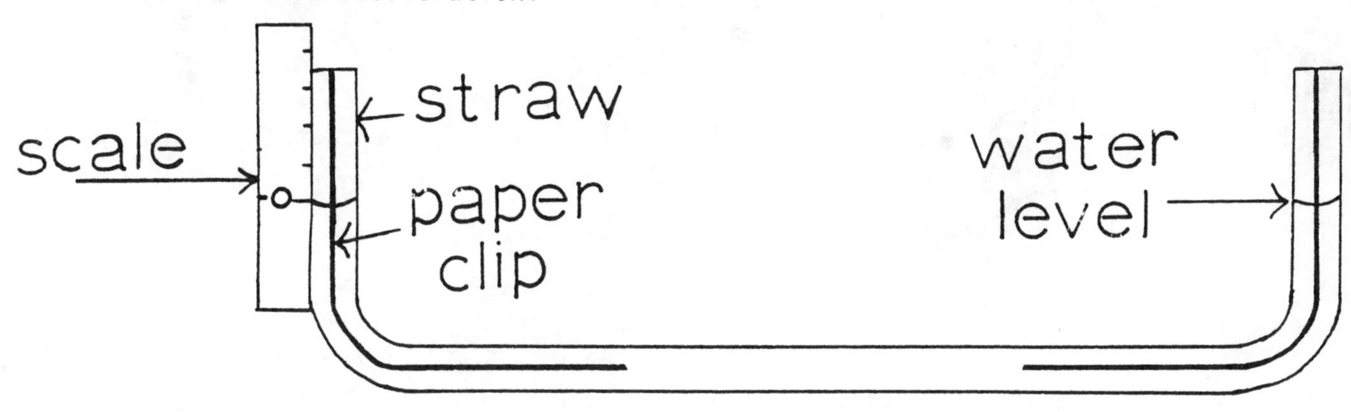

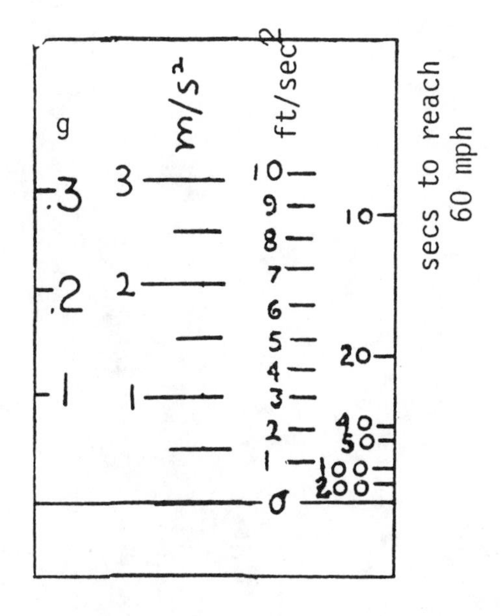

Experiment 1.45 The size of the Sun

Materials: Paper, card, pencil.

Procedure: Measuring the size of objects is always interesting, from the size of the nucleus (see experiment 8.02) to the size of the universe (no experiment included). The size of the sun, and of a tall nearby building present interesting experiments. Poke a large pinhole in a sheet of card or thick paper on a sunny day, and allow the sun's image to fall on a sheet of paper a fixed distance from the hole. Make pencil marks at the edge of the

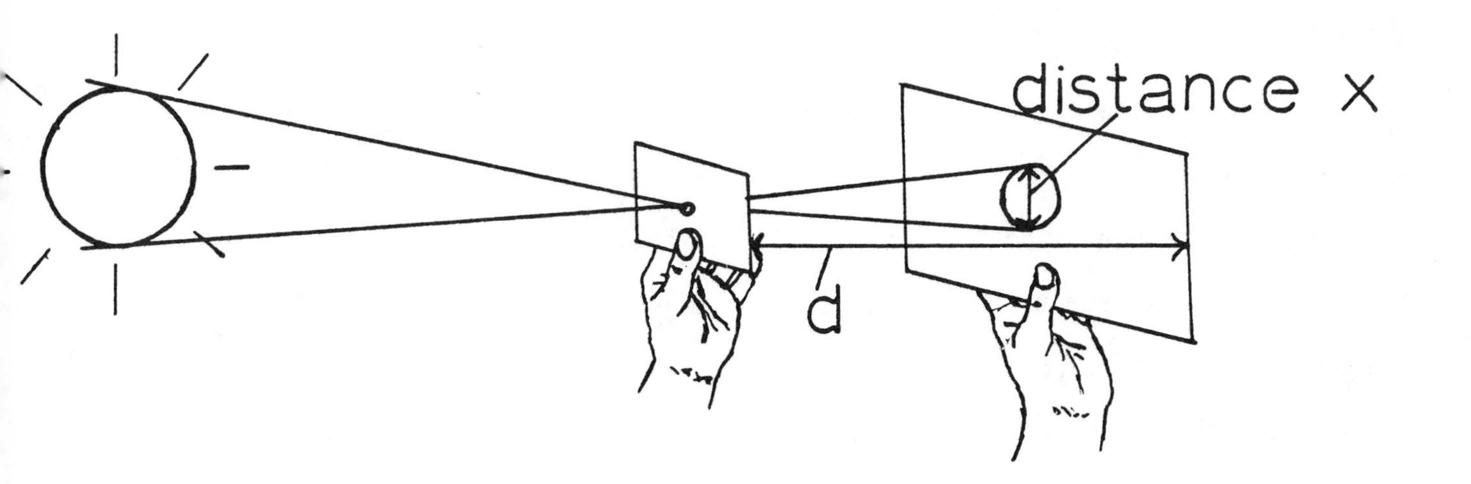

image, and measure the separation with a ruler to be X cm. Measure the distance of the pinhole from the paper to be d cm. If you look at the figure, you will see

$$\frac{\text{diameter of sun}}{\text{distance of sun from earth}} = \frac{\text{diameter of spot X}}{\text{distance of spot from pinhole}}$$

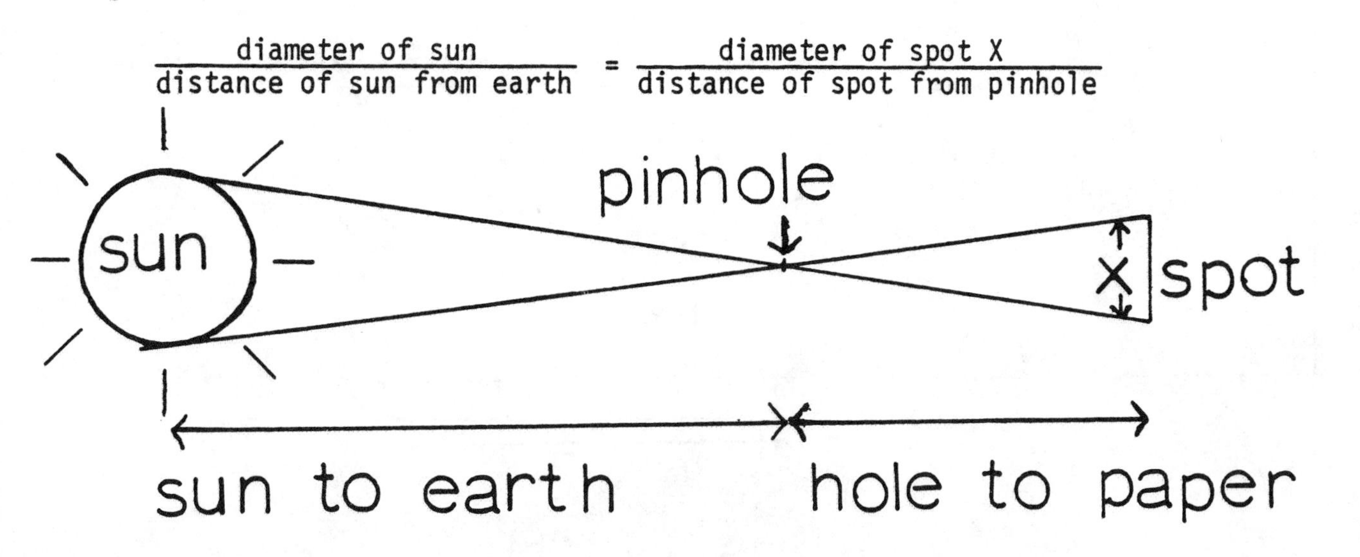

All we need to know here is the distance of the earth from the sun. This is a little more difficult, and has been obtained by using the earth as a base line by taking sights of one point on the sun simultaneously from opposite sides of the earth, to get the angle θ as shown.

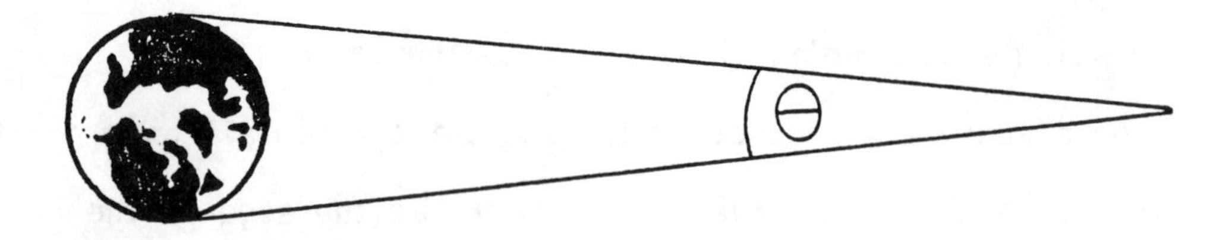

$$\text{The distance to sun} = \frac{\text{diameter of earth}}{\theta}$$

This gives the distance to the sun as 92.9 million miles, or 169 million Km. What is its diameter?

Now for a simple method for the heights of buildings. Pace out a suitable distance from the building, about equal to what you think is its height. Count the number of paces.

Cut out the figure on the following page, and fold it up, as shown. Aim the lower part at the base of the building, and hold a pencil in line with the top.

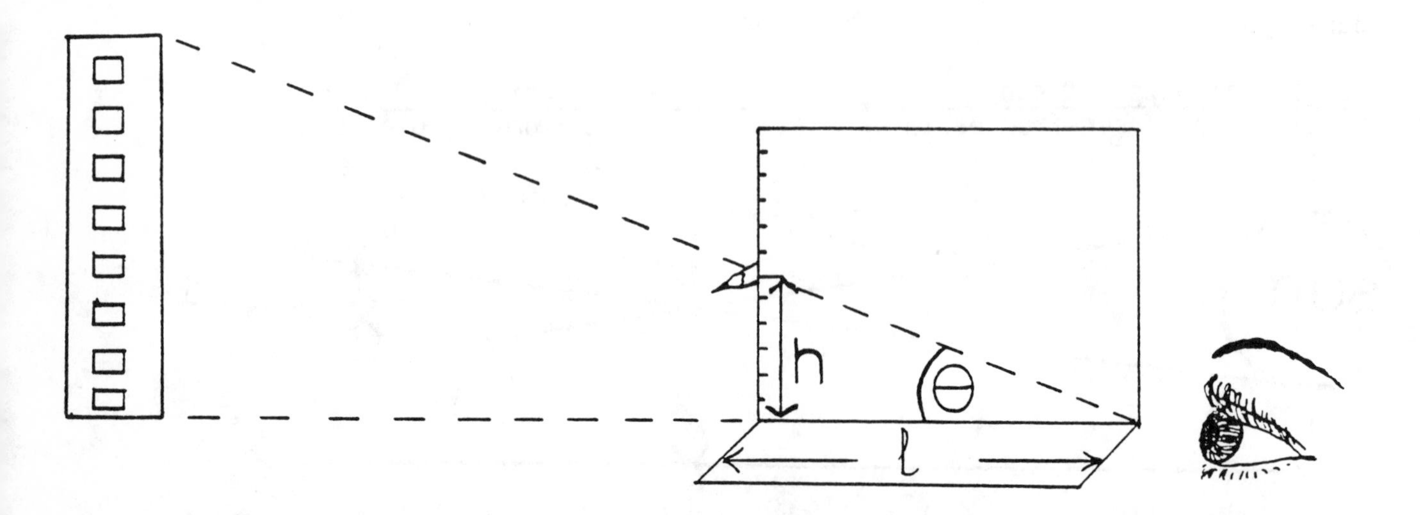

The edge of the sheet is graduated to give you the tangent of the angle

$\theta = \frac{h}{l}$.

The height of the building is then tangent of θ x number of paces x length of pace.

Measure ten paces to obtain the length of your pace. A pace is about two and a half feet, three quarters of a meter.

Length of your pace =

Number of paces =

Tan θ =

Height of building =

Repeat the whole process to get the building height.

How close together are your values?

How accurate do you think this experiment is?

How best could it be improved?

1
.9
.8
.7
.6
.5
.4
.3
.2
.1

sight from
this point

sight along this line

Experiment 1.46 Rolling Uphill or Antigravity

Materials: Three straws, paper clips, marble.

Procedure: Fasten two straws together at one end using paper clips as shown.

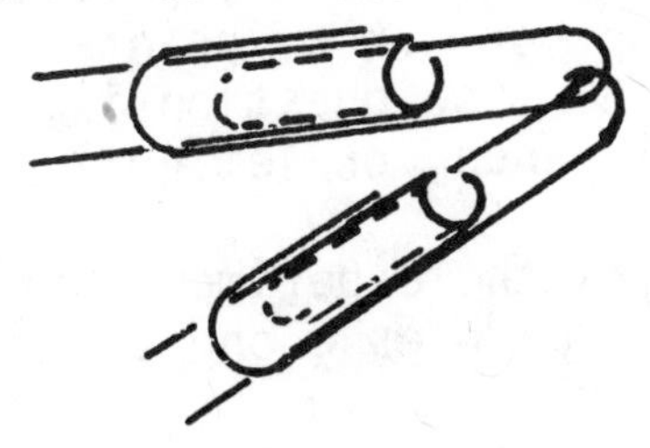

Place the straws together over a third straw or pencil.

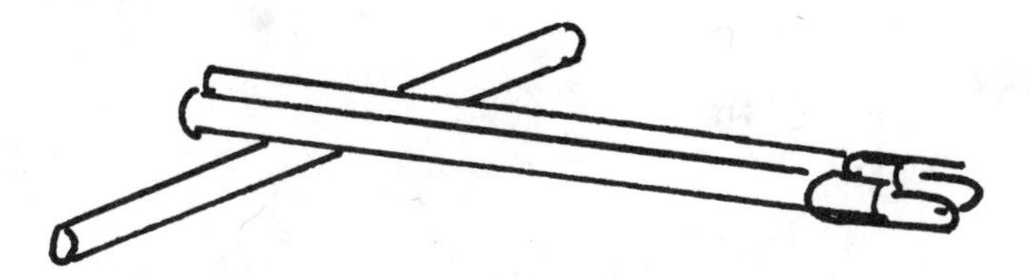

Now, balance a marble at the lower end and gradually separate the straws. the marble will ultimately roll toward the upper end. Why does the marble appear to roll uphill? By moving the straws apart and together it is possible to make the marble roll completely to the top end. Can you explain this? The solution lies in the fact that by separating the straws, the marble rolls horizontaly between them, the kinetic energy being provided by the center of mass falling slightly. Before the marble drops to the table the straws are brought rapidly together. This increases the potential energy, but does not appreciably affect the kinetic energy so the marble continues to roll to the upper end.

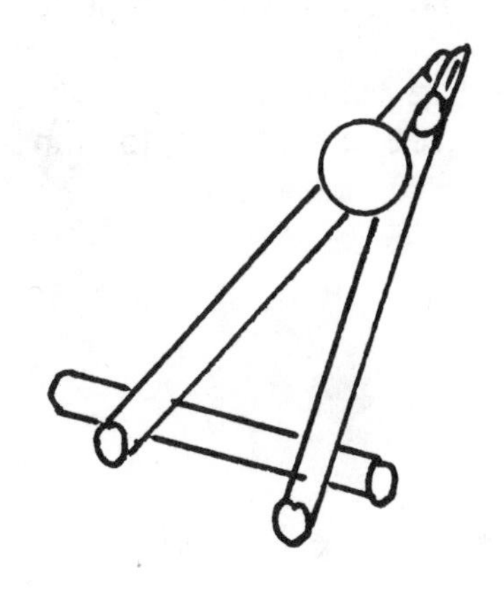

Experiment 1.47 The Monkey Puzzle

Materials: Marbles, rubber band, paper clips, string, tape.

Procedure: A famous problem in physics is the monkey (or squirrel) problem. A hunter aims a gun (or bow and arrow) so that the gun (or arrow) points directly at a monkey (or squirrel) hanging from a tree. The question is, as the hunter fires, should the monkey hang from his branch, or let go?

To see what the monkey should do, let us perform the experiment. Cut, or tear, out the monkey overpage, and attach five marbles along the bottom of the page,
to weight it. Attach two strings from the top of the picture, pass the strings through two paper clips attached to the back of a chair, or table, as shown below.

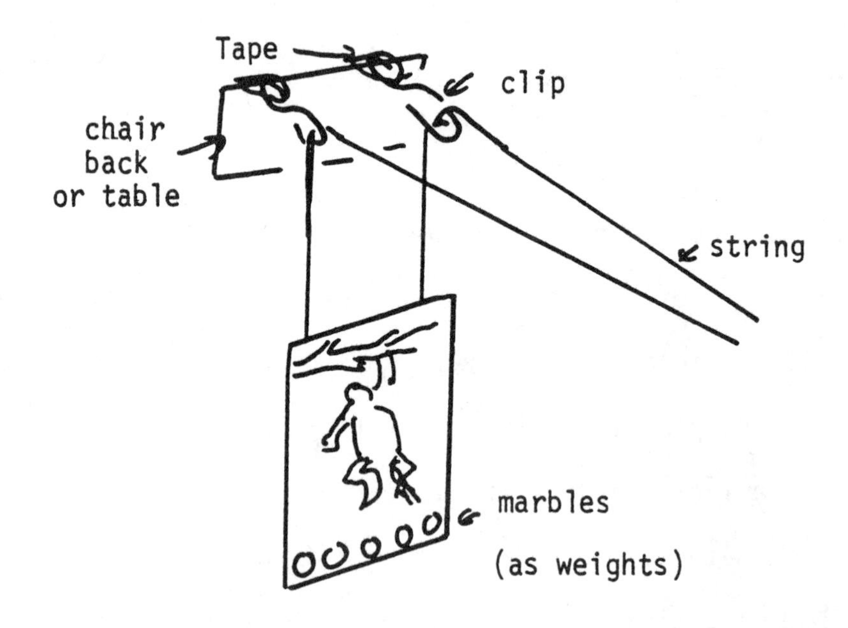

To make the gun, roll a sheet of paper around a pencil and tape it as shown in the figure. Tape a rubber band to the paper cylinder, as shown, and tape the cylinder to a book, as a support. Sight down the tube to aim at the monkey, about three or four feet away, which can be raised or lowered by its supporting strings. Stretch the rubber band over the end of the pencil in the tube, and hold pencil in the taut position, together with the strings supporting the monkey, with one finger. Release the strings and pencil simultaneously.

When does the pencil strike the monkey?
Should the monkey hold on or let go when the hunter fires? Why?

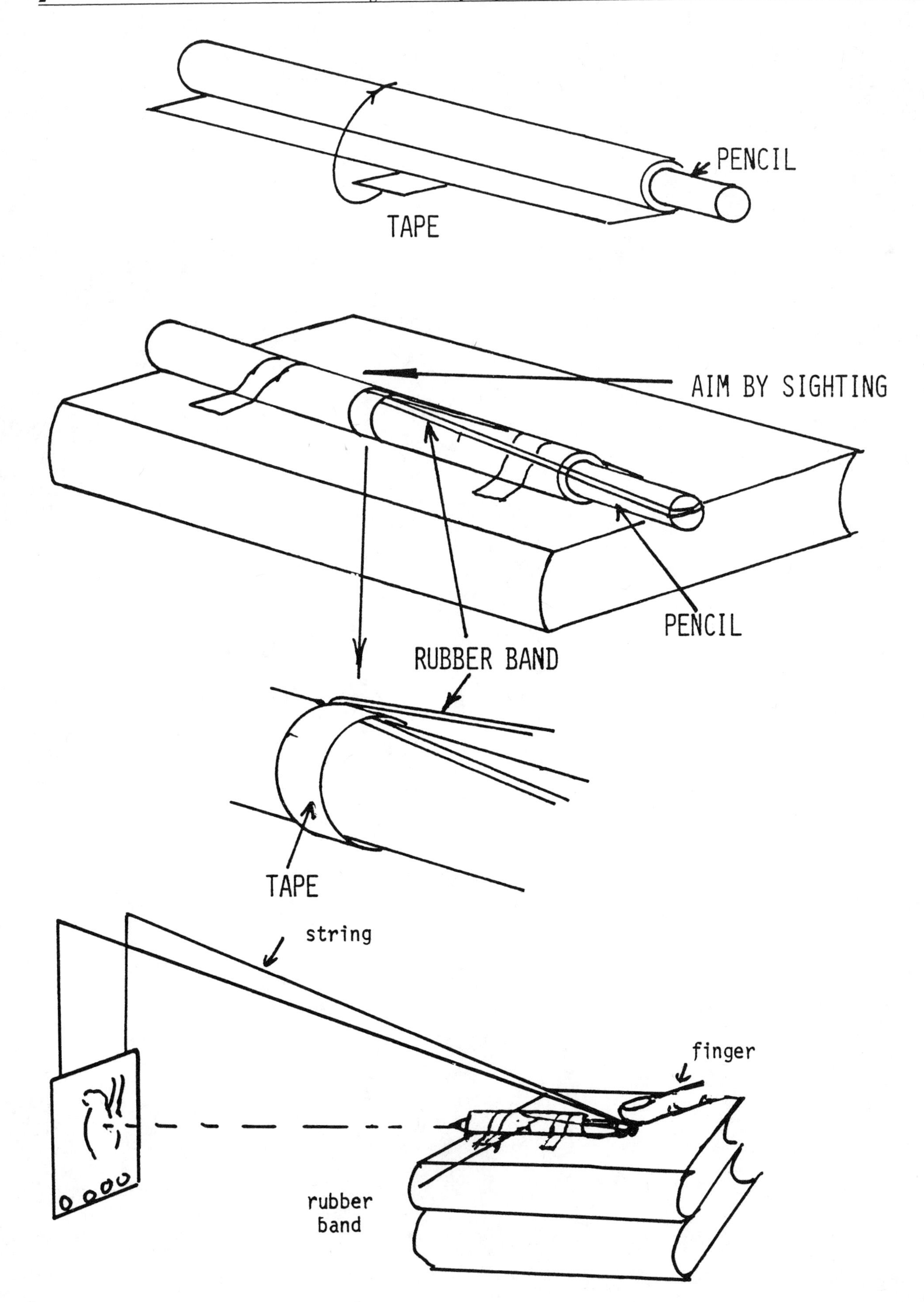
PENCIL
TAPE
AIM BY SIGHTING
PENCIL
RUBBER BAND
TAPE
string
finger
rubber
band

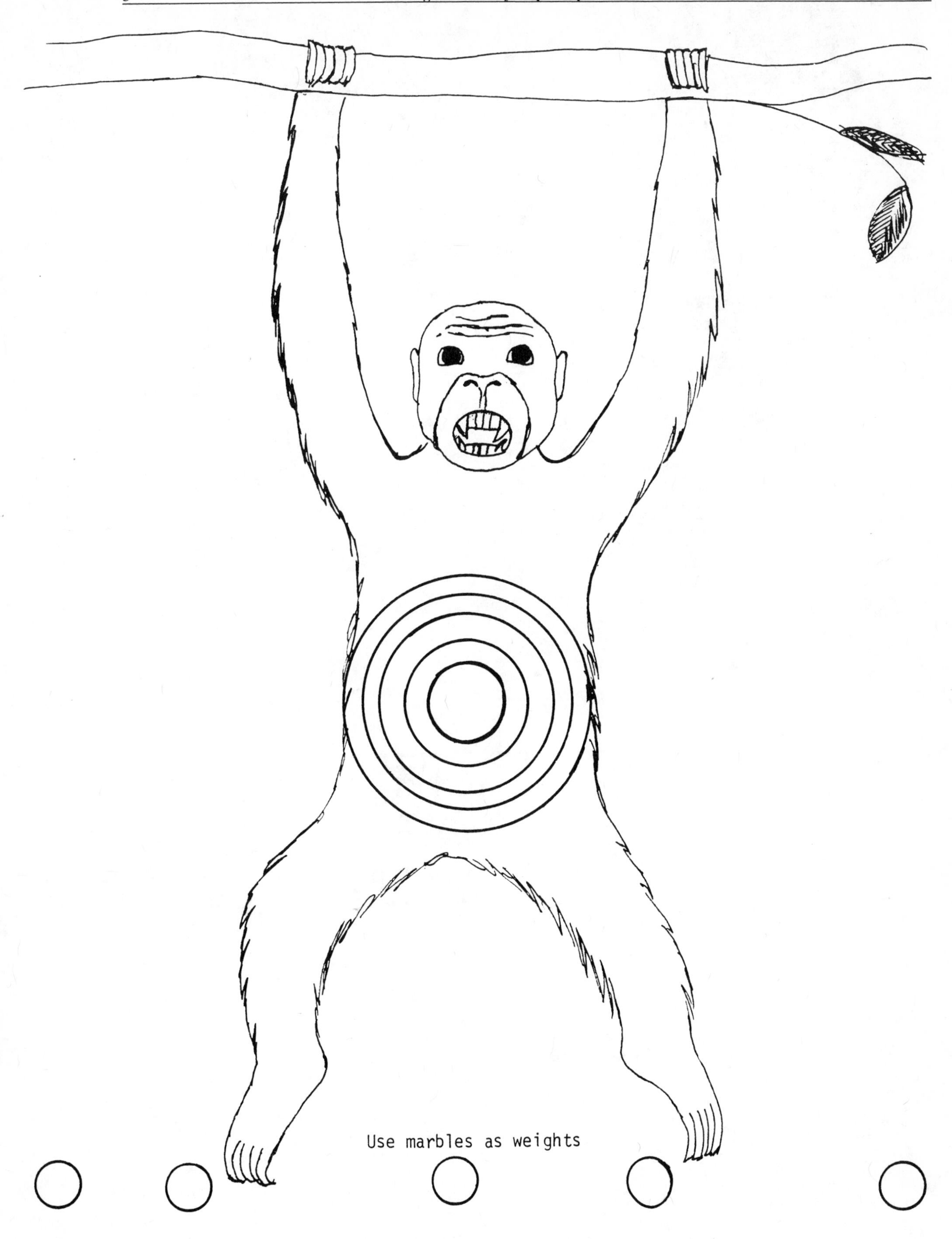
Use marbles as weights

Experiment 1.48 Angular momentum of coins

Materials: Two quarters and a nickel.

Procedure: Three coins are held as shown in figure 1, the nickel between the two quarters. The bottom coin is then released, holding on to the top coin. It is nearly impossible to release both sides of the quarter at the same instant, and the side first released starts at once to turn about the opposite side. This turning motion continues as the coin falls, and the nickel falls below the quarter, which is very odd to the uninformed observer. If it takes a fall of about ten inches for the coins to rotate 180°, many students believe it will take twenty inches for the coins to land with the nickel again above the quarter. In fact, of course, it is 40 inches. Once released, the coins rotate at a constant speed and so form a clock, since there is no torque, and the distance fallen will be proportional to the square of the angle of rotation. If we assume the quarter is 2.5 cm in diameter the velocity of the center must be about 15 cm/s at the point when the edge of the coin ceases to stick to the finger, to provide the correct angular velocity. Kinematically, this occurs after the center has fallen little more than a millimeter. In mentioning this to John Webb at Sussex University, he said "Oh yes - in fact you can win quite a lot of money in pubs with that experiment."

If, instead of simply dropping the coins you give them a slight twist as they are released, as shown in fig. 2, they do not turn over at all! The secret is in the gyroscopic forces involved. The twist makes the coins spin about a vertical axis. The small torque given the coins as they fall from the fingers, which previously turned them over, now merely gives a slight precession about the vertical axis.

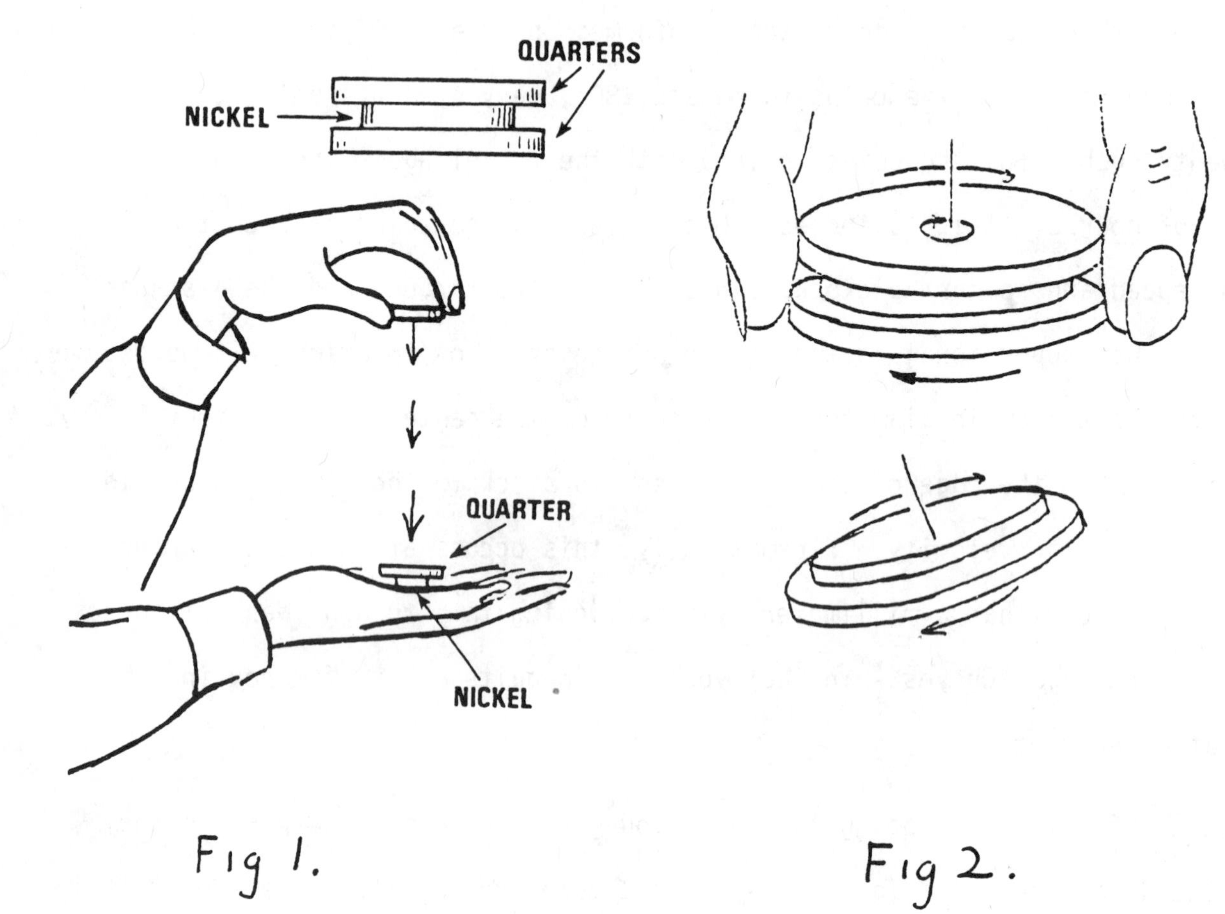

Fig 1.

Fig 2.

EXPERIMENT 1.49 Mobiles and Moments

Materials: marbles, straws, string

Method: The usual examples given for illustrating moments tend to be rather dull--balances of various kinds or levers lifting barrels. Mobiles provide an interesting, elegant and artistic way of demonstrating the principles of moments. The simplest way is merely to use combinations of marbles, attached to straws suspended by fine threads or strings. Paper clips can be used in place of marbles.

One example is shown in figure 1.

Moments give us that

$a \times 3 = b \times 1$ (Three marbles hung from a, one from b)
$c \times 1 = d \times 1$
$e \times 2 = f \times 4$
$h \times 6 = g \times 12$

Do experiments agree with theory?

Instead of marbles, we can use abstract cardboard cutouts, or other objects. The mass of the cutouts can be obtained from their area, for which one can count squares if they are drawn on squared paper--so the mass of the two objects drawn in figure 2 would be as 140 to 275 or $L \times 140 = M \times 276$. The mobile can be informative in other ways--one can make an atom, with a nucleus such as He^4, as shown in figure 3, or the planetary system (figure 4) -- though only the home planets can be displayed to scale, the others would be too far!

Furthermore, the abstract mobile shapes provide interesting experiments in the center of mass as they are supported from different positions.

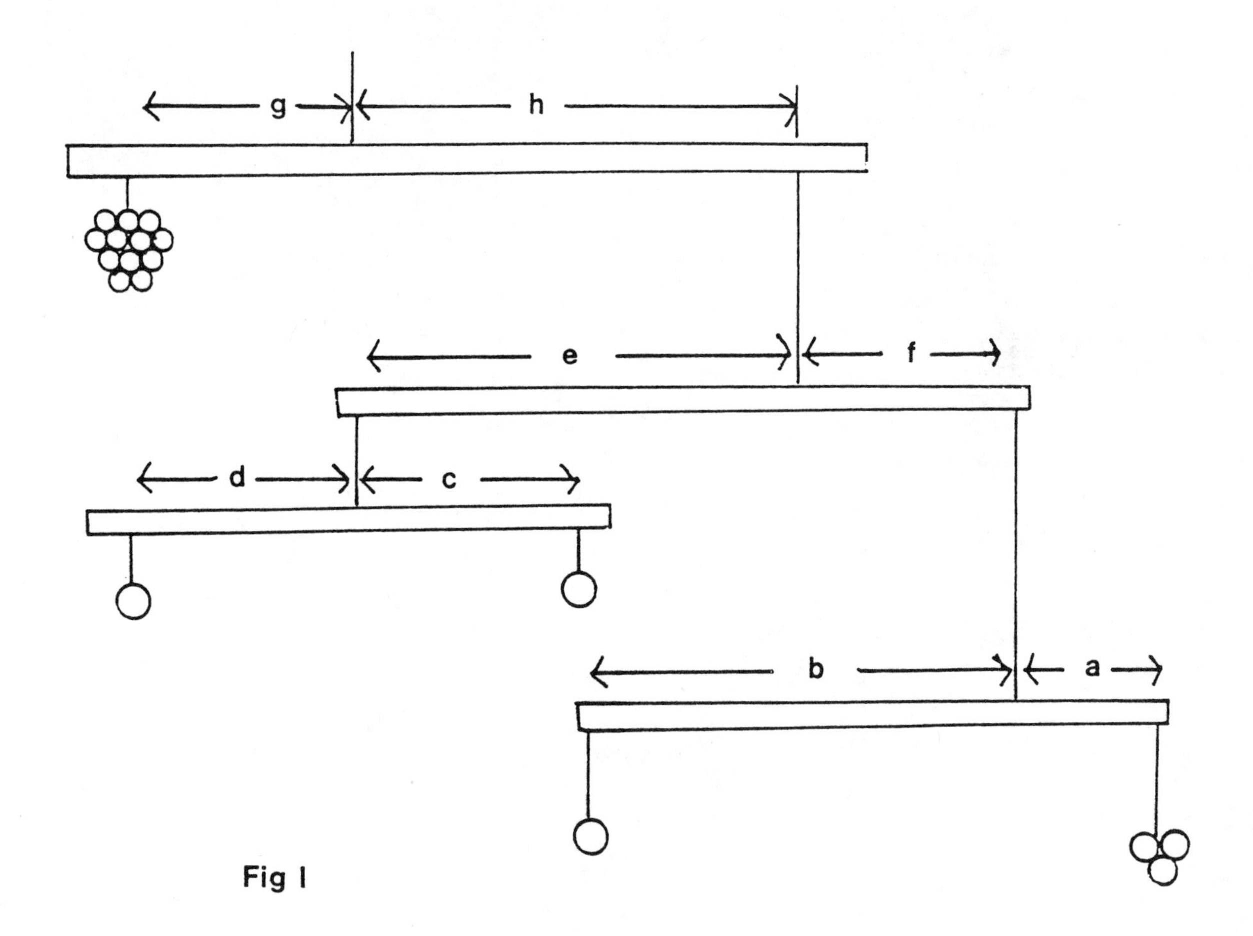

Fig 1

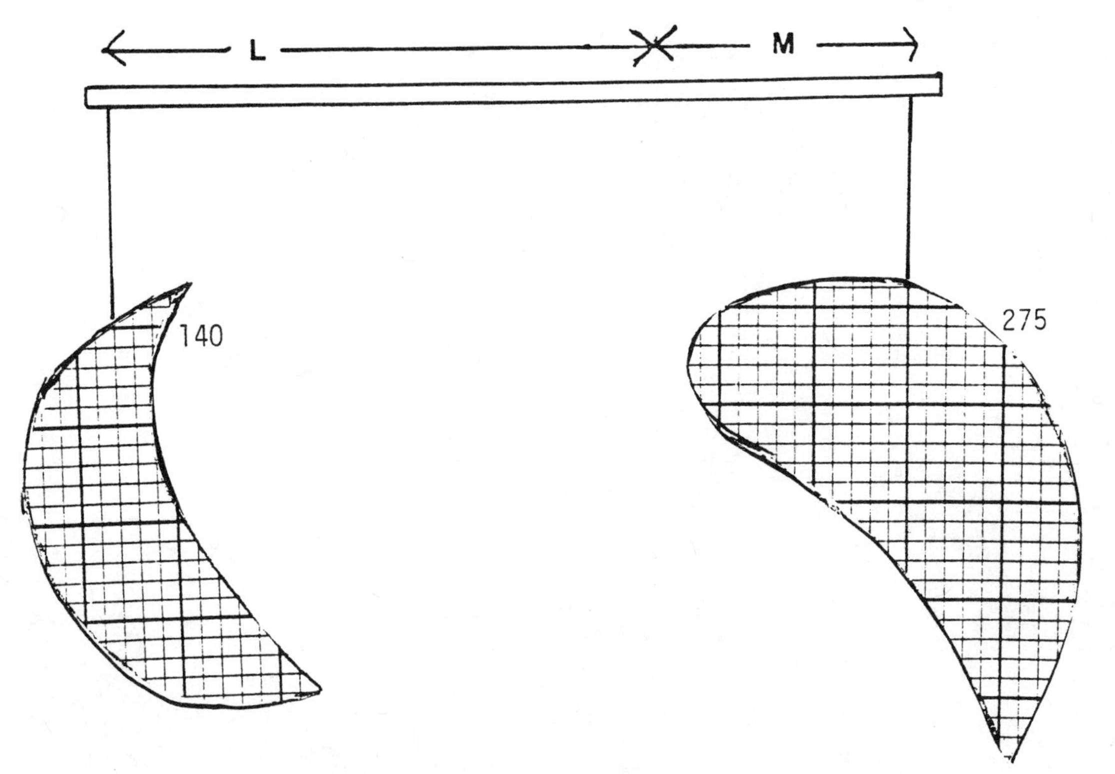

Fig 2

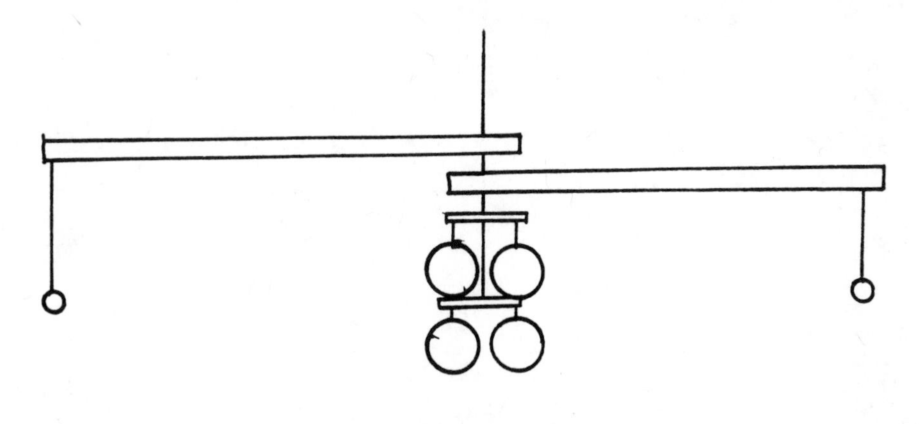

Fig 3

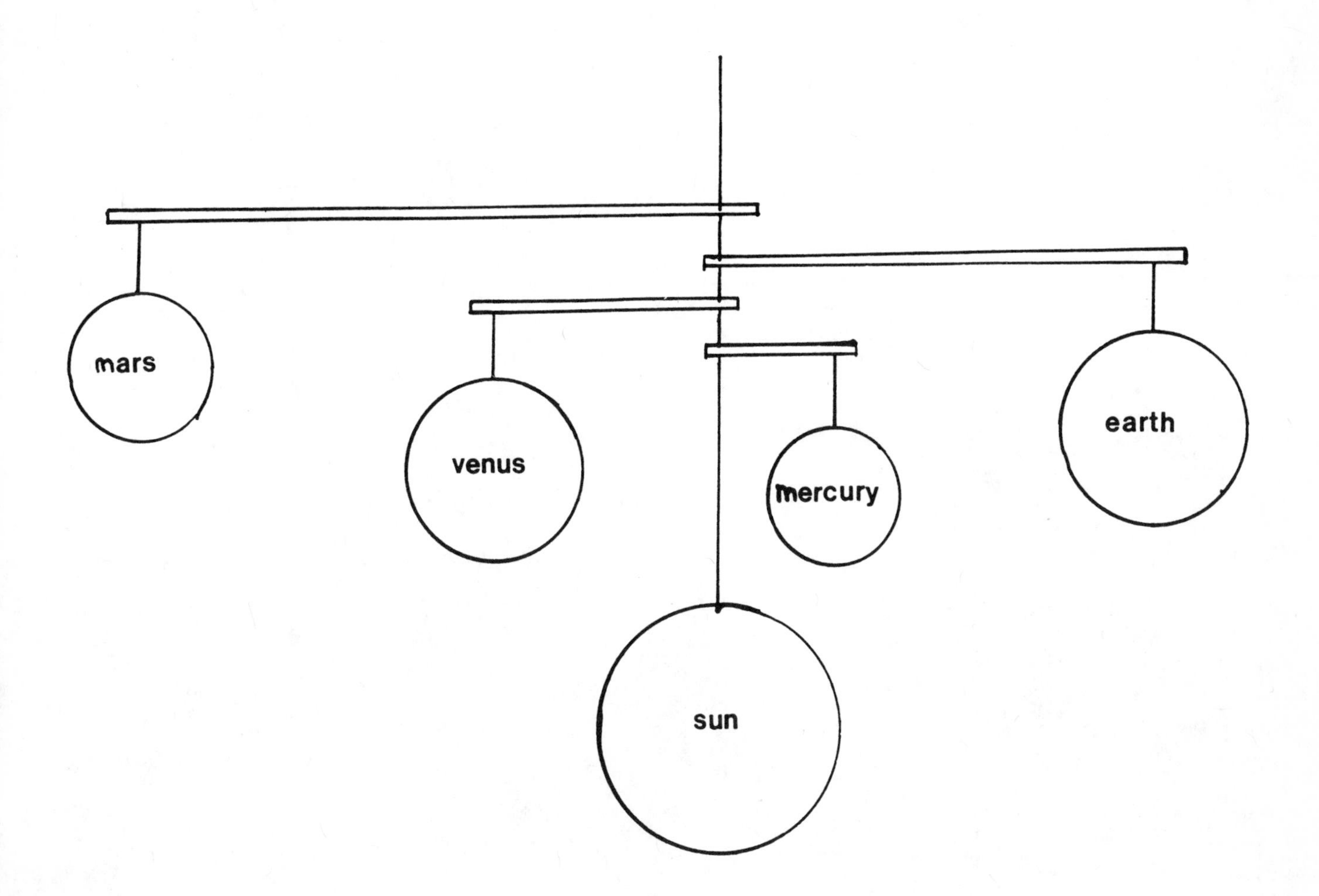

Fig. 4

EXPERIMENT 1.50 Angular Acceleration - the ball and ruler

Materials: ruler, tape, cup, marble

Method: Cut about 1" off the bottom of a styrofoam cup.

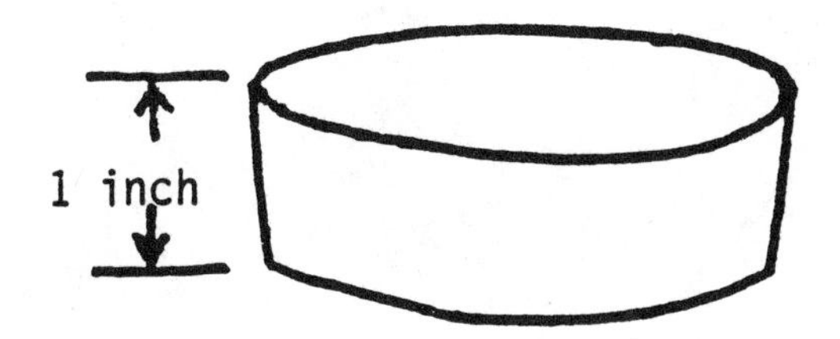

Push one side in so that it sits at about 30° to the vertical.

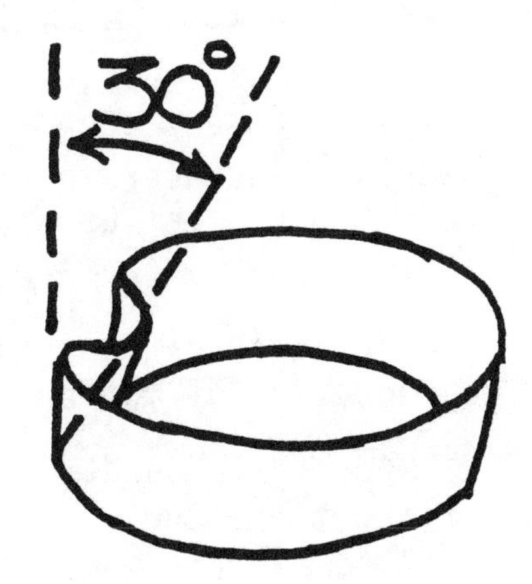

Now tape the cup one inch from the top of the ruler.

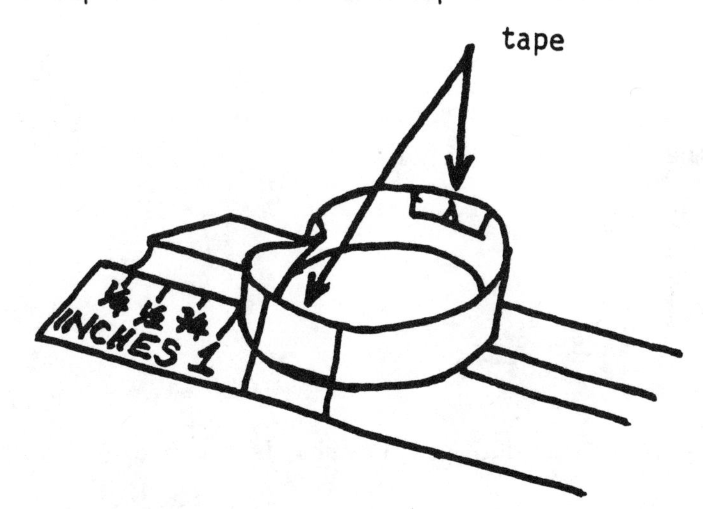

Hold the ruler at about 40° to the horizontal, with one end on the floor, as shown. Place a marble against the cup. Release the ruler, and the marble jumps into the cup. The device works best on a carpet which cushions the collision with the floor, which can make the marble jump out of the cup again. Why does the marble jump into the cup?

On a smooth table, it may be necessary to tape the end of the ruler to the table, to act as a hinge, and prevent slipping.

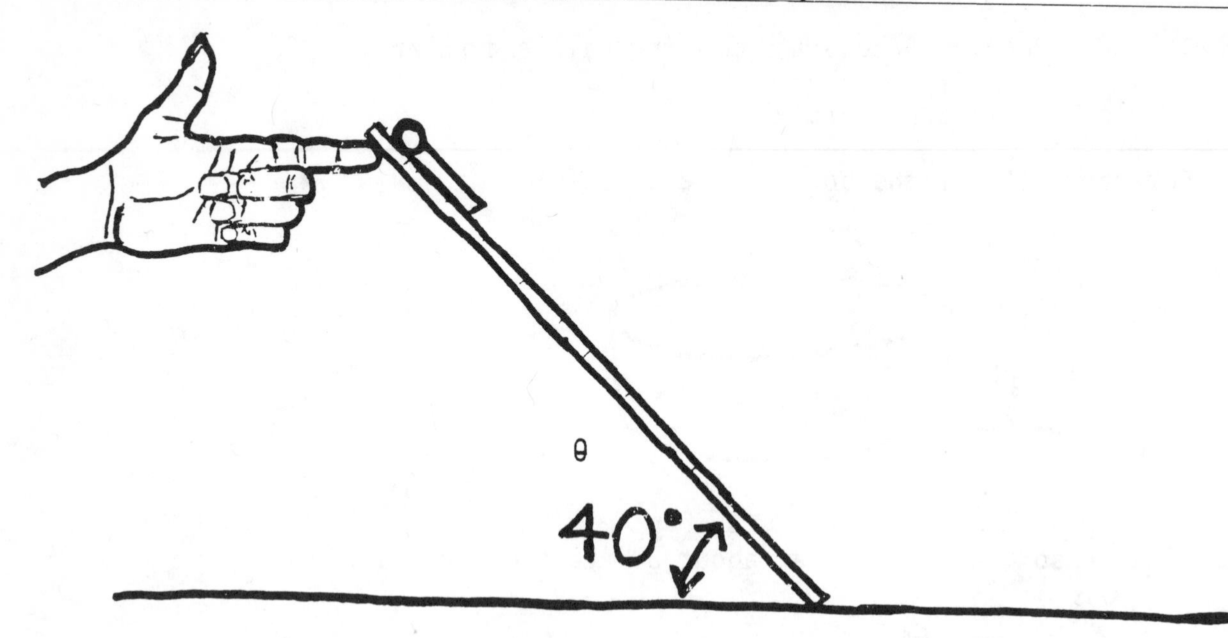

The cup accelerates more rapidly than g, and hence drops away from the marble. Roughly, we may say the center of the ruler wants to accelerate at g, and for this to be true, the end must accelerate more rapidly. Let us show how this works mathematically.

If the angle θ is varied, then only below an angle θ_c does the stick drop away from the ball. We can see this from the equation of motion of the ruler,

$$I\ddot{\theta} = -Mg\frac{L}{2}\cos\theta \qquad (\ddot{\theta} = d^2\theta/dt^2,\ \dot{\theta} = d\theta/dt)$$

where I is the moment of inertia of the ruler about one end since the rate of change of angular momentum Iθ equals the moment of the applied force. I is the moment of inertia of the ruler, and M its mass. Also, from geometry,

$$I = 1/3\ ML^2$$

Hence,

$$\ddot{\theta} = \frac{3g}{2L}\cos\theta$$

The end of the stick has a linear acceleration of

$$\ddot{S} = L\ddot{\theta} = -\frac{3g}{2}\cos\theta,$$

whose vertical component is

$$\ddot{Y} = \ddot{S}\cos\theta = -\frac{3g}{2}\cos^2\theta$$

Now, this only exceeds the acceleration due to gravity if $-3g/2\cos^2\theta < -g$

or $\cos^2\theta > 2/3$

which is true for angles smaller than about 35°. Miller, in his book "Physics Fun and Demonstrations," points out that this effect is related to the way in which tall chimneys buckle as they fall, since there is a torque trying to accelerate the end of the chimney faster than g when the angle is below 35°, which will break the chimney.

EXPERIMENT 1.51. Inertia, Momentum, and Blow Football

Materials: straws, paper, marble

Procedure: Most people have played blow football--a small light ball is placed between two contestants who endeavor, by blowing through tubes, to propel the ball over the opponents edge of the table. This note is to suggest the physical principles behind the game. Play one game using a marble and two straws. Now place a dampened piece of paper between the palms, and roll it into a ball about the same size as the marble. Play the game again and ponder the following questions:

In which game was it easier to get the ball moving?

The light paper ball is more mobile, and can easily be turned around--we say it has less inertial mass than the marble. The marble nicely exemplifies Newton's first law--a body continues in its state of rest or uniform motion in the absence of forces. The force you can deliver by blowing is the same in both cases, but you have to blow much longer to stop the marble, because for the same force, if the mass is higher, the acceleration (or deceleration) is slower, according to Newton's third law.

Experiment 1.52 Quantum Properties
(Suggested by Fream B. Minton)

It is often difficult to convey the idea of the quantum nature of the physical properties of substances. As an exercise in determining such properties, different numbers of marbles are placed in cups which are sealed with paper on top, so the number is not visible. Only marbles obtained from one source have a sufficiently constant mass to be used, and this should be checked. The number of marbles in each cup is arranged so that the differences between cups are both odd and even numbers of marbles, and no two cups are separated by a single marble. One empty cup is supplied, and the smallest number of marbles in a cup is five. The object is to determine the mass of one "quantum". Table I shows how the cups were prepared.

In the second part of this experiment the students are given a brief description of the famous experiment by Robert Millikan in which he determined the charge of a single electron. The students are not told his analytical method, only how data was obtained and what the numerical values mean. Pointing out that Millikan's problem was very similar to the "balls in the cup" problem, the instructor supplies a sample of data in which oil-drop charges are recorded. The students are asked to use this information to find the charge on a single electron. Table II is a sample of the kind of information supplied to the students. Although somewhat contrived, it serves the purpose.

By the time this part is reached, many of the students have discovered for themselves the principle that quantum units may be measured by determining differences between measured values and looking for the largest common factor. As a result, most of them are quickly successful in finding the charge on a single electron.

As a thought question, the instructor might ask if Millikan was justified in throwing out a small amount of data he obtained which indicated that the charge on an electron was one-third of the value he ultimately reported.

TABLE I

Can No.	No. of Balls
1	0
2	5
3	11
4	13
5	16
6	20
7	27

TABLE II

Oil Drop Event	Electric Charge x 10^{-19} Coul.
1	6.4
2	16.0
3	20.8
4	28.8
5	32.0
6	46.6
7	67.2
8	94.4

EXPERIMENT 1.53

The Brachistochrone - or, the longer way round may be the quickest way home

Students often find it difficult to distinguish between acceleration and velocity. If an object were not accelerated, the shortest path between two points in terms of both time and distance, would be straight line. The next simplest case occurs when a constant force acts on the body. Students find it hard to believe that the shortest time to travel between the two points does not now occur along a straight line unless the force acts along the line between the two points. The brachistochrone problem - to find the path producing the shortest travel time between two points for an object under constant gravitational force - exemplifies this difficulty. It was solved by John Bernoulli (1667-1748), and led to the formal foundation of the calculus of variations. In the 18th century, the Bernoulli family occupied a large number of mathematics chairs in the universities of Europe. John himself was an irascible individual - being "violent, abrasive, jealous, and when necessary, dishonest ... His son Daniel (who developed "Bernoulli's principle" in hydrodynamics) had the temerity to win a French Academy of Sciences prize which his father had sought. John gave him a special reward by throwing him out of the house."[1]

What would be the simplest way to attack the brachistrochrone problem? It should be noted that, although there is a constant vertical force, the force and hence the acceleration along the trajectory need not necessarily be constant. Experimentally, we could construct a series of curves joining two points considered. Then we could try each curve to see which path could be traveled in the shortest time. The calculus of variations performs such an iteration mathematically, varying the path to determine which provides the shortest time. The solution turned out to be a cycloid - the curve traced by a point on the circumference of a wheel rolling along (you may remember the question - which part of a fast-moving train is stationary? That part of the wheel in contact with the track. The subsequent motion of the point in contact with the track forms the cusp of the cycloid).

It is easy to draw a cycloid. Tape a piece of string to the end of a can, wrap it one or two times around the can, and tape the other end to a ruler, laid on a sheet of paper. Wrapping the string forces the can to roll along the ruler without slipping. Now, hold the tip of a pencil against the can (you can dent the can to make a groove for the pencil tip to sit in) and, as you roll the can along the ruler, the pencil traces the cycloid. The size of the can will determine the dimensions of the cycloid.

For students with little physics or mathematics, the mere words "brachistochrone" and "cycloid" are offputting, however, it is possible to devise a simple experiment to show that the cycloid (or a curve close to it) provides the path of shortest time. Common sense arguments can be used to reinforce this result. Since the time taken to travel between two points is the distance divided by the average velocity, it would be a big help to accelerate rapidly at first, and quickly gain a high velocity. The cycloid starts off almost vertically (Fig. 1), so we get the maximum acceleration at first. This means it takes a shorter time for the path than a straight line incline, which provides constant acceleration. However, for a curve having a more extended vertical portion, the increased distance from start to finish is not compensated by the larger vertical acceleration: Our experiment provides three paths, a straight line, a cycloid, and a path with a larger initial vertical drop.

The cycloid is outlined in a thick black line in Fig. 2. The figure may need to be enlarged - we found an 8½ x 11 sheet of paper or card works well, but anything larger is better. The simplest arrangement is to use sticky tape to stick straws over the section indicated in the figure. We approximate the curves by a series of straight lines, which works well for the purposes of timing the ball along the trajectory, and it is easy to enlarge the figure - relative distances are marked to help with this. It would be an improvement to tape a piece of narrow rubber hose, if available, along the cycloid curve.

Place the sheet on a large book or piece of stiff cardboard and tilt it, so that the bottom of the sheet remains horizontal. Start two marbles rolling along the cycloid and the straight line simultaneously. That rolling along the straight line clearly takes longer than that on the cycloid since it will arrive at the bottom behind the other. Now start one on the cycloid and one on the bottom path - again the cycloid wins. One can vary the speeds by tilting the book more or less. The principle is the same, with the curves vertical, although it is not possible to demonstrate it with this little experiment.

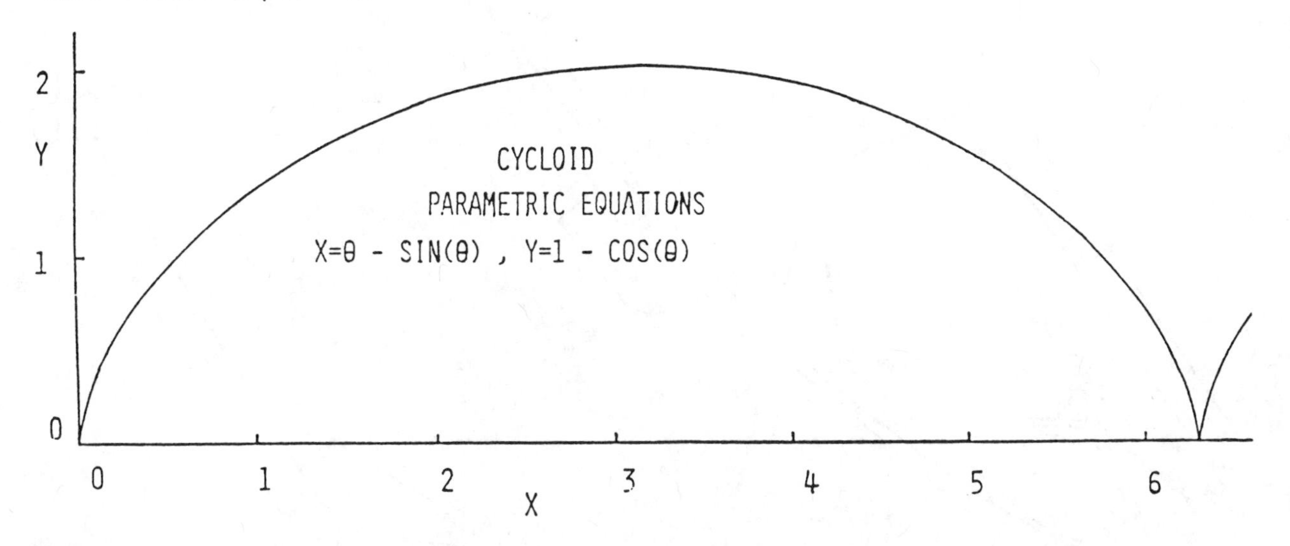

Reference

1. J. R. Newman, Vol. II World of Mathematics (Simon and Schuster, New York, 1956).

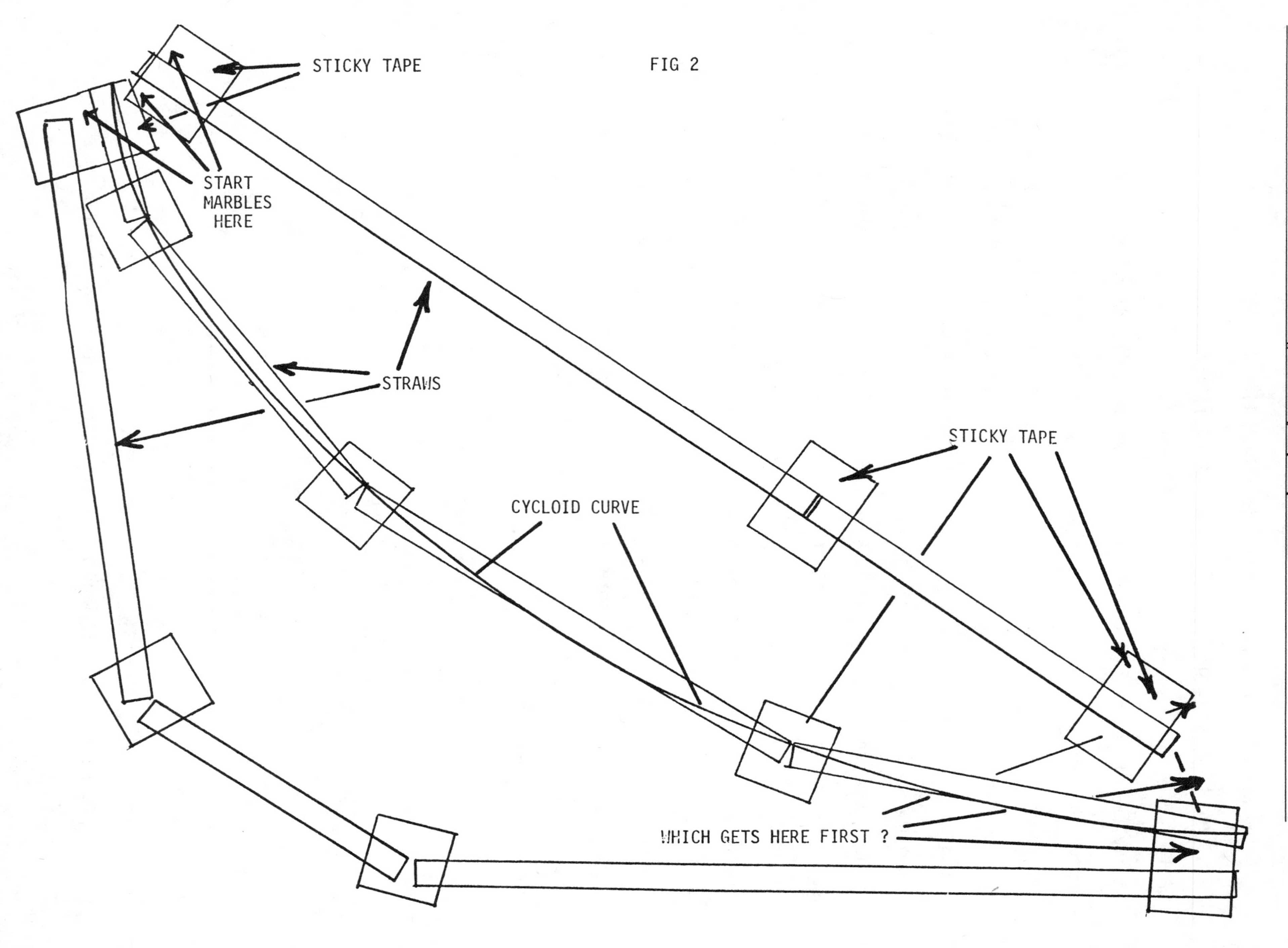
STICKY TAPE
FIG 2
START
MARBLES
HERE
STRAWS
STICKY TAPE
CYCLOID CURVE
WHICH GETS HERE FIRST ?

Experiment 1.54

Levitating a Marble - Motion in a Circle

If you cut the bottom off a styrofoam cup, can you use that cup to pick up a marble without touching the marble with your fingers? Place the cup over the marble, and swirl it round and round, as shown in figure 1. When the rotation speed is fast enough, the marble will rise up the wall, because of the reaction of the cup on the marble. The taper of the cup provides a vertical force which balances the gravitational attraction. The cup can then be safely lifted without the marble falling out. We can even measure the constant of gravitation g from this. Let the radius of the cup at which the marble rotates be r, the angle of the cup wall be $\Theta°$ to the vertical and the number of rotations per second be n Hz. The period, T is then 1/n sec. The velocity, v, of the marble is $2\pi r/T$, and the radial force is $mv^2/r = m(2\pi r/t)^2/r$. Since the reaction of the cup is normal to its surface, the ratio of the vertical force, mg, to the horizontal radial force must be

$$\tan \Theta = mg/mr(2\pi/T)^2 = g/r(2\pi/T)^2$$

Using a 6 oz. styrofoam cup, the internal diamter of the top is 7.4 cm, and of the bottom 4.8 cm. The cup (less the bottom) is 8.6 cm high, so

$\tan \Theta = (7.4 - 4.8)/2x8.6 = 0.15$

Now, from the previous formula

$$T = 2\pi\sqrt{r \tan \Theta/g} \qquad (1)$$

If we take it half way up the cup, r is 3.05 cm, from which we subtract the radius of the marble 0.8 cm. With these values, n = 8.54 Hz. Experimentally, it lies between 7 and 9 Hz, because the marble moves up and down the cup. If you quit after getting the marble to spin round the top of the cup the marble loses energy to friction, descends the cup, and kinetic and gravitational potential energy is converted to heat. Since the velocity is given by $v = g\sqrt{r/\tan\Theta}$, and r is decreasing the marble will slow down, but equation (1) shows the period decreases with r, so it will make more revolutions per second. (This experiment is a product of the Minnix & Carpenter Summer Institute at V.M.I.)

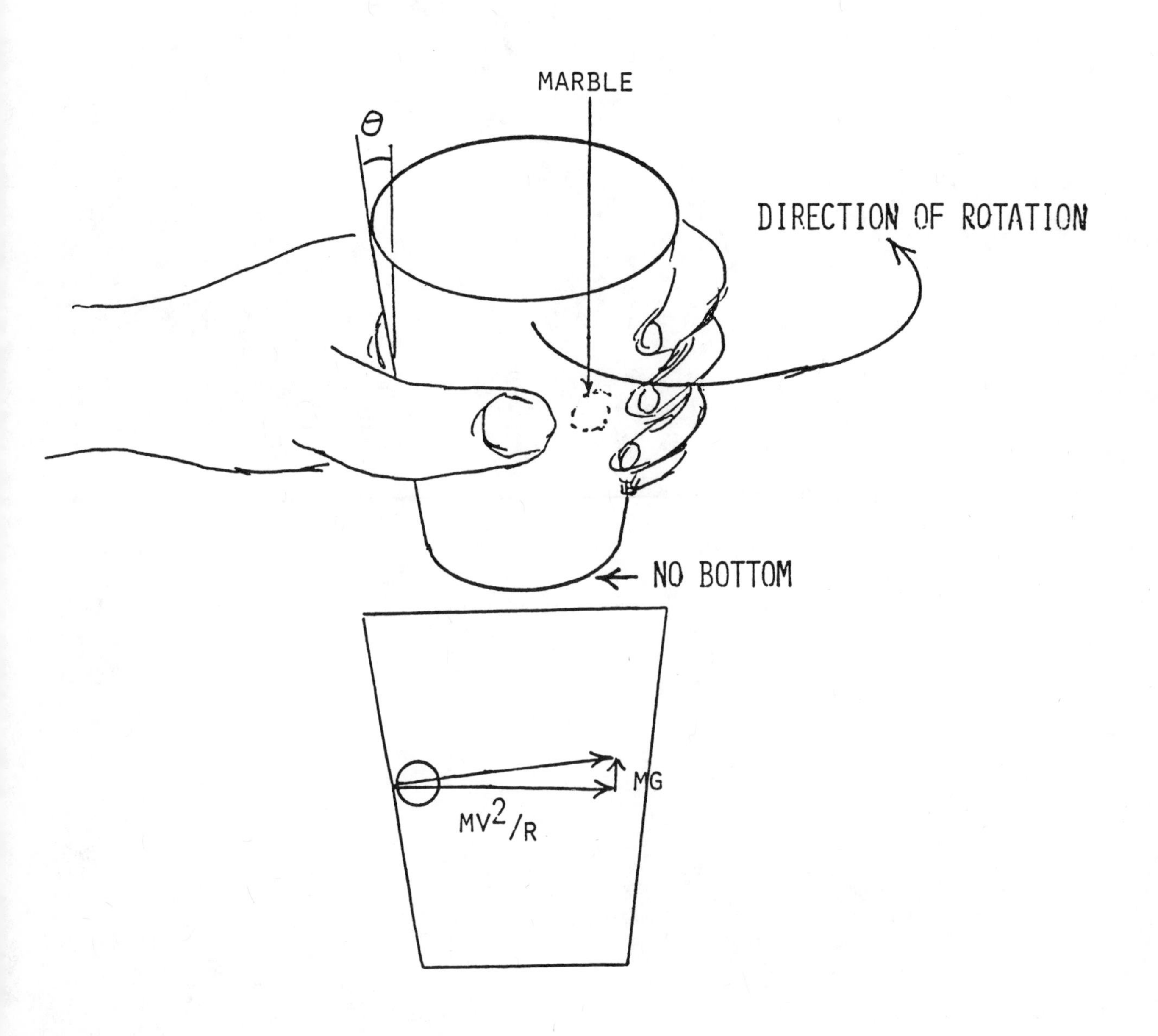

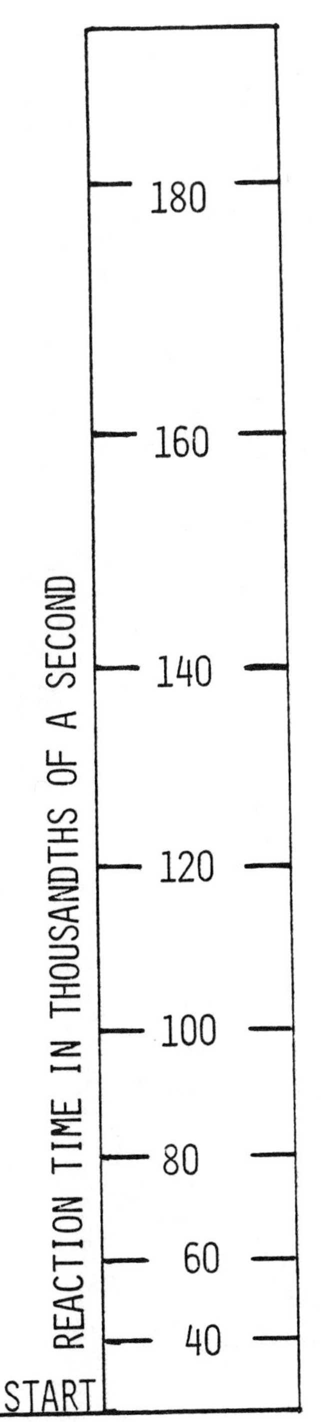

EXPERIMENT 1.55

An application of g - the reflex tester

The test of someone's reflexes whereby one person hold a dollar bill vertically and the other holds thumb and fore finger on opposite sides of the bill. He (or she) is told they can have the bill if they can catch it by pinching it. In fact, it is virtually impossible to catch the note - one's reactions are too slow. Using $S = \frac{1}{2} gt^2$, it is easy to mark on a piece of paper -- as shown here, the time it takes the sheet to fall that distance. Hence, we have a calibrated value for our reaction time.

Experiment 1.56 Trajectories, Drops and the Strobe

In addition to psychophysical experiments simulating motion using persistence of vision, a stroboscopic disk is a handy device for many other occasions - for example, although a fluorescent light looks white, in fact, for part of the A.C. cycle it is blue, and for part red, averaging out to white. Let's see how we can observe this.

Figure 1 shows how to construct a stroboscope with a wide range of speed. We need this because the speed with which we can spin the disk is restricted. It is not possible to spin the disk very rapidly - about 2-3 Hz is as fast as convenient - nor can one spin it too slowly - variable friction causes a variable speed. Cut out the disks in figures 2 and 3. It is best to glue them on to thick card or thin plywood and attach them to a wooden support with a pin. However, more simply, after cutting out, they can be attached back to back with four paper clips round the edge. Push a thumb tack through the center of both, with the black disk outward - then into the eraser end of a pencil, as shown. The disks can now be spun with one finger and the fluorescent light observed. At a suitable speed, a stationary pattern as shown will be seen. One can vary the number of slits on the disk from one to twelve by rotating one disk with respect to the other. The alternating stripes of blue and red occur because the mercury discharge part of the light occurs over less than half a cycle, and is bluish, the fluorescent part of the light over the full cycle, and is reddish. The spectrum is shown in figure 4.

Now, adjust a water tap so that a fine stream of water falls from it, breaking into drops just before striking the sink. You can adjust the speed of the strobe to the point where a drop appears stationary. Actually, of course, you are seeing successive drops separated by the time it takes the strobe to move the angle between two slits. Notice how the distance between the drops increases as they fall because of the acceleration due to gravity. The way in which the stream breaks up into drops is interesting - there is one tiny drop (called "Plateau's spherule") between any two big drops.

What other things can we observe through the strobe? Anything which is periodic - a tuning fork, a ruler held to the table at one end and weakened to vibrate. One interesting object acoustically is a bowed violin. A cello is easier to see, being bigger. Put the cello on the table and have someone bow it. Look along the length of the string through the strobe. One thinks of the violin string as having a uniform vibration too and fro. What you will see is a spike (called the Helmholtz spike) running up and down the string. A guitar string, when plucked, has two spikes running in opposite directions, which die out rapidly. An electric fan can be made to appear stationary - but it can also be made to look as though it was moving backwards. Which is moving a little faster, the fan or the slots in the disk, to make the fan appear to move backward, assuming they move at the same rate when the fan appears stationary?

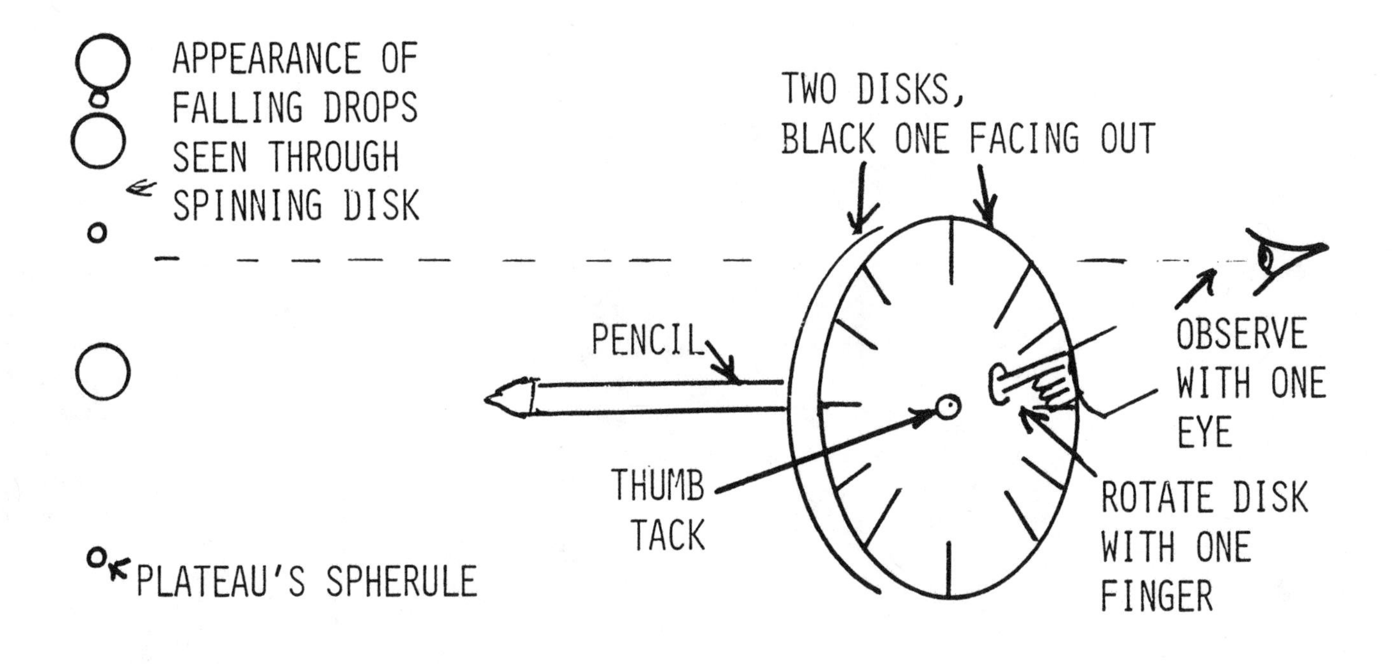

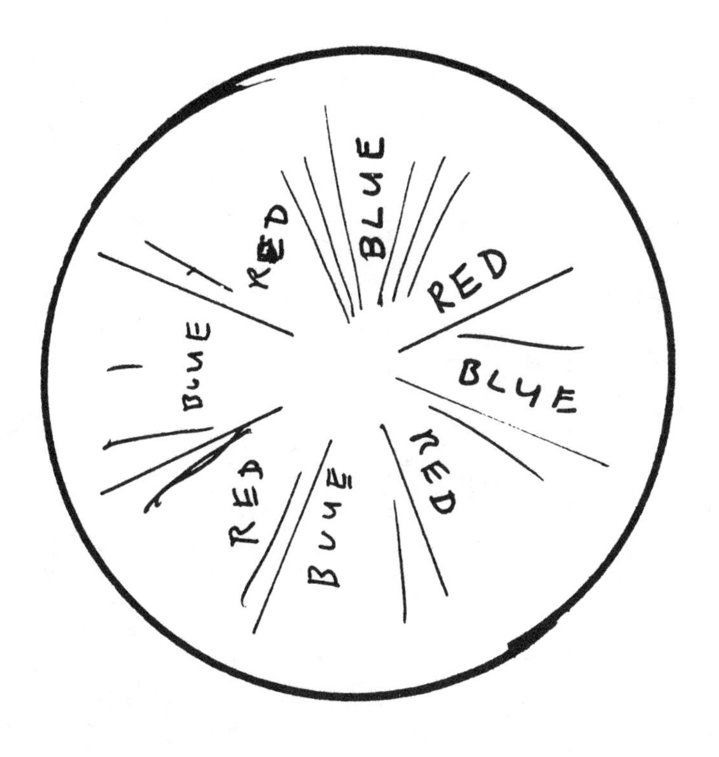

APPEARANCE OF
FLUORESCENT LIGHT
ON LOOKING THROUGH
ROTATING STROBE

Fig 1

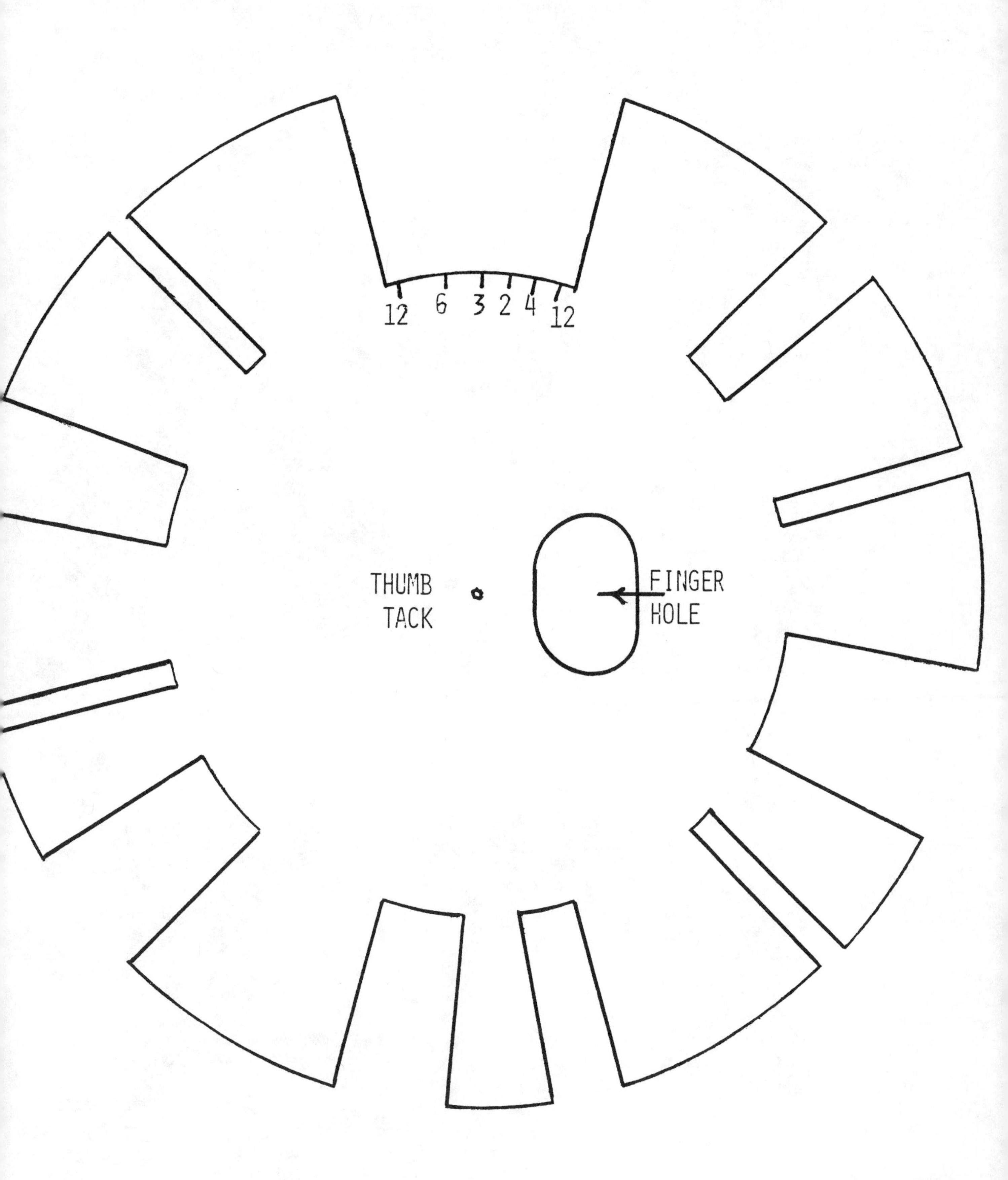
12
6
3
2
4
12
THUMB
TACK
FINGER
HOLE

Fig 2

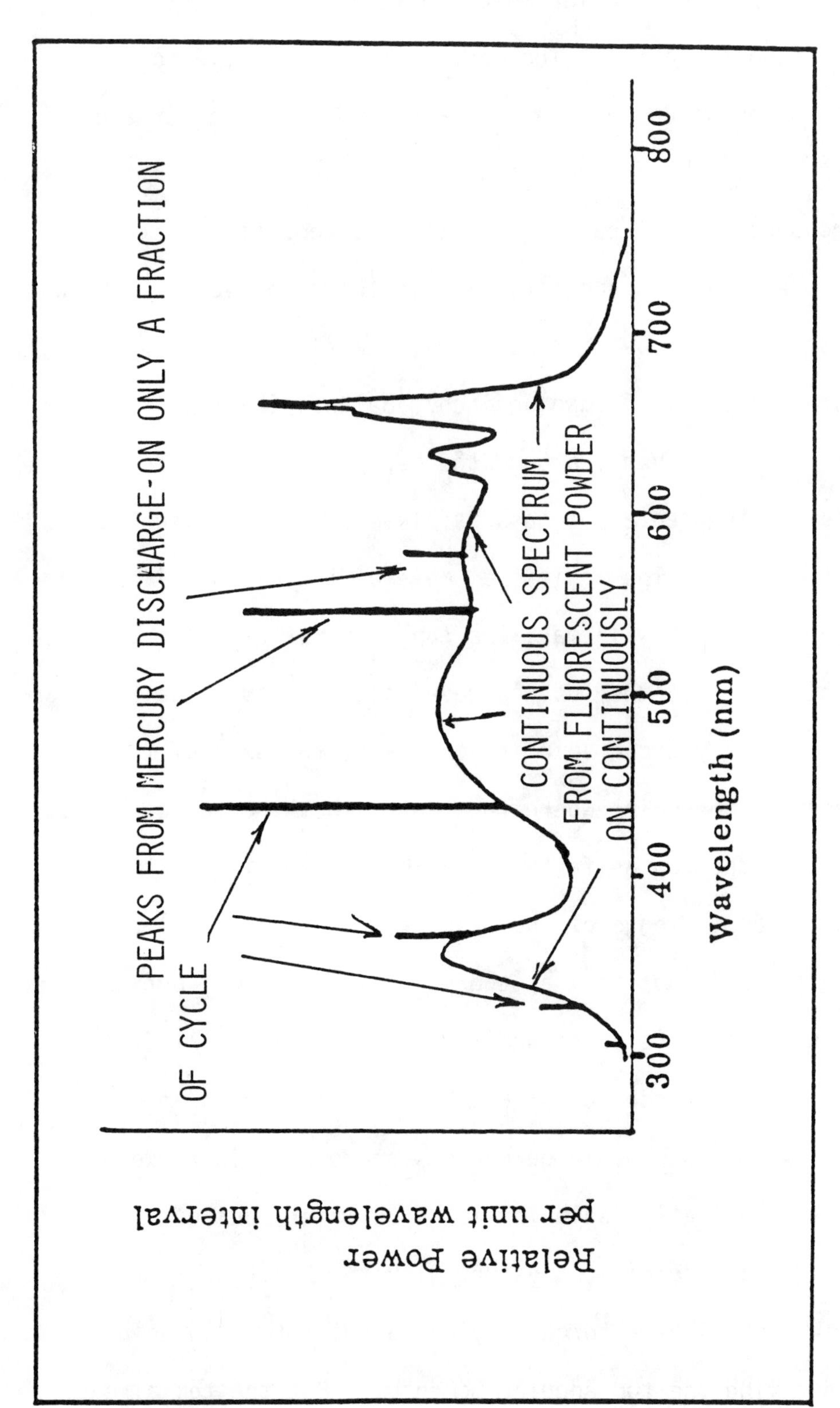

Spectral power distribution of a fluorescent tube (Artificial daylight, Thorn Lighting Ltd.)

Fig 4

Experiment 1.57 - SPINNING TOPS

The most intriguing aspect of angular momentum is the spinning top - and its many relations (tippetop etc.). The most obvious properties of the top are precession and nutation, and we must see how these are started by giving an impulse to the top. To make a top, cut a disk about four inches in diameter out of cardboard (or you can use the plastic top off a nut tin etc.), sharpen your shortest pencil (about two inches long) and push it through a hole, which you can start with a pair of scissors, exactly through the center of the disk but it must protrude less than an inch. Balance the top, if it needs it, by pushing paper clips evenly around the periphery of the disk, as shown. (Fig. 1) Try to get the center of mass as close as possible to the point about which the top spins. The additional mass of the clips will provide a greater moment of inertia, which will make the top spin better. The clips can be struck on with a little tape so they do not move once the top is balanced.

Now spin the top - unless you can spin it very fast, the axis of the top will slowly rotate about the point where the top touches the table. This is called precession. Spin the top as fast as possible, then blow vertically through a straw at a point on the edge of the disk, as hard as possible, as shown in figure 2. The top will not fall at the point where blown, but one side will slowly rise at 90° to this point and the opposite side will fall as shown in the figure. This gyroscopic torque is the basis for the toy gyroscope, the gyroscopic compass, and in guidance systems used in rockets. What makes it work? - A quantitative solution is given later, but qualitatively, blowing on the edge of the disk, together with the reaction where the top axis touches the table provides a torque trying to twist the top down where the air presses it, and giving the top angular momentum. But the top already has a large angular momentum about its axis - and the vector representing this, which is like an arrow pointing along the axis rotates to put a larger component of angular momentum along the arrow representing the torqe. Strangely enough, this moves the top in a direction perpendicular to that we would expect - i.e. to move down where the puff of air hits it - instead rotating the axis of the spining top at right angles to the direction the air would appear push it.

Gravity acting on the center of mass of the top provides a similar torque - and again, the top moves at right angles to the direction we would naively expect - this is precession - the faster the top spins, and the larger the axis to the vertical, the faster the precessional rotation. Lastly, if we give a sharp puff of air - an impulse - to the edge of the top, it will bounce up and down as it precesses - as shown in figure 3 - this is called nutation. Even objects as large as the earth precess and nutate - but very slowly - in the precession of the earth, the axis take 20,000 years to make one rotation. Effects such as the growth of large mountains, which change the moment of inertia of the earth, can affect the rate of precession.

Quantitative -

The quantitative "why" of the spinning top depends on realizing that angular mometnum can be represented as a vector - so a top spinning right handed seen from above is represented by a vector pointing downward along the axis, of magnitude $I\omega$ where I is the moment of inertia and ω is the angular velocity (2π x the number of rotations per second). Correspondingly, the torque, due to gravity, of magntiude $mg\ell$, shown in Figure 4, can also be represented as a vector pointing as shown. Now this torque must accelerate the top, and it does so by increasing the angular momentum by $G\delta t$ in time δt -

however, this change in momentum is actually accomplished by rotating the axis perpendicular to its length. Since the angular momentum is $I\omega$, the change in $I\omega$ in δt must equal $G\delta t$ - and this must occur in the direction of G - the y direction in the diagram.

So $\delta(I\omega)_y = G\delta t$

From the diagram $\delta(I\omega) = (I\omega)\theta = (I\omega)\Omega\delta t$

and $I\omega\Omega\delta t = G\delta t$

$mg\ell = I\omega\Omega$

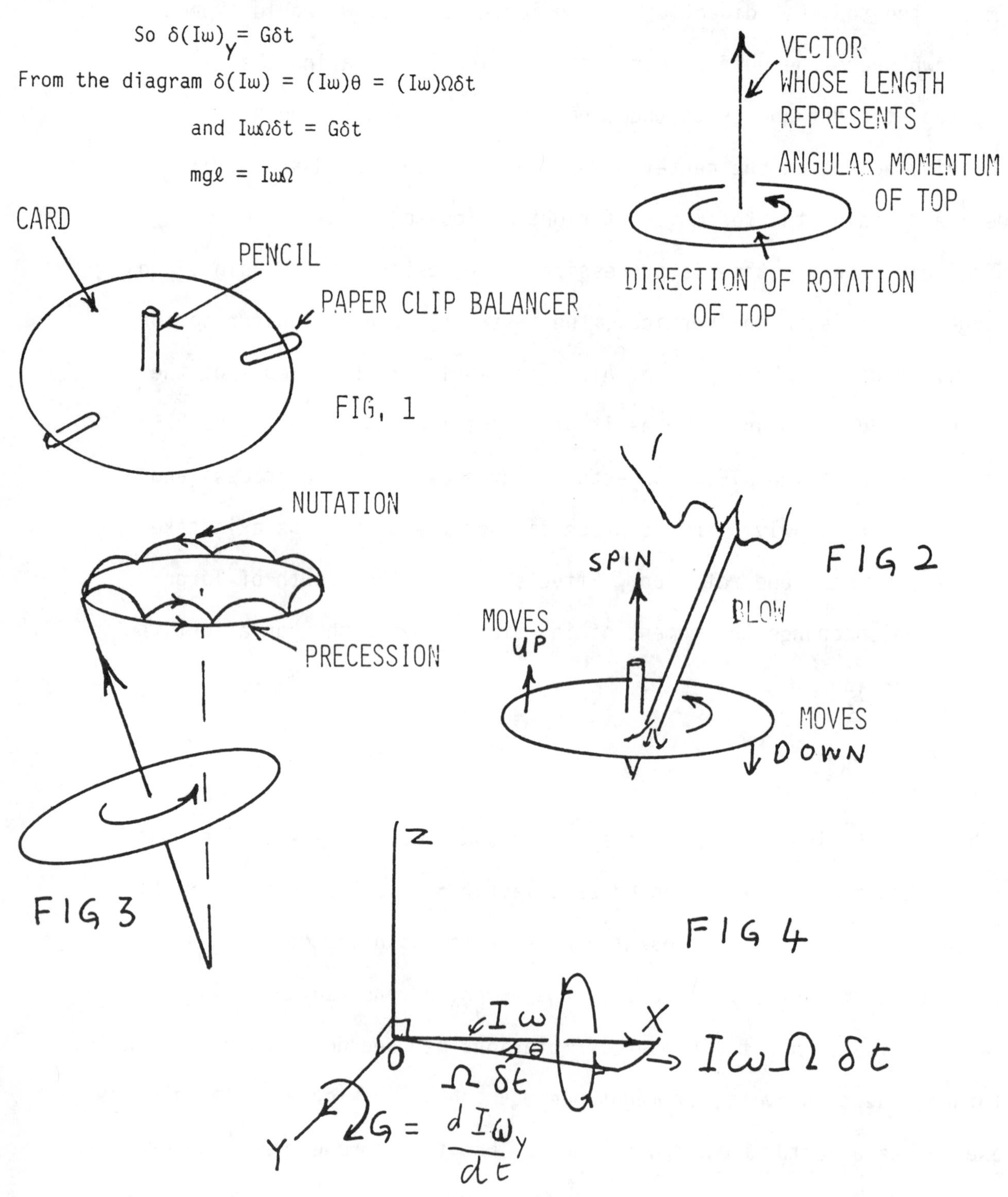

Experiment 2.01 Properties of matter-stress and tension in chalk

Materials: - an unbroken piece of chalk

Take an unbroken piece of chalk, Grasp the opposite ends and pull, taking care not to twist or bend it.

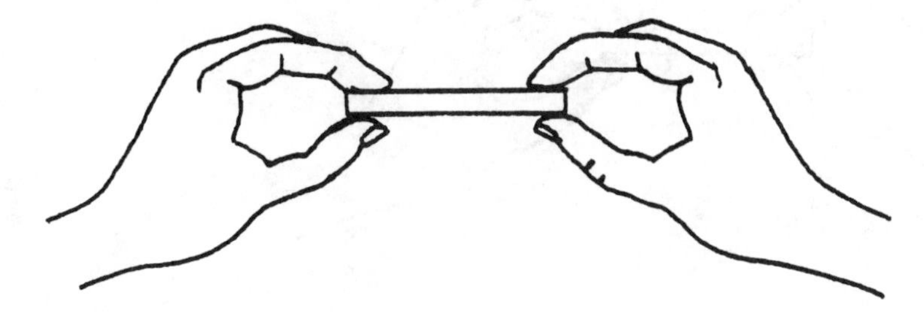

The chalk will break cleanly across

This is because chalk, being a compacted dust, can easily withstand compression, but very little tension. However, a crack, once started in the chalk, propagates all the way across, as a sheet of paper tears: -

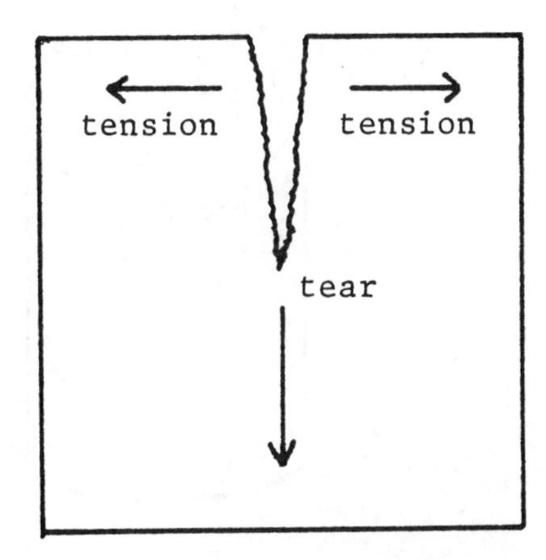

See Feynmann Lectures in Physics Vol. II p. 39-9.

Experiment 2.02 Properties of matter - torsional stress in chalk

Materials: - an unbroken piece of chalk

This is a continuation of the last experiment. Grasp the opposite ends of the piece of chalk, and twist without pulling or bending.

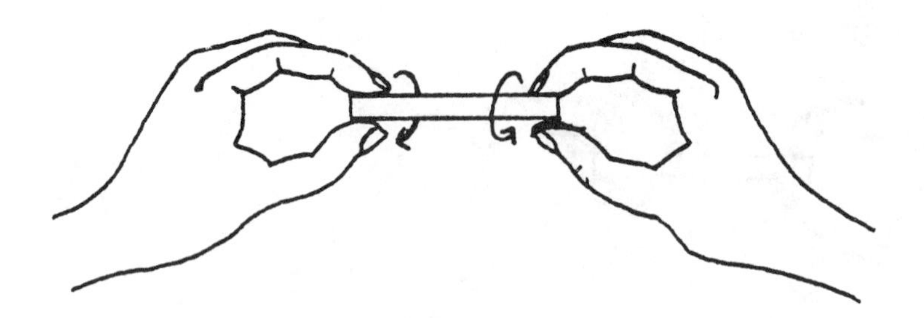

The chalk will break, with at least one portion showing a helical edge which is at 45° to the axis of the chalk.

This arises because the torsional stress strains the chalk in two directions at right angles.

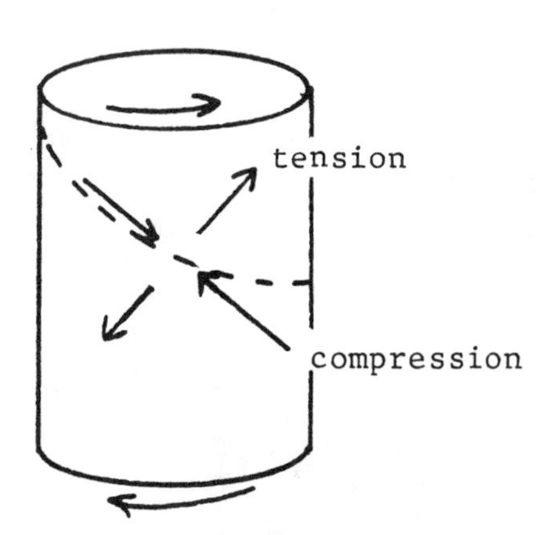

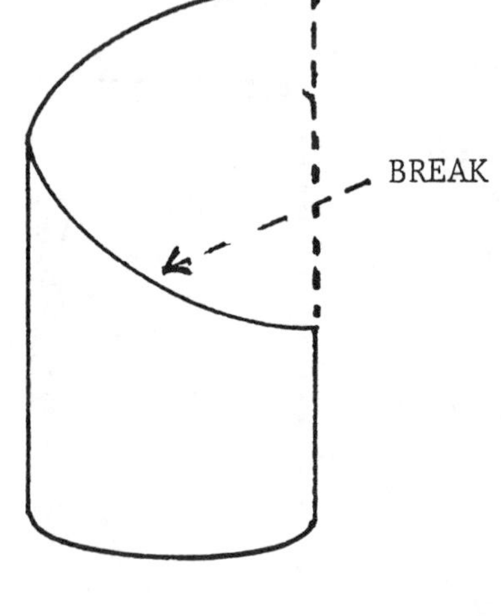

One strain is a compression at 45° to the axis of its torque, the other is a tension, also at 45° to the axis. The chalk cannot withstand the tension, and breaks perpendicular to the tensile force.

Experiment 2.03 Surface Tension

Materials: - cup, two paper clips

Procedure: - Bend one paper clip as shown

Lay the other one across it

Fill the cup with water. Now, slowly lower the clip into the water, and, if done carefully enough, and provided the clip is perfectly flat, it will float. Rubbing the clip down the side of one's nose, to grease it, helps repel the water. The way in which the clip lies on the surface "skin" is clear to see - the surface dimples under the clip. Sometimes one or two clips may have to be tried, until one works, because of sharp edges.

Qualitative Questions:

(1) Why must you lay the clip gently on the surface?

(2) Why would an aluminum clip work better than a steel one?

In the body of the liquid, the attractive forces between molecules act on the molecules in all directions.

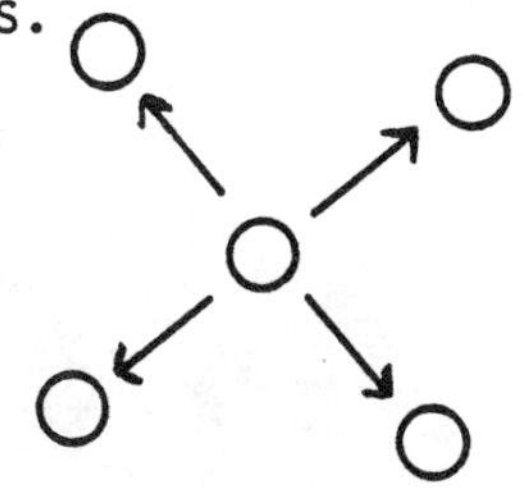

However, in the surface, they can only act inward.

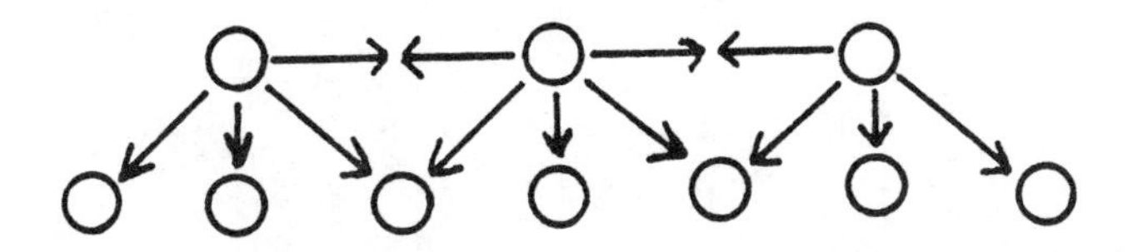

This forms a surface "skin", and the attraction between molecules in the surface pulls inward to hold the liquid in drops. If a group of people join hands in a circle, and pull together, they move inwards, in the same sort of way.

Experiment 2.04 Surface Tension in a liquid stream

Materials: Styrofoam (or paper) cup, sharp pencil

Procedure: Poke three small holes in the side of the cup, as close together and near the bottom as possible. Three streams of water will shoot out. Now run one finger close to the cup through the streams and note how they join together even after you've removed your finger.

Qualitative: What causes the streams of water to join together? Notice how they spiral about one another.

Quantitative: Can you estimate the surface tension force from the point where the streams join the size of the holes and the rate at which the cup empties?

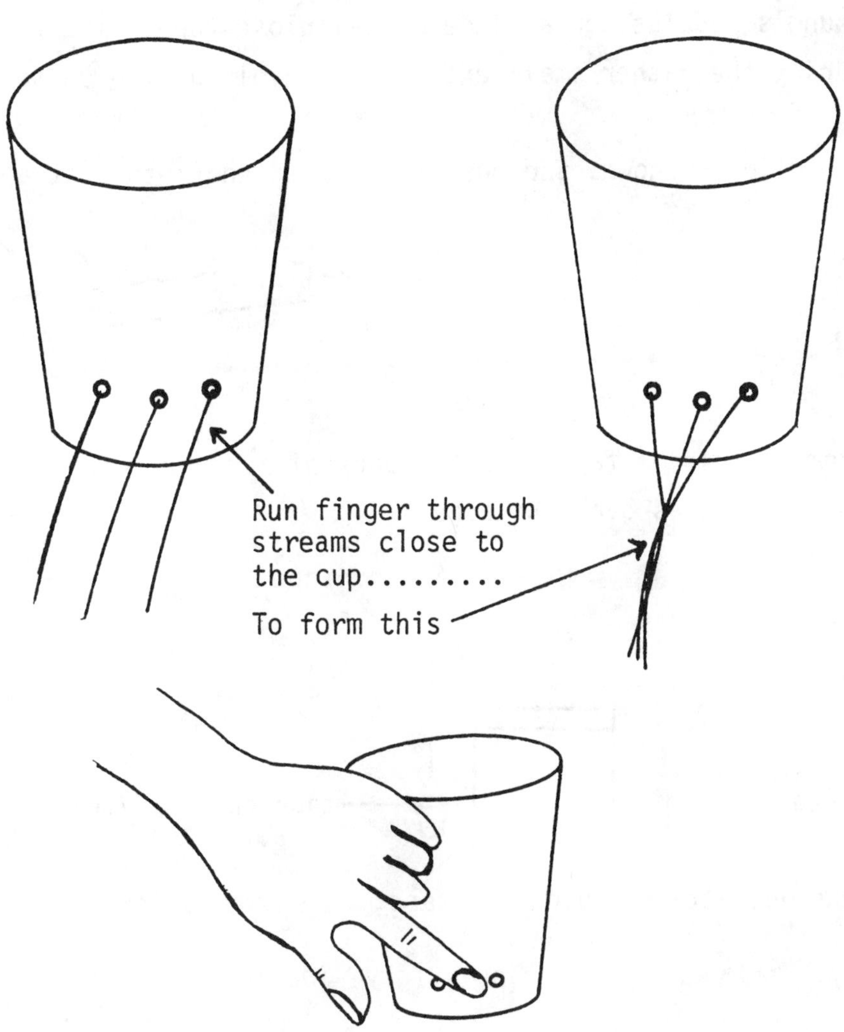
Run finger through streams close to the cup.........
To form this

Experiment 2.05 Youngs Modulus for Cellophane Tape

Materials: Cellophane tape, paper clip, cup, straw, marbles, paper clip

Procedure: The modulus of elasticity determines the change in shape of a body when a force acts on it - for example, if we hang a weight on a wire, it stretches, and the stretch is given by Young's modulus of elasticity. The modulus is defined as the stress, or fractional change in shape, divided by the strain, or force per unit area. So, for Young's modulus $Y = \frac{\text{stress}}{\text{strain}} = \frac{\text{fractional change in length}}{\text{force per unit area}}$

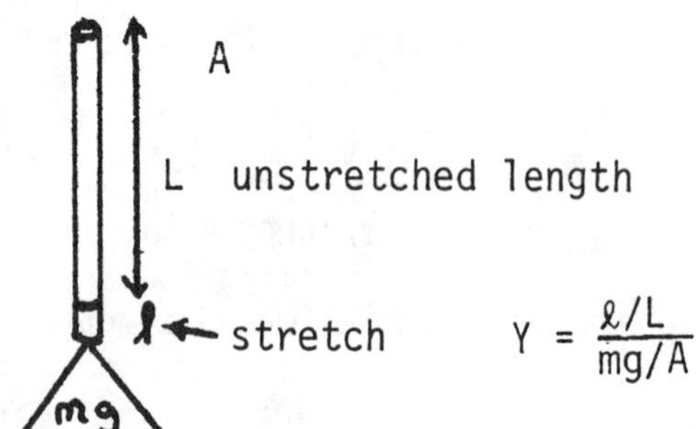

$$Y = \frac{\ell/L}{mg/A}$$

where L = unstretched length of wire
ℓ = extension of wire
Mg = force acting on wire
A = area of wire

We shall measure Young's Modulus for a piece of cellulose tape. The principal difficulty is measuring the rather small extensions involved. To do so we shall use a paper clip as a roller.

Open up a paper clip as shown, and push the closed end into a straw.

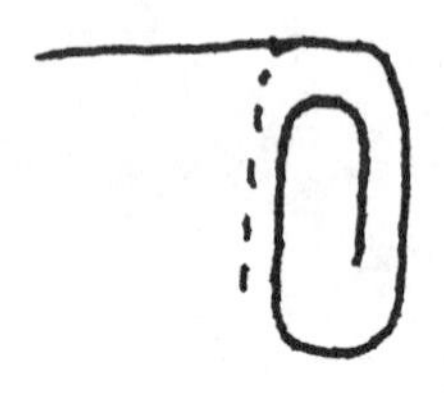

Hang a paper cup from a strip of tape over the back of a chair, edge of a table, etc. Make sure the tape passes over the edge

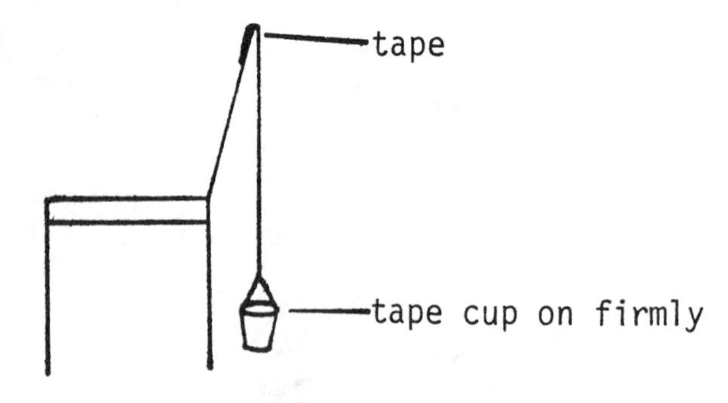

To measure the extension, attach a piece of tape, as shown, sticky side out, over the end of a ruler

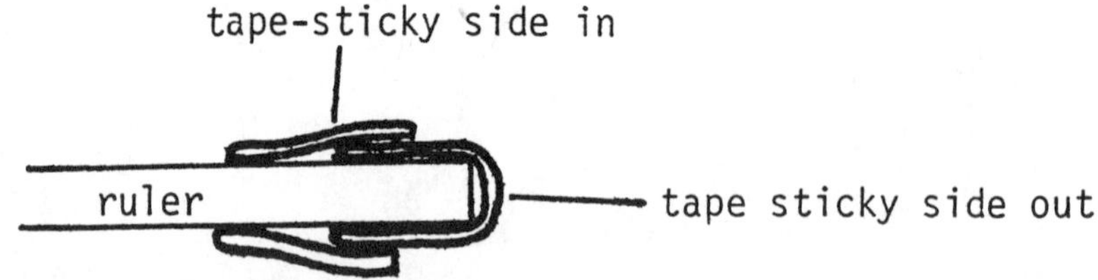

Attach the opened paper clip to this, so

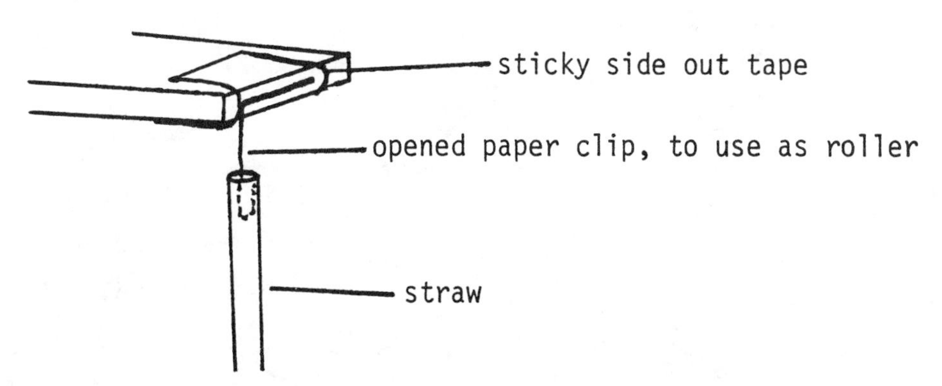

Now, place the ruler on a pile of books on the floor, and put some books on top to hold it steady. Then move the whole set-up so the roller paper clip sticks to the vertical hanging tape. One side of the roller should be on the tape on the ruler, the other side on the vertical tape. The two sticky tapes must not stick to one another

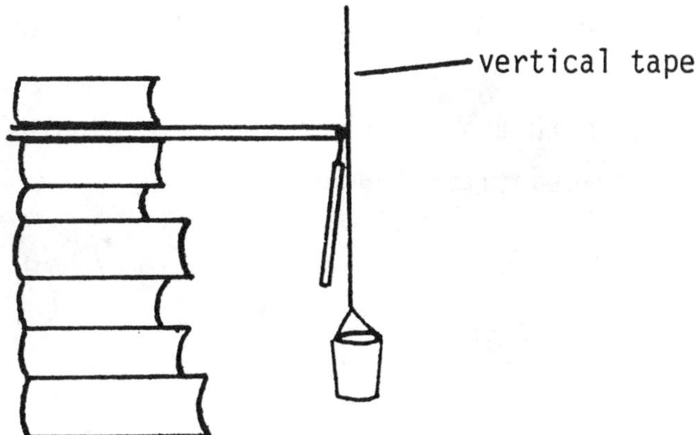

A thumb nail may be run across the roller to make sure the tape sticks. Now, as marbles are placed in the cup attached to the vertical tape, it stretches, rolls the paper clip, and the motion is magnified, so that the motion of the end of the straw is very large. Measure the movement of the end of the straw, using a ruler, or mark it on a sheet of paper to measure later.

To find the modulus of elasticity, we need to know the thickness of the tape. To find this, stick ten or twenty small pieces on top of one another, cut across, and measure the thickness with the ruler.

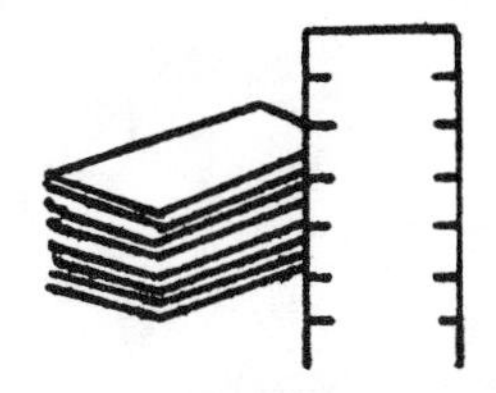

The area of the cross section of the tape is this thickness multiplied by the width.

The extension of the tape, ℓ, is the distance moved by the end of the straw x multiplied by the ratio of the diameter of the paper clip wire forming the roller, 2r to the length of straw, $\ell = \frac{2rx}{S}$ (S is the straw length)

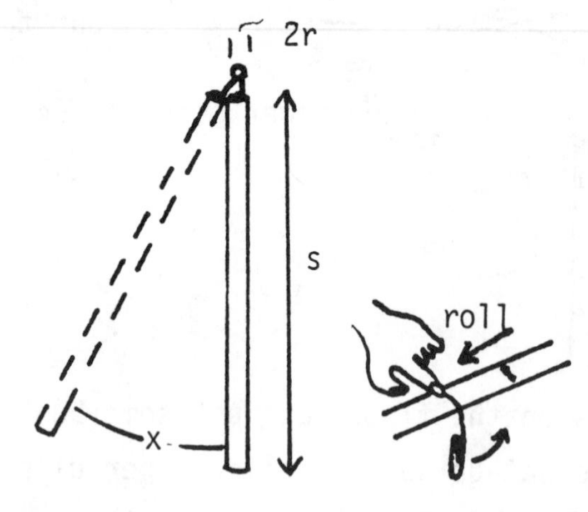

The diameter of the wire may be measured roughly with a ruler. Otherwise, roll the paper clip, without the straw, along a strip of sticky tape, four or five times, and note how far it rolls. The distance is $2\pi r\, n$ where n is the number of rolls.

The length of tape L used extends from the support to the roller.

The mass of the marble can be found by weighing against a known quantity of water. The density of glass P_g is about 2.5, so the mass can be calculated roughly as $\frac{4}{3}\pi r^3 P_g$, measuring r with the ruler. (Generally marbles weigh about 5.4 gm.)

We now have all the quantities required to measure Y.

$$Y = \frac{\ell/L}{mg/A} = \frac{2rx/sL}{mg/Wt} = \frac{2rxwt}{sLmg}$$

W = width of tape

t = thickness of tape

Y is generally of the order 10^{11} dynes/cm^2. (10^{10} N/m^2)

Experiment 3.01 Hydrostatics and Hydrodynamics - Hydrostatic Pressure

Materials: - Styrofoam (or paper) cup.

Procedure: - Poke small holes of the same size with a pencil in the cup at various levels, and around the cup at the same level as shown

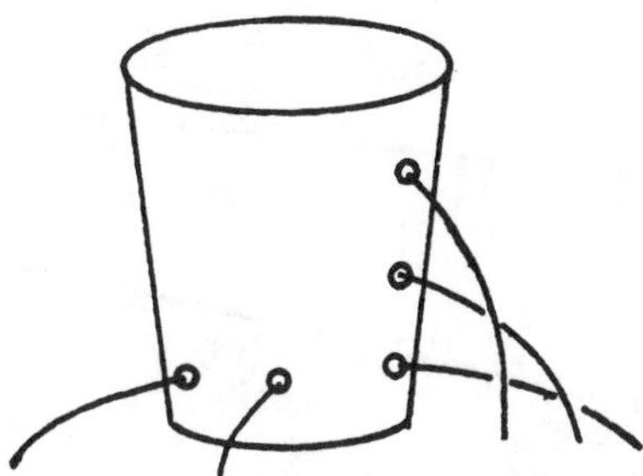

Fill the cup with water. Note how the stream from the bottom hole shoots out much faster than that from the top. Also note how the streams from the holes at the same level always shoot the same distance, even though that distance changes as the water level in the cup drops.

Qualitative: - The pressure forces the water through the holes. Since it goes further, the pressure is largest at the bottom, smallest at the top, and since, at the same level, the holes all shoot the same distance, the pressure is constant a constant distance from the top. Pressure acts in all directions at any one point in the liquid

Quantitative: - The pressure is the force per unit area, and is given by

$p = \rho gh$ where ρ = density of liquid

g = acceleration due to gravity, L = the length to the point where the pressure is measured.

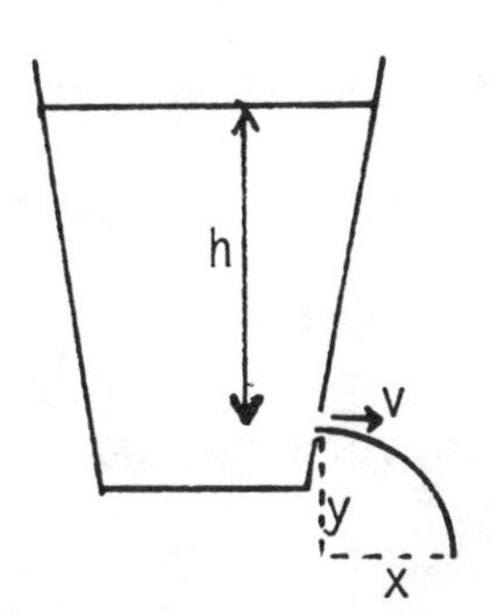

From the equation for the velocity of a particle falling a distance h, the velocity of the stream is given by $2\,gh = v^2$

If the stream projects horizontally

$$x = vt,\ y = \frac{1}{2} gt^2 = \frac{1}{2} g \frac{x^2}{v^2} = \frac{1}{2} \frac{x^2}{2h} = \frac{x^2}{4h}$$

You can measure x, y and h. Is the formula correct?

Experiment 3.02 The syphon

Materials: soda straw, cup

Procedure: The difficulty is to bend the straw, as shown, without breaking or kinking it too much. Unfold three paper clips, and push them as far as possible into the straw.

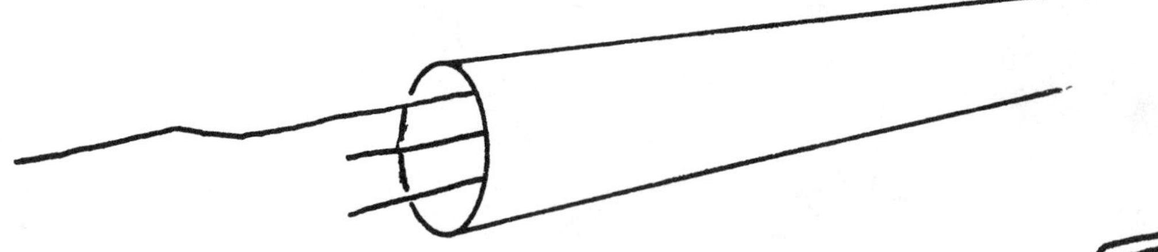

Now bend the straw until it is curved through 180°.

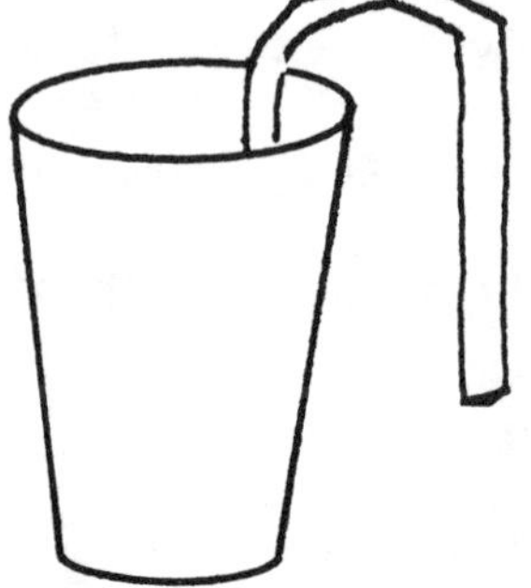

Fill the cup with water, place the syphon in as shown and suck the open end. The water will flow uphill, down the far side, and continue to run until the cup is empty.

Qualitative: Why does the water flow uphill? Why is the flow so slow?

Quantitative: How high could the syphon go? Could you syphon over a mountain? Because the syphon is very narrow, it can be caused to start on its own by tipping until the water is just below the lip. What makes the water flow uphill without suction?

Note: Running very hot water over the straw helps it to bend.

Experiment 3.03 Bernoulli Effect

Materials: - Styrofoam cup, and paper

Procedure: - Bore a hole about 1/4" in diameter in the bottom of the cup. Place over the sheet of paper. Place the cup over your mouth, and blow into it. You should feel a stream of air coming out of the hole. Place the cup over a sheet of paper, and blow. The sheet may easily be picked up, rather than being blown away, because of the low pressure rapidly moving air.

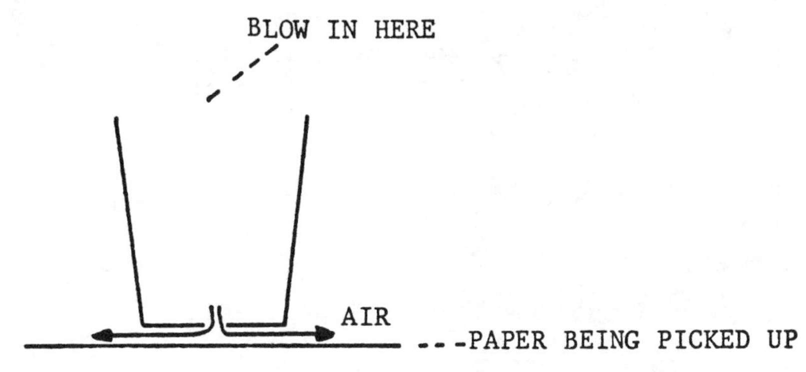

A straw through a cotton reel will do the same thing, but a Dixie cup with its raised bottom will not work as well. The reason being air must flow through a narrow space over a large area between the paper and the flat bottom of the cup, to provide sufficient force to pick up the paper.

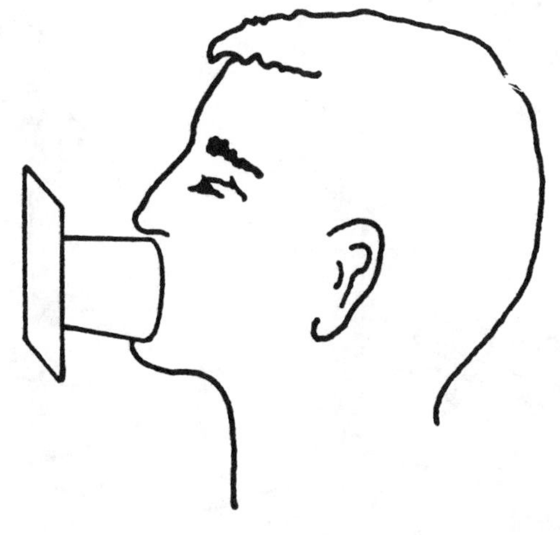

Experiment 3.04 Bernoulli's Principle

Materials: - A sheet of paper, and a straw

Instructions: - Place the paper (8 1/2 x 11) over the straw at the edge of the table. Blow. Does the paper blow away, as you might expect? The end of the straw under the paper should be flattened.

Try peeling the paper off the table as you blow. Does it tend to stick?

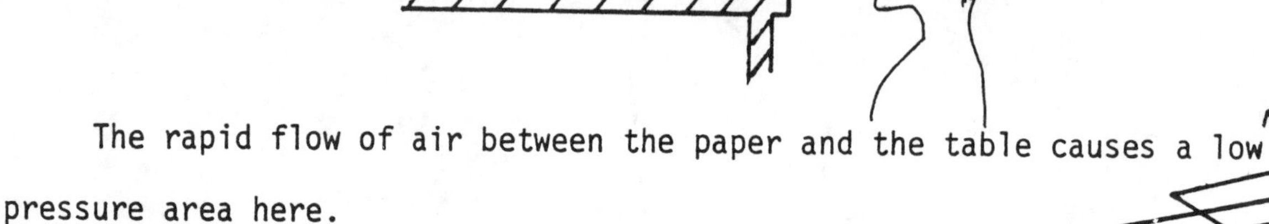

The rapid flow of air between the paper and the table causes a low pressure area here.

Note: If you have trouble with this experiment, stick a piece of tape across the end of the straw, and try again.

A slight variation on this experiment is to blow under the folded paper, as shown below, which sucks down onto the table.

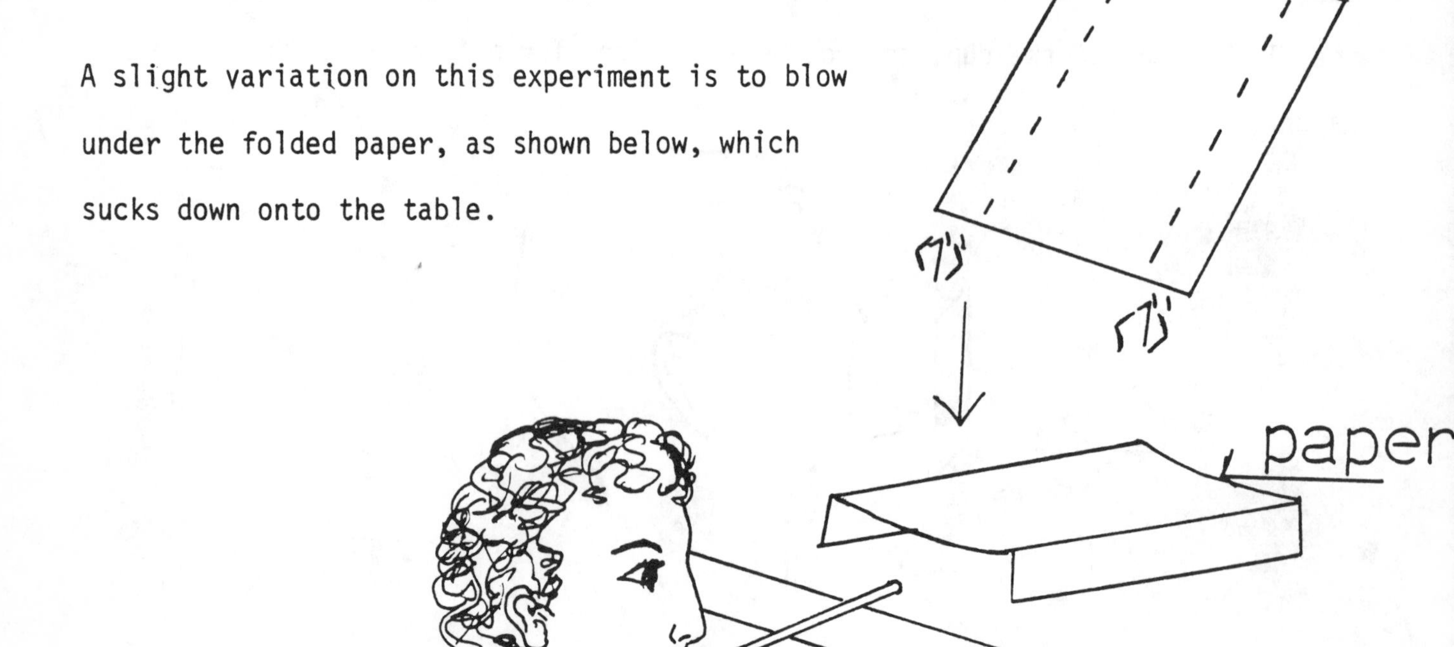

EXPERIMENT 3.05 The Atomiser

Materials: Straw, cup

Procedure: Either cut the straw partially through, as shown, or completely through, and tape the two halves to a sheet of card.

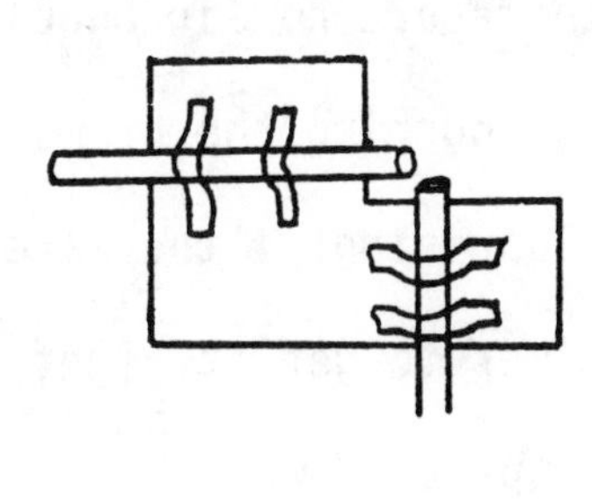

Put one end into a cup of water, and blow through the other end. The suction is small, so fill the cup quite full of water, and ensure the top of the vertical straw is near the surface.

Qualitative Question: What happens?

Bernoulli's principle tells us that, when the speed of the air is high, the pressure is low. Here, the speed is high just above the straw dipping into the water, and the low pressure sucks it up the straw, allowing it to be sprayed "atomized". Sometimes this suction is not very great and the straw has to go deep into the liquid.

Experiment 3.06 Atmospheric Pressure

Materials: - cup, paper, water.

Procedure: - Fill the cup to the brim with water, so that the meniscus stands high. Place a sheet of paper or card on top, and smooth it down so that there are no air bubbles, and the paper lies smoothly across the top, touching the brim. Smoothly, but rapidly turn the cup upside down. Suprisingly, the water does not run out. You may require two or three times to get it right, and it may be necessary to hold the paper on until upside down.

Qualitative: - The atmosphere presses down on us from all sides, not merely from above. When the water presses down on the paper, the pressure

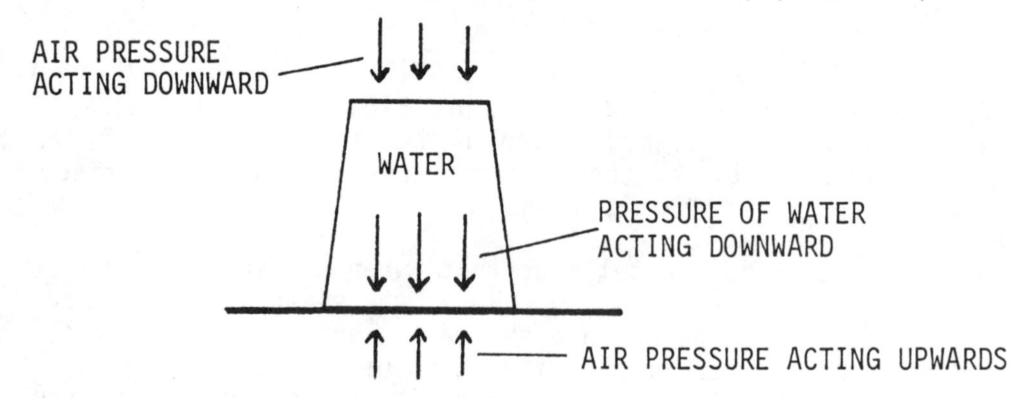

of the atmosphere pushes it back, and it cannot run out.

Quantitative: If we built a cup so high the pressure of water when inverted were greater than air pressure, the water would run out. How tall must such a cup be?

Experiment 3.07 Vortex Rings

Materials: - Cup, colored water or ink, (Grape Soda works Fine), soda straw.

Instructions: - Fill the Cup with water, suck up a little ink in the soda straw, and pinch it off. Place the open end of the straw about one inch above the water in the cup, and give a sharp squeeze in the middle, rapidly expelling some ink.

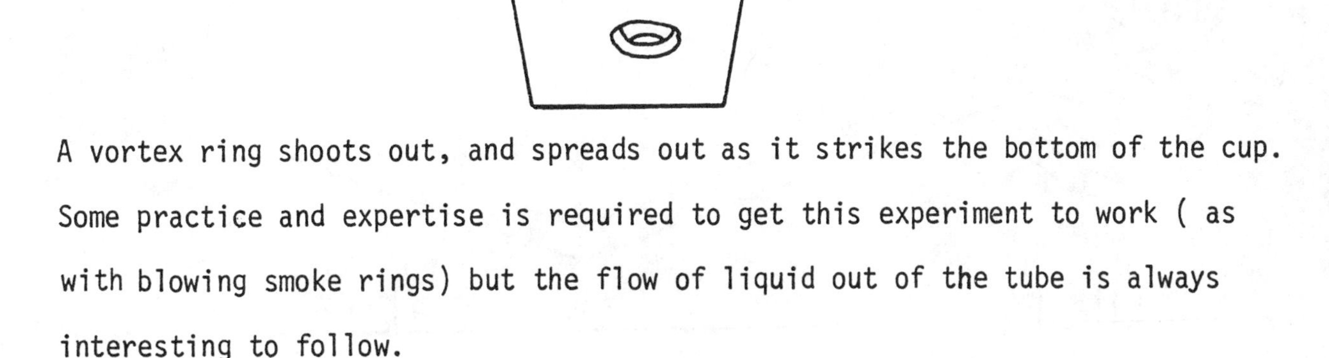

A vortex ring shoots out, and spreads out as it strikes the bottom of the cup. Some practice and expertise is required to get this experiment to work (as with blowing smoke rings) but the flow of liquid out of the tube is always interesting to follow.

Qualitative: - A vortex involves circulation of an annulus of fluid. It cannot be generated in a non-viscous fluid, but once formed, is highly stable, persisting for long periods of time.

Note: If no ink is handy, coke, tea, coffee or even milk will do.

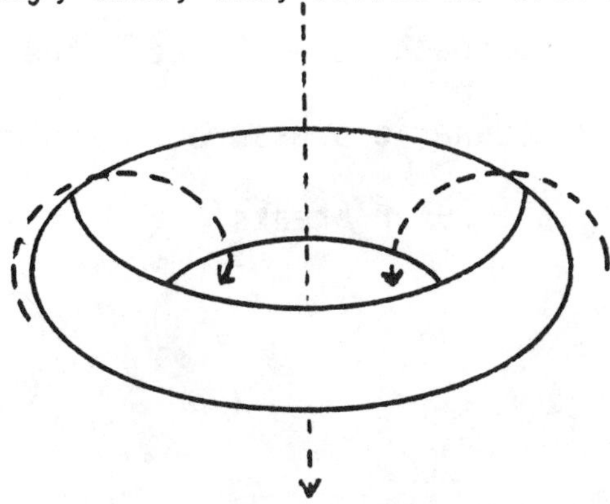

See: "The Flying Circus of Physics", 4.74, Jearl Walker (Wiley).

Experiment 3.08 Pressure of the Air

Materials: Sheet of paper, ruler.

Procedure: Lay the ruler over the edge of the table, and place the sheet of paper on top. The ruler may easily be dealt a blow capable of breaking it (either use an old ruler, or don't hit so hard!) even though held down only by the paper.

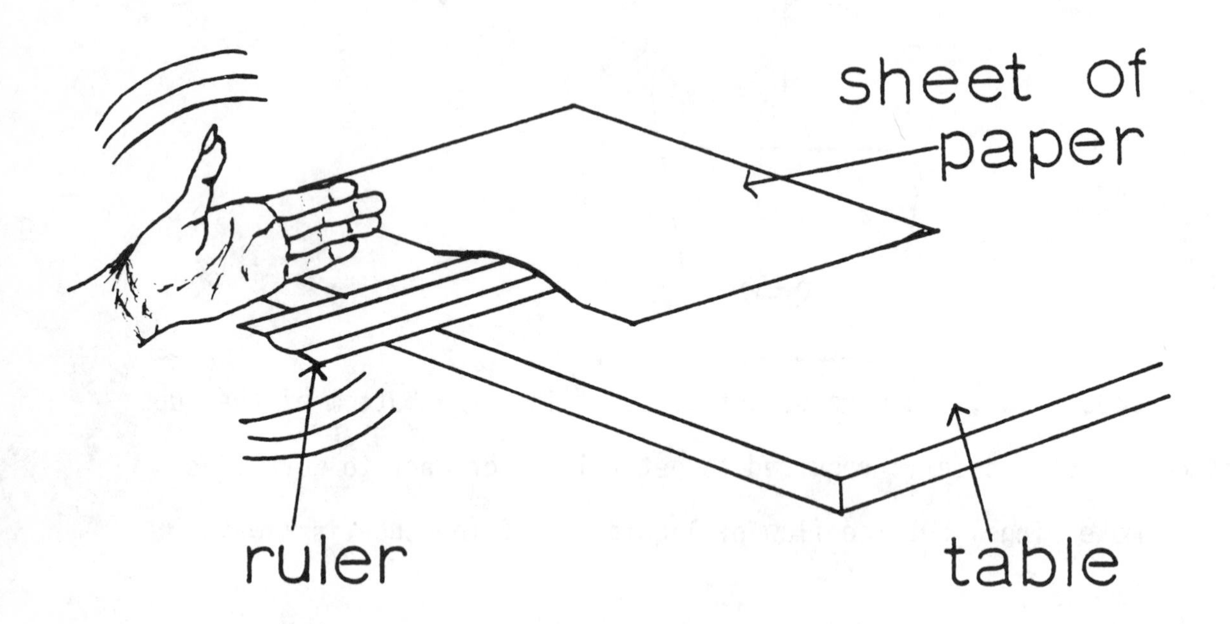

Why is this? The reason is that the ruler pushes up on the paper when struck, attempting to create a vacuum underneath it. Since there is no time for air to rush into this vacuum, the atmospheric pressure on top of the paper holds the ruler down so strongly that the ruler breaks.

Experiment 3.09 Sail boat log

Materials: Soda straw, paper clip.

Procedure: Straighten a paper clip, and insert it into a straw as far as it will go. Bend the combined clip and straw through a right angle, and cut out and attach the scale over as shown.

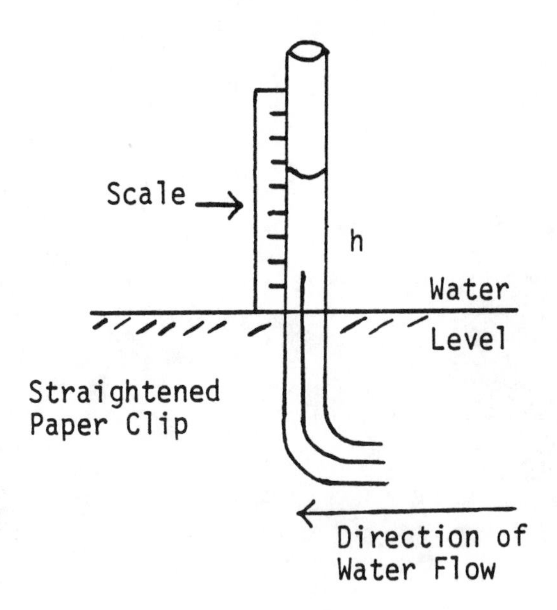

Insert the log in the water over the side of the boat, so the open end faces forward, and the zero of the scale lies in the surface of the water. As the boat moves forward, water will rise inside the straw. Bernoullis theorem (or simple conservation of energy) tells us that

$$gh = \frac{1}{2} v^2$$

where h is the height the water rises in the tube, and v is the speed of the water past the open end of the tube. The scale calibration can be checked by a conventional old fashioned log - a piece of wood attached to a string.

One knot is tied in the string every 50 ft. The log is thrown overboard and the number of knots in 30 seconds counted. A knot is a nautical mile per hour; or 1.15 mph.

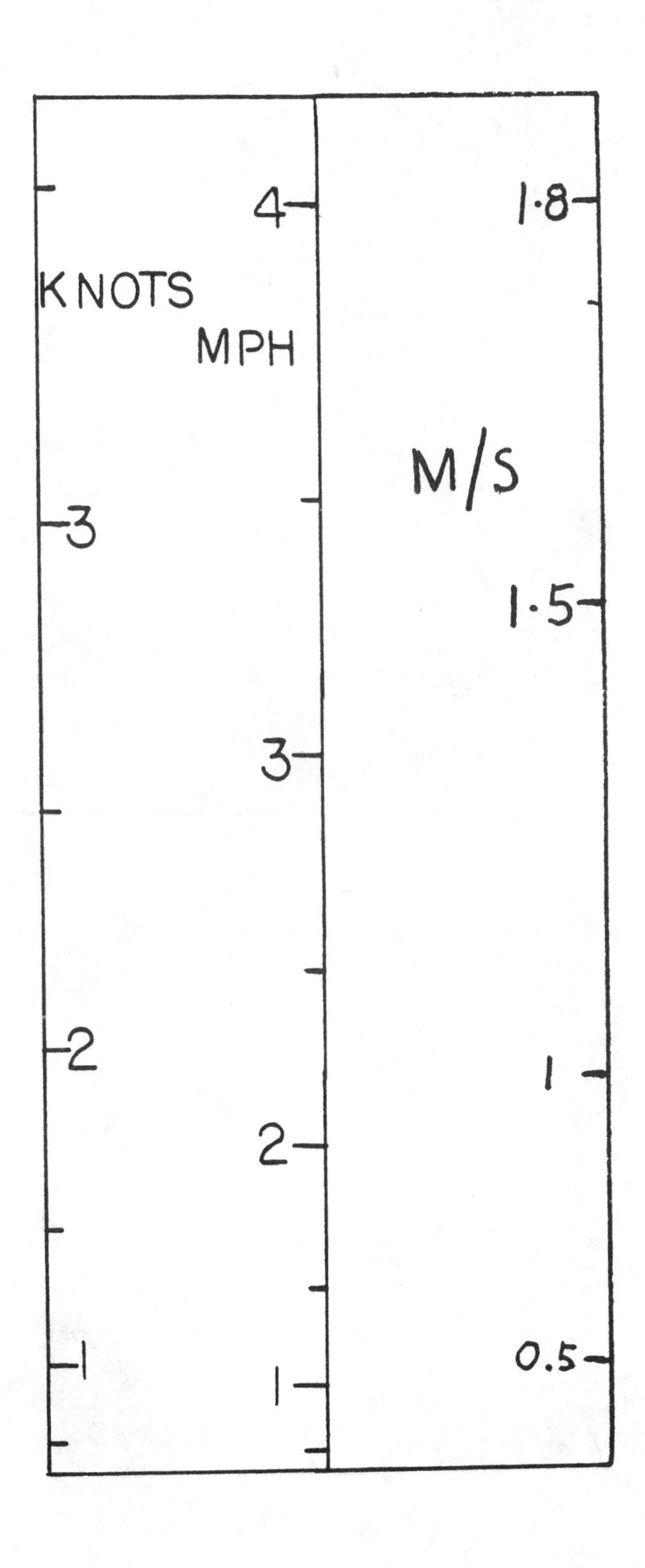

Experiment 3.10 Bernoulli Effect

Materials: Paper cups, straw, sheet of paper, string

Procedure: Hang two cups on strings, as shown, close to one another. You can hold the strings by hand, or off a table. Blow between them, and notice how they are drawn together.

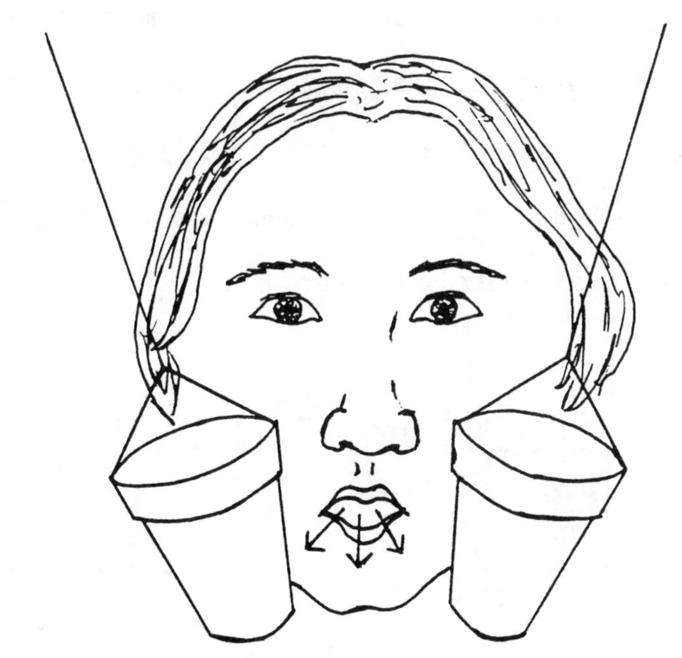

Try to explain why.

Explanation: The air moves rapidly between the two cups, and Bernoulli's principle tells us that where there is a high velocity, the pressure is low, sucking the cups together.

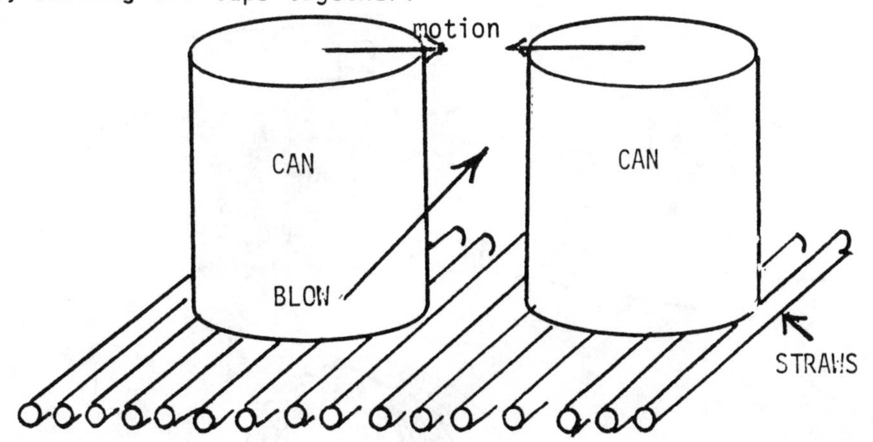

In his book "Invitations to science enquiry",* Tik L. Liem has a nice variation on this experiment, where you blow between two empty aluminum cans supported, as shown, on drinking straw rollers. The cans roll together.

* Obtainable from :- Dr. Tik L. Liem, St. Francis Xavier University, Antigonish, Nova Scotia, Canada, B2G 1C0

Experiment 3.11 Streamlining of Aerofoils

Materials: Paper, tape, string.

Procedure: The flow of air over objects is of vital importance - not only for aircraft, but to stop windows being sucked out of tall buildings, etc. Fold a sheet of paper into an aerofoil shape by taping the ends together as shown. Blow over it, and use a small piece of tape on a thread or piece of thin string to show the flow of air. Now replace the airfoil section with a sheet of paper or card the same width. Notice your detector does not move, or moves towards the card showing the motion of the air is as shown below

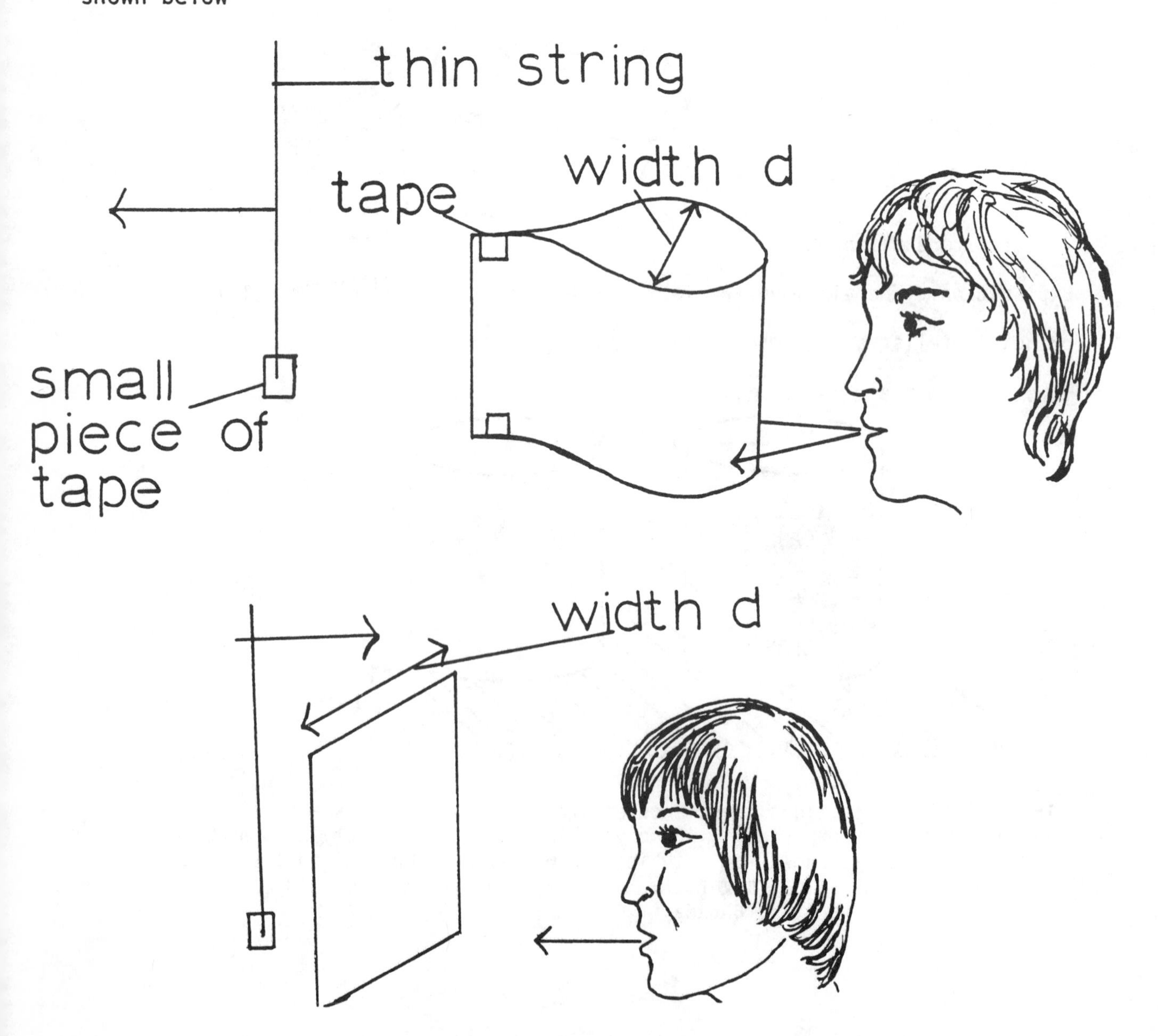

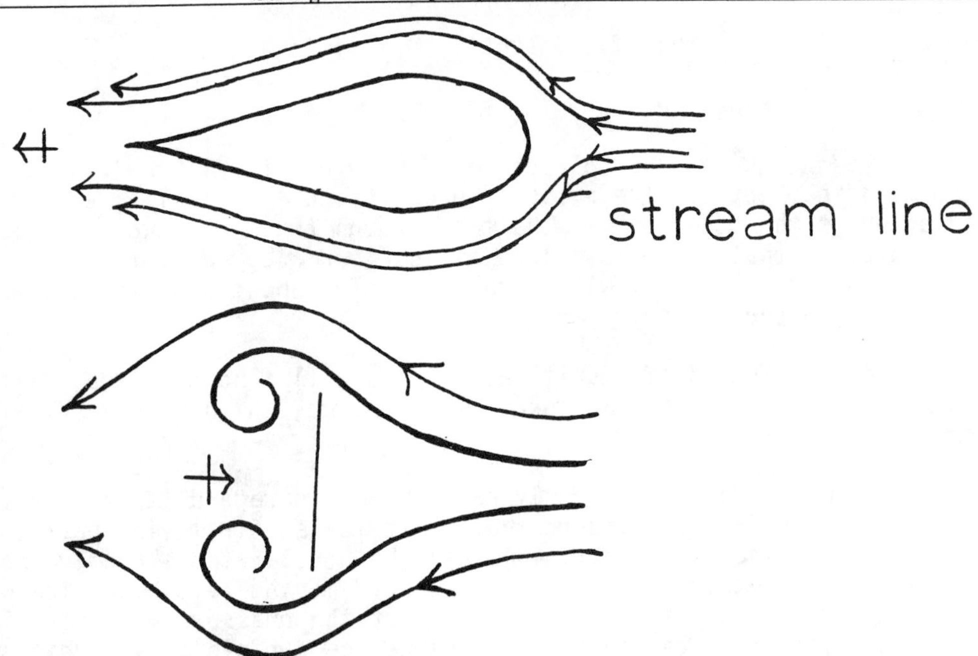

The airfoil is a wing shape, and you can use it to investigate stalling, which used to be a prime cause of aircraft crashes. Turn the foil as you blow over it, as shown. You will notice a lift produced by the faster motion of the air on top of the wing. But if you twist further, very

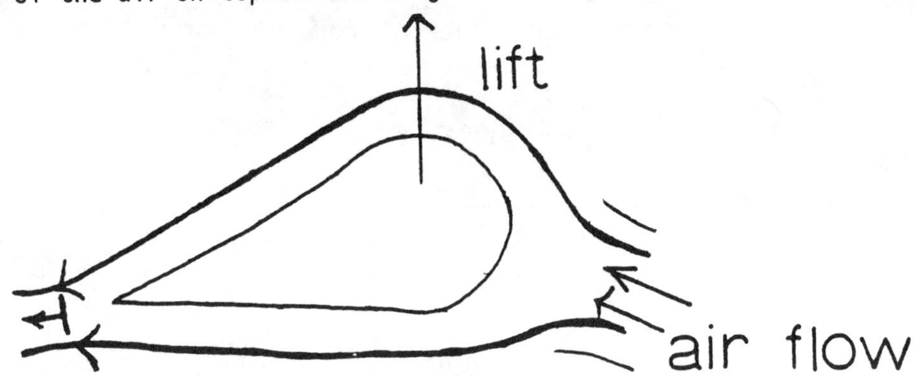

suddenly the flow detector will be sucked on top of the wing, as the air ceases to flow smoothly over the trailing-edge, the lift ceases, and, if it were a plane, it would suddenly dive.

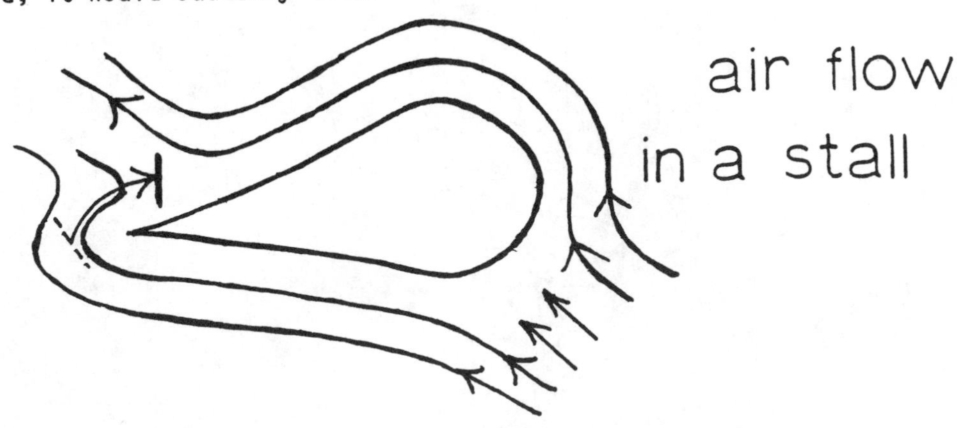

Experiment 3.12 The Flight of a Baseball or Golf Ball

Equipment: Cup, string, marbles, scissors

Procedure: We can use a paper dirigible to show how the flight of a ball is affected by spin. Cut out the strip of paper (figure 1a) making a slot at one end and a tab at the other. Slip the tab through the slot, and hold the strip in the middle, so that it forms a loop as in Figure 1b. Hold the loop as high as possible, and drop it. It will rotate rapidly, about a horizontal axis, falling slowly to the ground.

Which way does the loop drift as it falls? You will find the loop drifts bodily in the direction the lower part is traveling, which ever way that is, as shown in Figure 2.

If the dirigible rotates sufficiently rapidly we can regard it as a continuous surface, and we see from the figure that as it falls, air moving past the rising right hand side of the dirigible will be accelerated slightly, moving faster, whereas that traveling past the downward moving left hand side will be slowed. Bernoulli's principle tells us that the pressure will be slightly less in the region of faster moving air, hence the dirigible, in addition to falling, will be sucked to the right. Were it to revolve the other way, it would move to the left. It is for a similar reason, of course, that we put spin on a golf ball to encourage it to stay aloft. We are assuming the effect of air inside the rotating loop can be neglected.

Another way of demonstrating the same principle is to attach a fairly long piece of string to a cup as shown,

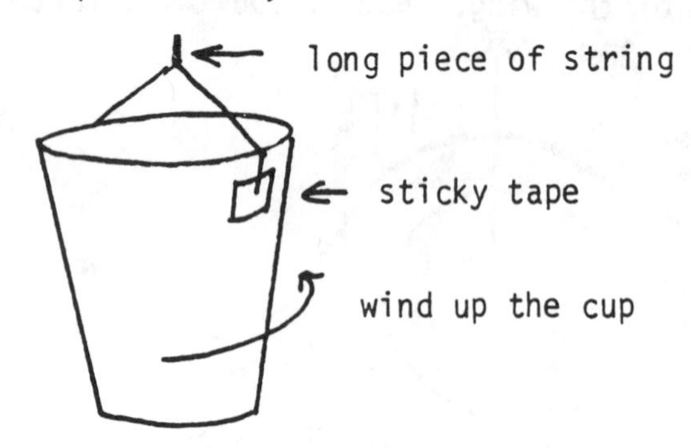

and hang it from some fairly high place, or have a friend hold it. Now, weight the cup with one or two marbles, and wind it up, so that it spins on release. When it is spinning as fast as possible, blow directly at the cup.

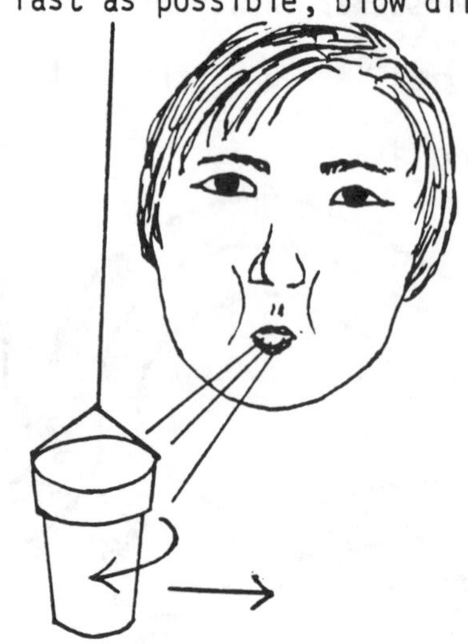

TAB →

FIG 1A

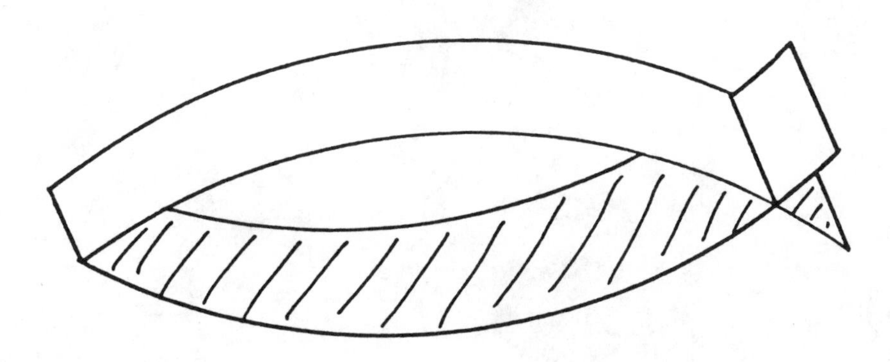

FIG. 1b

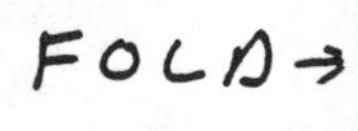

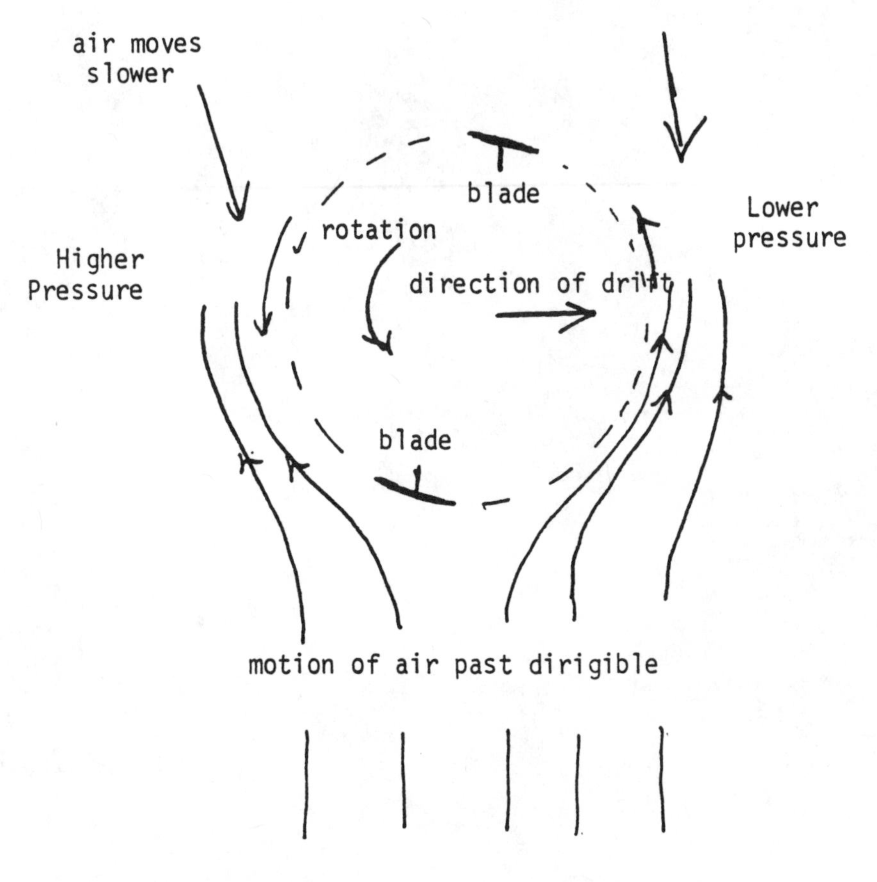

Fig. 2

SLOT →

If the cup is spinning clockwise, seen from above, as shown, it will not be blown directly backwards, but should also move towards the left. If the cup spins anticlockwise, it should move to the right. Do you find this is so? Can you explain this in terms of Bernoulli's principle?

Explanation: Looking down from above, the spinning cup drags the air with it on one side, and the high air velocity produces a lower pressure, according

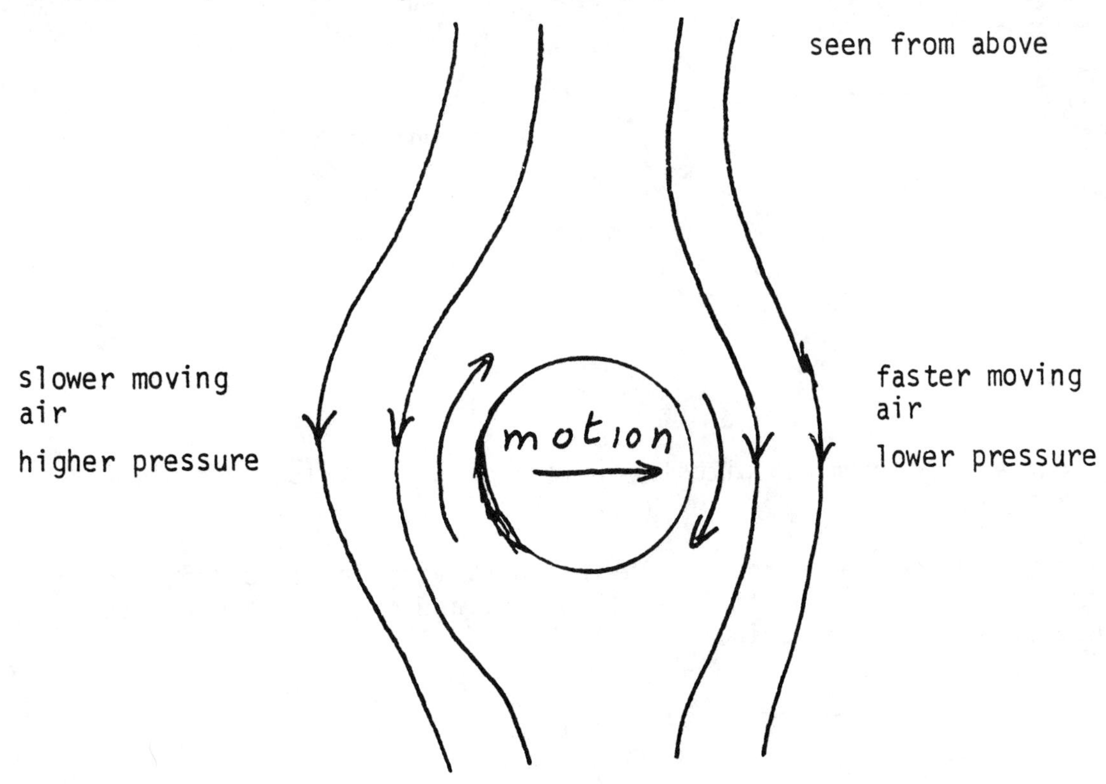

to Bernoulli's principle. Contrariwise, on the opposite side of the cup the air is slowed down by the spinning cup, producing a higher pressure. The cup will move towards the region of low pressure.

This effect finds application in the flight of a golf ball, which is given bottom spin to make it stay aloft. The dimples on the ball help drag the air round with it.

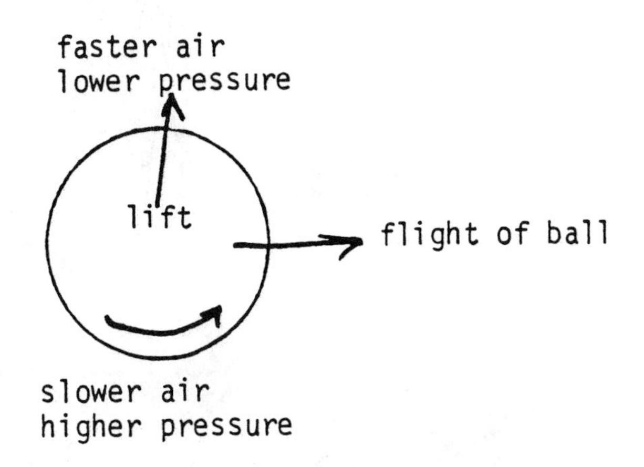

Correspondingly, if the ball is "sliced", i.e. given a clockwise spin seen from above, by a right handed player, it will curve away from the player. The situation where the ball curves away from the player is shown below.

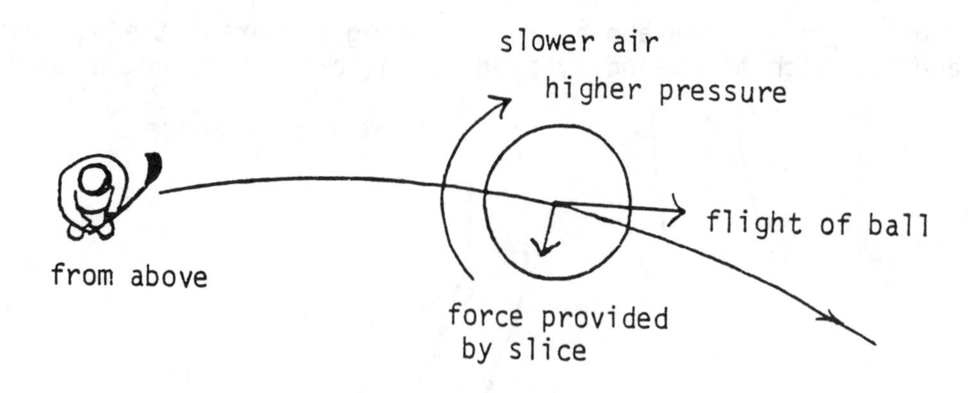

A baseball will curve away from the pitcher, if spinning clockwise seen from above, for the same reason.

Even more simply, take a flat card - a playing card, visiting card 4x5 card or computer card will do, and drop it. It will rotate, and drift in a specific direction because of Bernouilli's principle, as shown in the figure, taken with a strobe light.

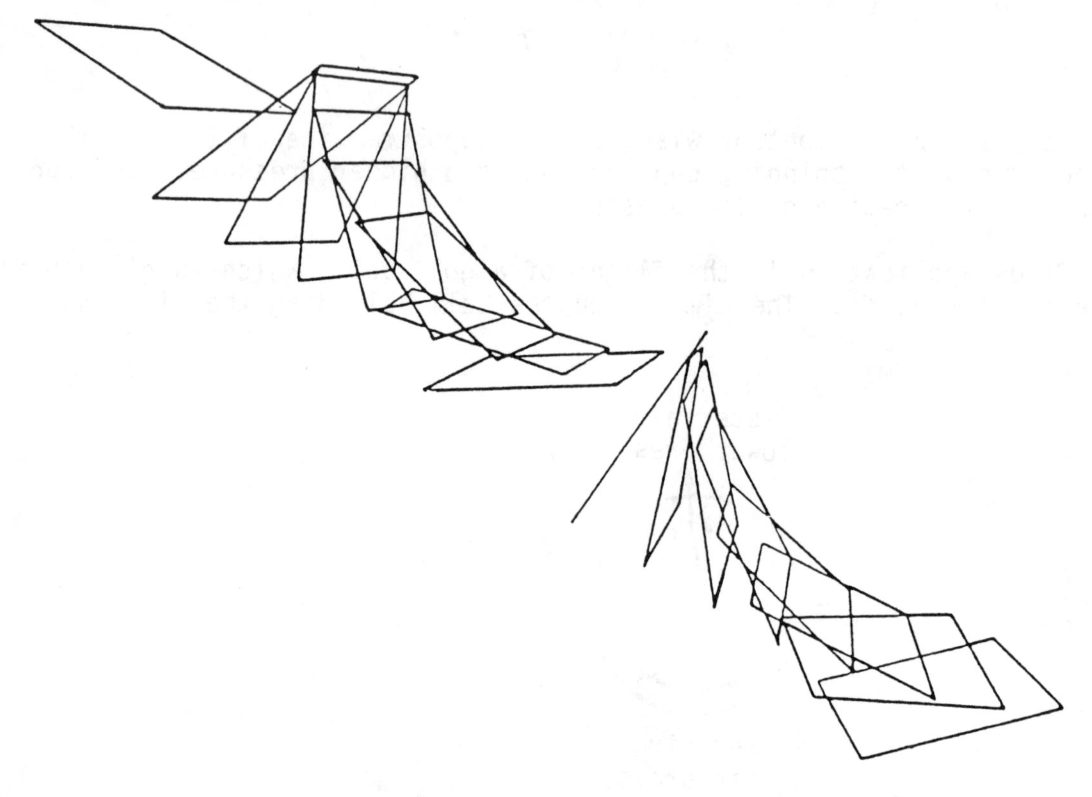

Stroboscopic photograph of a falling computer card, 50 flashes/sec. Note the rapid rotation to the position beyond the horizontal.

Experiment 3.13. Levitating a Dime

It is truly amazing how many qualitative and quantitative physics experiments can be done with very simple equipment if you put your mind to it. There are many curious and paradoxical problems which can be solved using Bernoulli's principle. One is to get a dime off a table into a cup on the table without touching the coin. The cup must be shallow, or tilted so that the lip is about 2 cm off the top of the table, and about 2-3 cm from the edge of the table, as shown. It looks impossible for the dime to get into the cup, but if you blow hard and suddenly, parallel to the table top, the dime hops in. As the air is blown rapidly over the top of the dime, Bernoulli's principle tells us that the pressure is lowered there, and the pressure differential between the top and bottom of the dime raises it off the table, and allows it to be blown into the cup. So much for the qualitative explanation, which will satisfy most people, but suppose we apply a little mathematics. Bernoulli's principle states that the pressure differential, p, between the top and bottom of the dime is given by

$$p = \frac{1}{2}\rho V^2$$

where ρ is the density of air (1 Kg/m^3) and V is the velocity of the air blown over the dime. Now, the area of the dime A ($2.5 \times 10^{-4} m^2$) multiplied by this pressure differential must equal the gravitational attraction on the dime, mg, if it is to rise off the table. Since the mass of the dime is 2.24 gm, the gravitational force is about 0.0224N.

High Speed

Low Pressure

So

$$mg = A\frac{1}{2}\rho V^2 \text{ and, putting in numbers,}$$

$$0.0224 = 2.5 \times 10^{-4} \times \frac{1}{2} \times 1 \times V^2$$

This gives V = 13.4 m/sec which is 48 km/hr or 30 mph. If we perform the same trick with a quarter, we require 16 m/sec or 37 mph, and for a nickel it is 17 m/sec or 38 mph. In fact, it must be much more than this to get the coin into the cup. Our little calculation tells us a surprising fact--we can blow at lease 30 to 40 mph--something you can tell any blowhards you know!

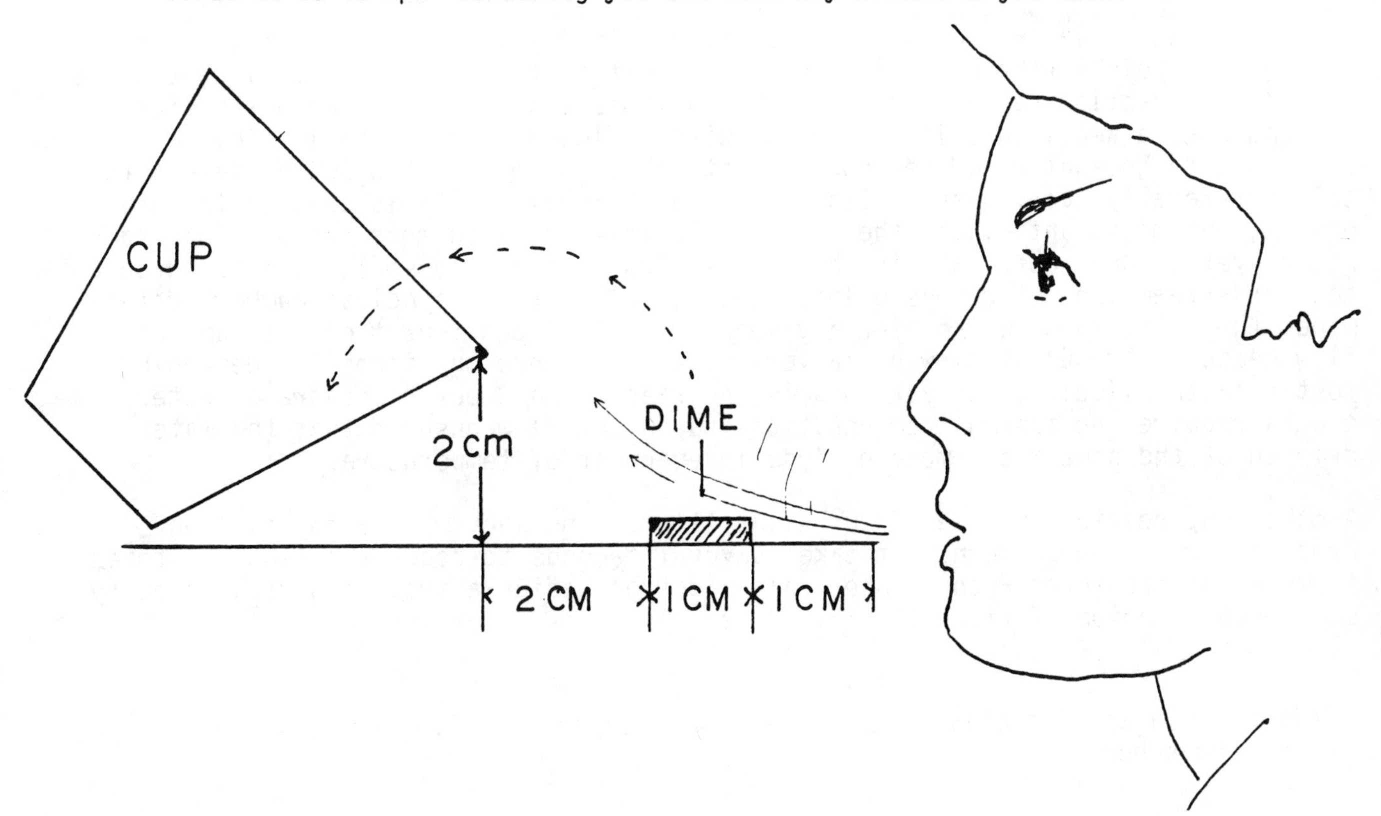

Experiment 3.14: Viscosity

Materials required: Styrofoam cups, hot and cold water, drinking straw.

Procedure: Viscosity is an interesting subject for simple experiments. When a liquid flows over a fixed surface S (fig. 1) a layer of the solution in a plane parallel to, and a distance from S flows with a velocity greater than layers close r to S. As a result of this relative motion

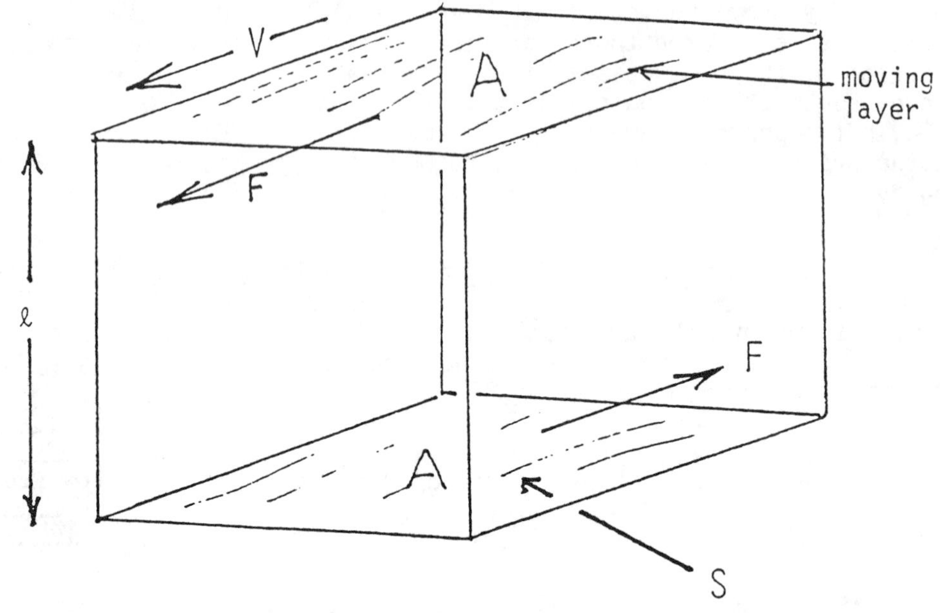

Figure 1

of layers, internal friction or viscosity arises, and the coefficient of viscosity η is given by $\eta = \frac{F}{A}\frac{\ell}{v}$ where F is the tangential viscous force between layers of area A distance ℓ apart moving with relative velocity v. The viscosity of water is 1.8 centipoises (the poise is the c.g.s. unit-gm/(sec Cm)) at 0°C, 1 centipoise at 20°C, .5 centipoises at 50°C and .28 centipoises at 100°C. Take a styrofoam cup and poke a small hole in the bottom with a fine needle or pencil tip. Fill the cup with hot water and see how long it takes to empty. Now do the same with cold (preferably iced) water. Since the change in viscosity is about a factor of six, one might anticipate the water would leak out much more rapidly for the hot water--yet it does not. Why is this? The flow through the small hole is turbulent and not streamline. A dimensionless number, called the "Reynolds' number" determines the velocity at which, for a given geometrical configuration, streamline flow ceases. Turbulent flow has a very weak dependence on viscosity, depending mostly on the liquid density. This is the reason the Greek Klepsydra or water clocks required no thermal compensation--turbulent flow ensured that the water dripped at the same rate, more or less independent of temperature.

Another interesting example of turbulent flow can be seen with a toy balloon. Drop an air filled balloon. It takes several seconds to reach the floor. Stokes' formula for the force F on a spherical object of radius a traveling with velocity v through a medium of viscosity η is given by

$$F = 6\,\pi\eta a v$$

This formula applies only if the flow is non-turbulent, i.e. streamlined around the sphere.

It is familiar to those who have performed Millikan's oil drop experiment. For the balloon of radius 30 cm, F is the gravitational attraction, mg. Now, m is, lets say 10 gm, so mg = 10 x 981 dynes = 6π x 0.1 x 30 x v
(.01kg.) (.9 N)

giving $v = 1.735$ m/sec.

The balloon certainly does not fall at a rate of 1.7 m/sec. The critical velocity is given by $V_c = \frac{K\eta}{\rho a}$ where K, the Reynolds number is about 1000, η, the viscosity of air is 182μ poises ρ, the density of the medium (air) is .0012 gm/cc so

$$V_c = 15 \text{ cm/sec}$$

The flow will be turbulent above this velocity, which is quite low. Only for very tiny spheres will the flow be streamline, unless very low speeds are used. However, try dropping a marble, radius .8 cm, mass 5 gm, in a cup of glycerine (viscosity about 4000 centipoises at 10°C). The density of glycerine is 1.26. (Syrup will also work, if sufficiently concentrated).

$$V_c = \frac{1000 \times (4000 \times .01)}{1.26 \times .8}$$

$$= 3.9 \times 10^4 \text{ cm/sec}$$

whereas $6\pi(4000 \times .01) \times .8 \times v = 5 \times 981$

The terminal velocity of the marble through the glycerine is only 8 cm/sec. So this is a very easy way to measure the viscosity of viscous liquids--simply drop a marble in and find how long it takes to reach the bottom. For less viscous liquids such as water, we must employ a very low velocity, to ensure streamline flow occurs.

One way of doing this is shown in the fig. A hole is poked at X in a styrofoam cup with a pencil point, so that a straw can be pushed through, making a tight fit. The end of the straw rests on the top of a second cup which has had only about 1/2 cm torn from the lip. Water is poured constantly overflowing the first cup until the second is filled. The flow, being slow, is now streamlined, and it will be found it takes much longer to fill the second cup with cold than hot water, because of its lower viscosity. However, the head of water must not exceed 1 cm.

The quantitative aspects of this experiment show (Poiseuille's formula) that the quantity of liquid Q flowing per second through a tube of radius a and length ℓ (2 mm and 20 cm for a straw propelled by a pressure p) is given by

$$Q = \frac{\pi P a^4}{8\ell\eta}$$

If the height differential is h, p = $\rho g h$ = 1 x 981 x h. The volume of the cup is 6 oz or 203 cc, so if h ~ .5 cm and η = .01

$$Q = \frac{\pi \times 1 \times 981 \times .5 \times (.2)^4}{8 \times 20 \times .01} = 1.5 \text{ cc/sec}$$

so it should take 132 sec (about 2 minutes) to fill the cup, which it does.

The dramatic difference in flow time filling the cup with hot or cold water arises because of the large temperature dependence of viscosity which is even more noticeable for the glycerine. This presents a severe problem with automobiles, since the cold oil does not flow and lubricate on starting the engine. This has lead to the development of motor oils whose viscosity remains relatively constant with temperature. It is interesting to measure the viscosity of different motor oils as a function of temperature, using the technique above, but qualitatively the difference may easily be seen as one pours oil into one's car from the can in mid winter and summer.

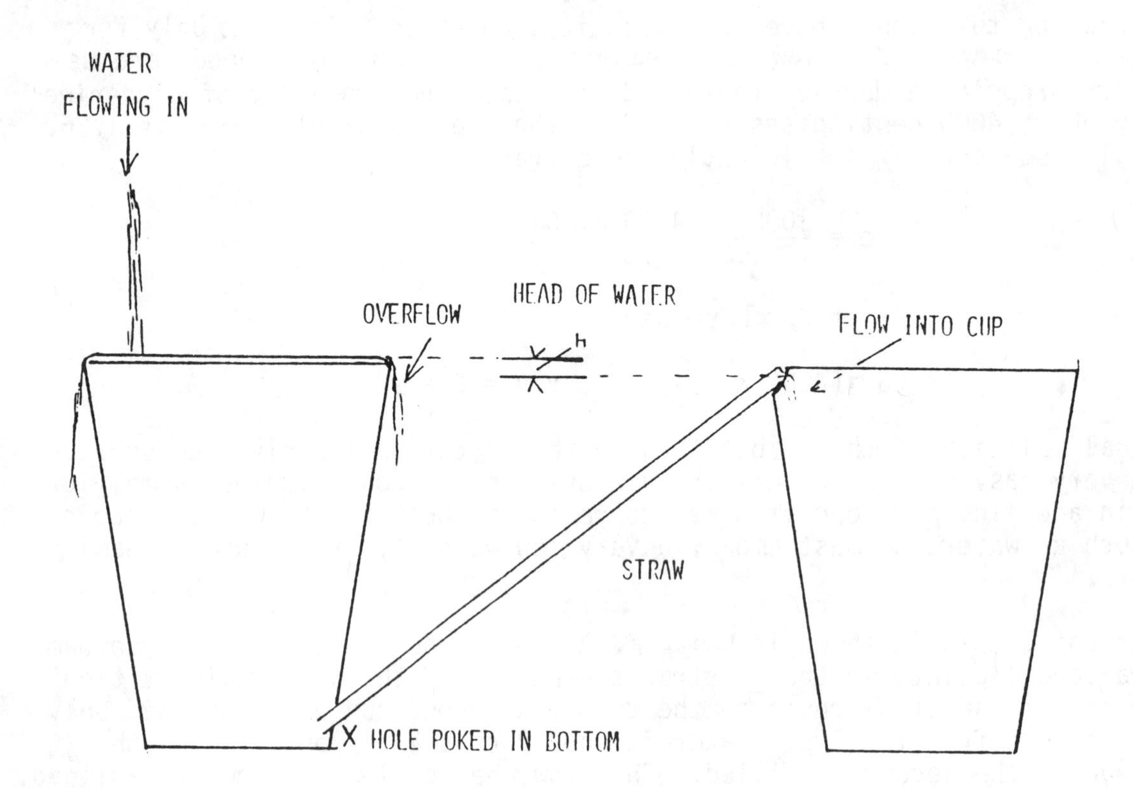

Experiment 3.15 Effects of pressure

Materials: garbage bag, vacuum cleaner which blows

Method: This experiment was suggested by Rae Carpenter and Dick Minnix, and is outside our other experiments because it required more equipment--which is, nevertheless, simple and fun. It requires a vacuum cleaner with an outlet that can blow rather than suck. You take a plastic garbage bag and tape the top end with duct tape so that it is air tight, except for the vacuum hose, which is sealed poking into the bag through the opening, as shown. Put a piece of thin plywood on the bag, then sit someone on the bag, and turn on the cleaner, (Fig. 1). The bag blows up, lifting quite a heavy individual--and tipping him over unless he is steadied! If the board is, say 50 cm by 50 cm (2500 cm^2, 387 $in.^2$) and if the individual weighs 75 kg (165 pounds) the pressure is 2940 pascals (4.26 pounds/sq. in.) which even a relatively inefficient vacuum cleaner can supply.

A variation of this is to seal the garbage bag completely with tape, then poke eight to ten straws in along the edges (they may need sealing to the bag).Eight or ten people blow through the straws, and can easily lift anyone sitting on the bag.

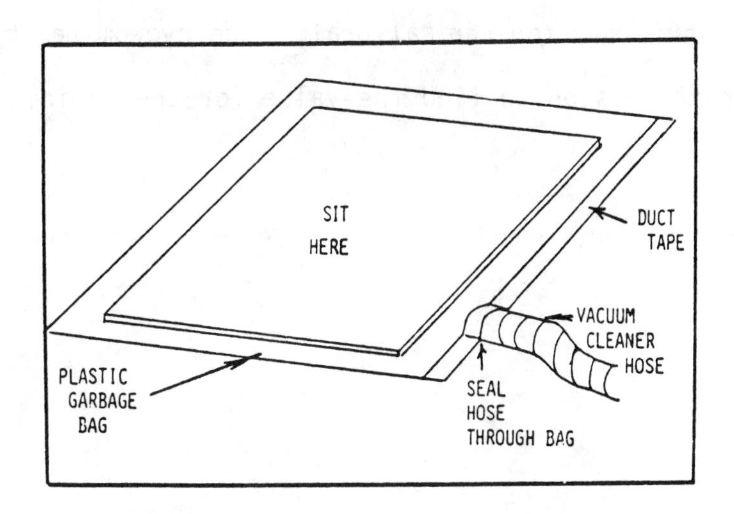

Experiment 4.01 The hygrometer

Materials: straws, card, rubber band, tape, hair.

Instructions :-

Attach a long hair (blonde works best!) to one end of the straw, and keep it taut with a rubber band on the other end.

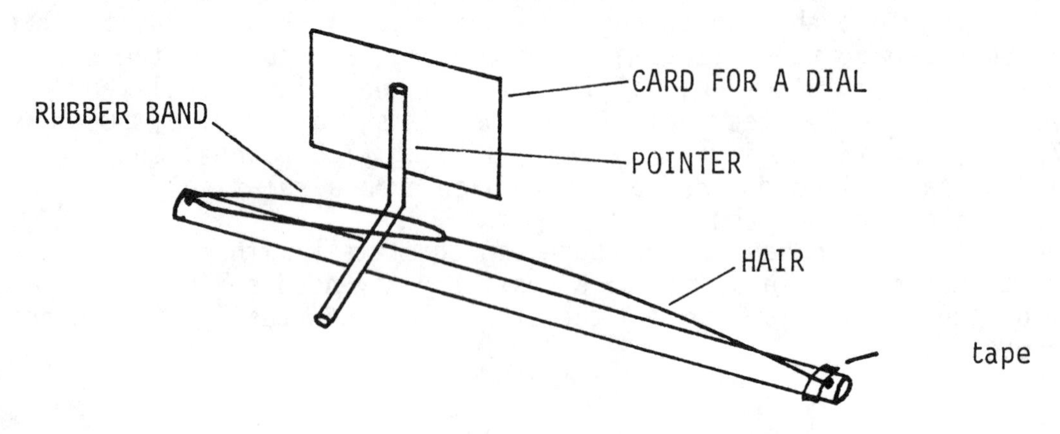

On moist days, hair absorbs water vapor and becomes longer, rotating the pointer. On dry days it shrinks. You can calibrate the hygrometer by listening to the radio or television to find the value for the humidity.

Experiment 4.02 Thermal Expansion of a Soda Straw

Experiments on heat requiring only very simple equipment are difficult to find. This experiment on expansion requires only three plastic drinking straws, sticky tape, very hot water, a pencil, a piece of card (or paper) and a Styrofoam cup. Bind two straws together very tightly along their length with sticky tape, as shown in Fig. 1. Fasten the top end to a sheet of card or paper using tape. Mark the position of the bottom end very carefully. Now, you must use the cup to pour the hottest water available through the lower straw. To do so, make a little funnel by folding one end of the third straw, as shown in the figure, so that it will fit into the top end of the lower straw. Slice the top end off the third straw diagonally to make it easier to pour through. Record how much the lower end of the straw shifts by making a pencil mark. Qualitatively, it is easy to see how the hot straw expands against the cold one, as shown in Fig. 2, pushing both straws into a bow shape.

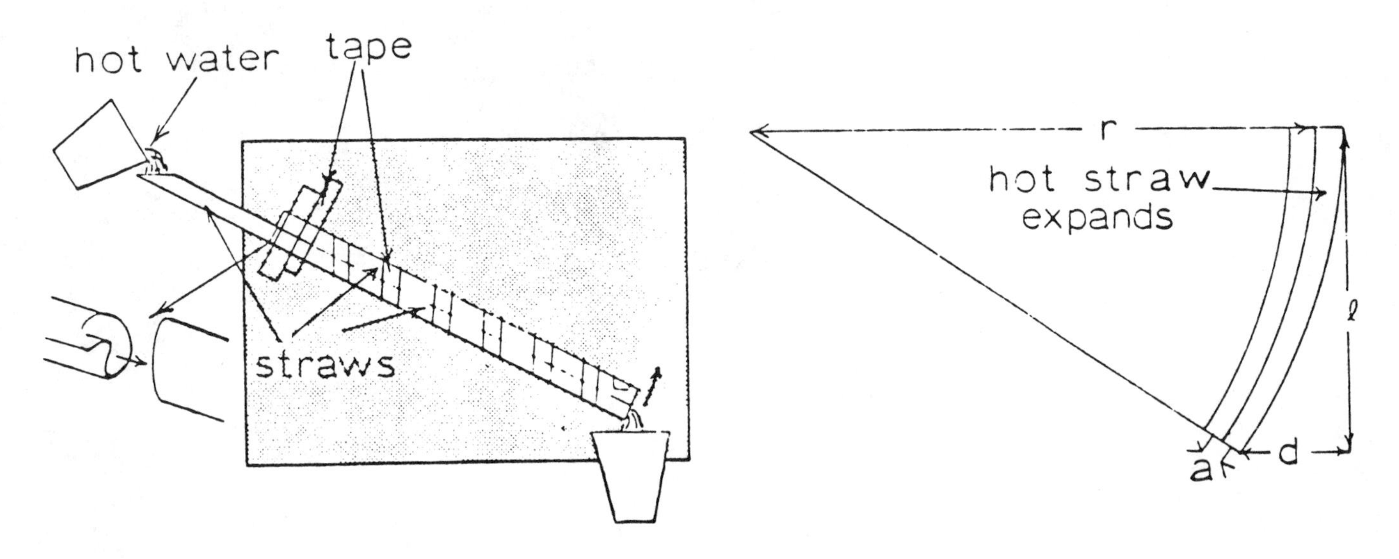

Fig. 1 Fig. 2

It is also interesting to study the result quantitatively. Boiling water will be at 100°C but if it is not available, measure the temperature using a thermometer (we shall show in a later issue how to make a thermometer out of a soda straw). Measure the length of the straw ℓ, and the distance it moves, d. Then, the radius to which the straws bow, r, is given by $2rd = \ell^2$. If the center of the straws is separated by a distance a, and the hot straw expands an amount x, then

$$x = (r + a)\,\theta - r\,\theta$$

$$\text{where } \theta = \ell/r$$

$$x = a\ell/r = 2\,ad/\ell$$

The coefficient of linear expansion α is given by the ratio

$$\alpha = \text{expansion/original length x increase in temperature}$$

$$= x/(\ell t)$$

where t is the temperature increase.

Hence,

$$d = 1/2\ \alpha \ell^2\ t/a$$

The coefficient is roughly 10^{-4}/°C for the type of plastic of which straws are made, so a temperature rise of 50°C where a is 0.5 cm and ℓ is 20 cm moves the bottom end of the straw 2 cm which is easily measurable. Heat loss and other problems generally give rise to a low measured value for the coefficient.

Experiment 4.03 The effect of heat on a rubber band

Materials : - Rubber band, two straws, match

Instructions: -

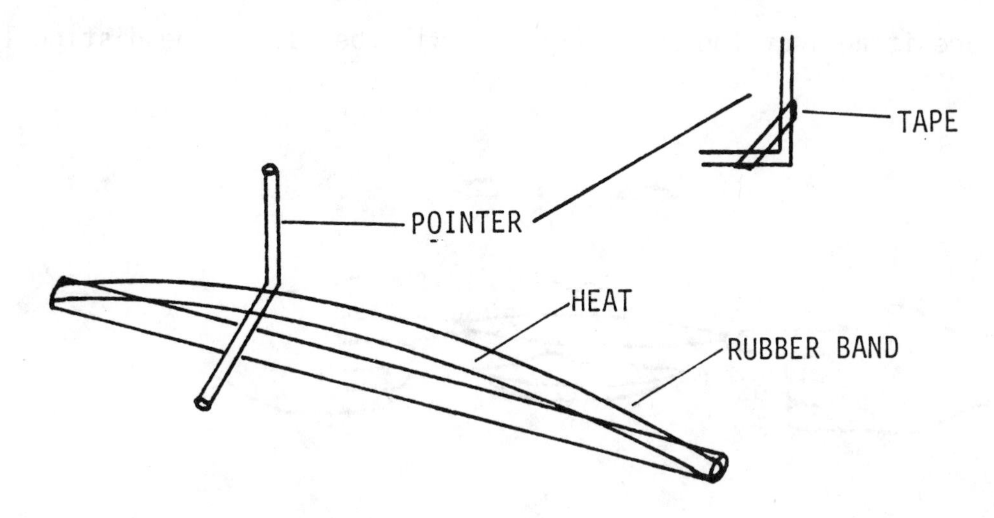

Stretch the rubber band over the ends of the straw (cut the straw to a suitable length and notch the ends if necessary. Bend about one inch at right angles for the second straw, to act as a pointer. Tape it to stay in that position. Place the pointer under the band. Heat one side of the band, and notice from the pointer that side contracts. Breathing on it may provide enough heat - if not, use a match.

Qualitative: - Most materials expand on heating - however the molecular structure of rubber is such that it contracts on heating, or rather, its elasticity decreases.

Experiment 4.04 Heat and Work

Materials: - one rubber band (reasonably large)

Instructions: - Place the rubber band between two fingers as shown, stretch it, and place it against the upper lip. It will be felt to be distinctly warm.

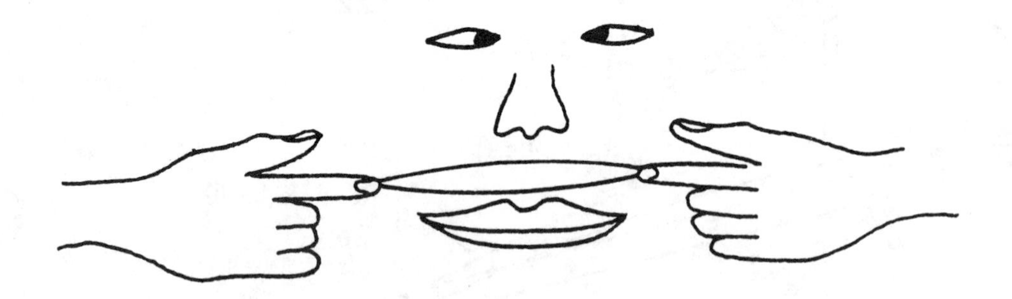

Hold it away from the face in this stretched position for a few seconds, release the tension and again place against the upper lip. It will be felt to be distinctly cooler than the surroundings.

Qualitative: - In stretching the rubber band, work was done on the rubber band. Part of this work went to heating the rubber band. On relaxing, the rubber band did work on the fingers, and drew on its heat energy to provide the work required. In the same way a gas, when compressed, rise in temperature, and cools on expansion against a piston.

Quantitative: -

This is an experiment in thermodynamics, and is described by Feynmann, Volume I 44-1.

Experiment 4.05 A soda-straw thermometer

It is difficult for students to understand the concept of temperature without a thermometer. Here is a simple experiment requiring only sticky tape, a soda straw, and a little water, which demonstrates Charles' law as well as giving the temperature.

Fold the end of a straw over two or three times as shown, and fasten it with sticky tape, (Fig. 1).

Fill the open end of the straw with about 5 cm of water (it may be easier to put the water in first, before sealing the other end). If you place the closed end in your mouth, you can see that the expanding hot air forces the water out. Remove the straw from the mouth, and notice how the air moves the water back up the tube (to its original position) as it cools. Now squirt cold water from the drinking fountain over the straw, or place it in a cold drink. The water will move way back.

The thermometer may be used quantitatively by marking the position of the water meniscus (with a pen) on the side away from the open end, first at room temperature, then for the ice cold water, for your mouth, and for very hot (preferably boiling) water. You can calibrate your thermometer on the basis of Charles' law, which states that the volume of air, or length of the air column, is proportional to its absolute temperature (temperature in °C + 273).written °K where K stands for Kelvin.

$$\frac{V_1}{V_2} = \frac{T_1 + 273}{T_2 + 273} = \frac{L_1}{L_2}$$

where V_1 is the volume and L_1 the length of the air column of temperature T_1°C, and V_2 and L_2 the corresponding values at T_2°C. Figure 2 shows typical measurements. The length of air at room temperature in this example T_r is 12.4 cm, for boiling water it is 16 cm, and for ice water 11.5 cm, and for body temperature T_B it is 13.2 cm.

$$\frac{\text{Temp °K}}{\text{Length of air}} = \frac{373}{16} = \frac{273}{11.5} = 23.5 = \frac{T_r}{12.4} = \frac{T_B}{13.2}$$

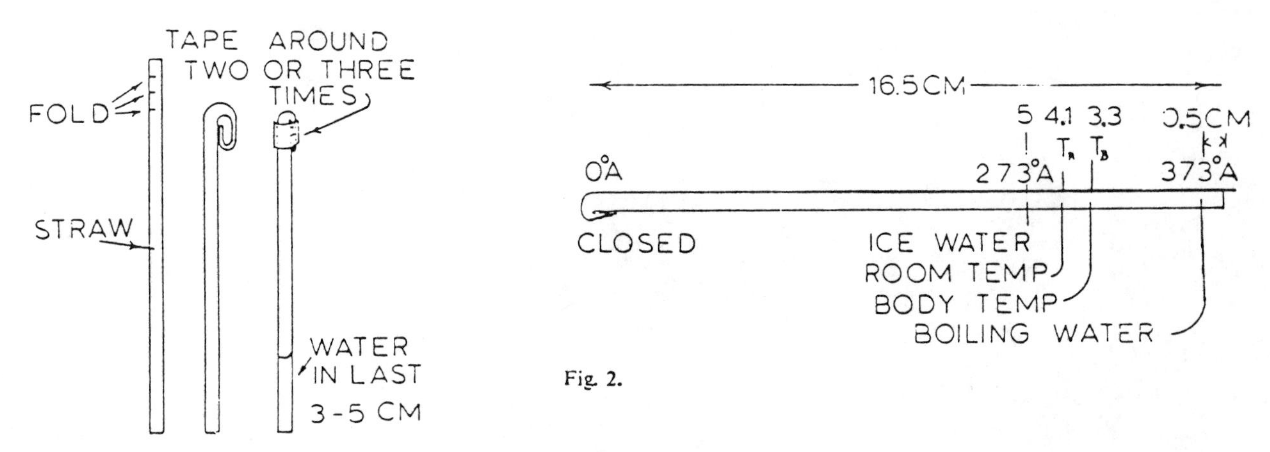

Fig. 1.

Fig. 2.

This gives room temperature as 291° K or 18°C, and body temperature as 310°A or 37°C. You can mark a linear scale (from 0 to 100) on the side of the thermometer if you can obtain the fixed points at 0°C and 100°C as described.

It is difficult to place the whole length of the straw in the mouth or cup. Unfold a paper clip, and drop it in the straw as shown in Fig. 3. The paper clip, within the straw, may then be bent to reduce the overall length.

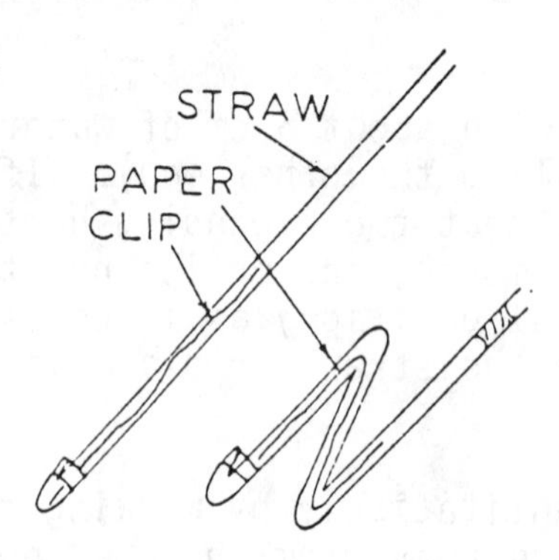

Fig. 3.

Experiment 4.06 The pressure-volume relationship in a gas

Materials: - Soda straw, water, tape

Procedure: - Fold over the end of the straw, twice, and wrap around with tape. Fill about half full with water. The straw will be about seven inches long, with about three and a half inches of water, as shown.

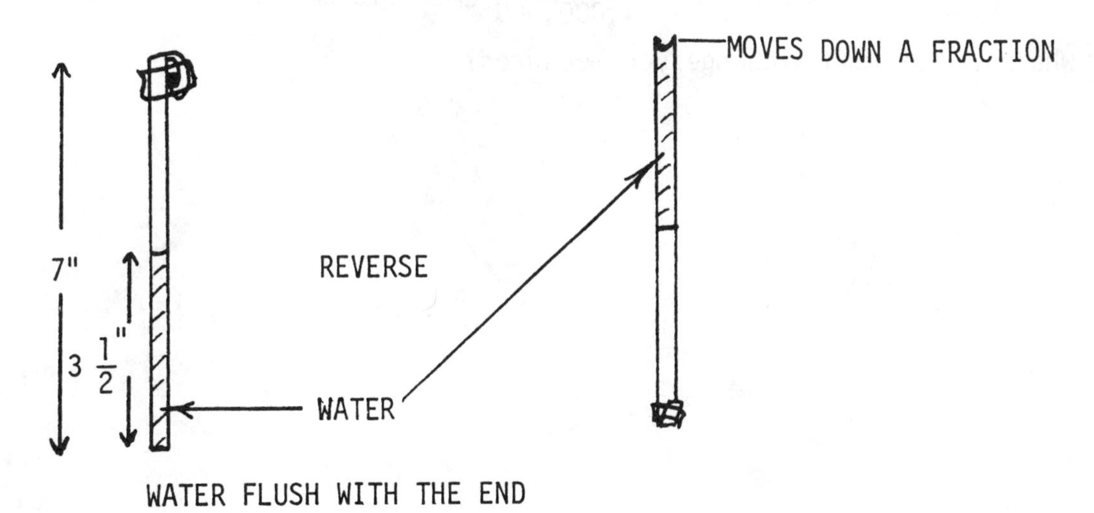

The water should be flush with the end of the straw. Now invert the straw. Notice how the water compresses the air, and moves a fraction down the tube. Estimate how far down. On reversing the tube, the water should again be flush with the end.

Qualitative meaning: -

The air in the tube is under atmospheric pressure. When held with the open end upward, to the atmospheric pressure is added the pressure of the water in the tube, reducing the volume of air. Held upside down, the water pulls downward, reducing the pressure, and enlarging the volume of air, so the water runs toward the mouth of the tube a little.

Quantitative: -

Boyle's law states

$$pV = \text{constant}$$

When p is the pressure and V the volume. Atmospheric pressure is approximately 1,000,000 dynes per cm^2. A height of 1cm of water, since it gives 1 gm per cm^2, provides a pressure of 981 dynes/cm^2. If the straw has

9 cm of water in it, and the volume of air is proportioned to the length, say 9 cm, the fractional compression of the air or reversing the straw will correspond to a change from

$$1{,}000{,}000 + 9 \times 981, \text{ to } 1{,}000{,}000 - 9 \times 981$$

a change in length of $9 \times \frac{18 \times 981}{1{,}000{,}000}$ cm = .15 cm

What is the length change you measured?

Experiment 4.07 Convection

Materials required: - Candle (a small birthday candle works best)

cup - sheet of cardboard, slightly damp paper - or, paper to make smoke.

Procedure : Cut the cardboard as shown to fit tightly in the dixie cup. Fasten a candle to one side of the cup and light it. After about two minutes, to allow the convection to build up, light the damp paper (paper towel works well) and observe the way the smoke is sucked down one side and goes up the other side with the candle.

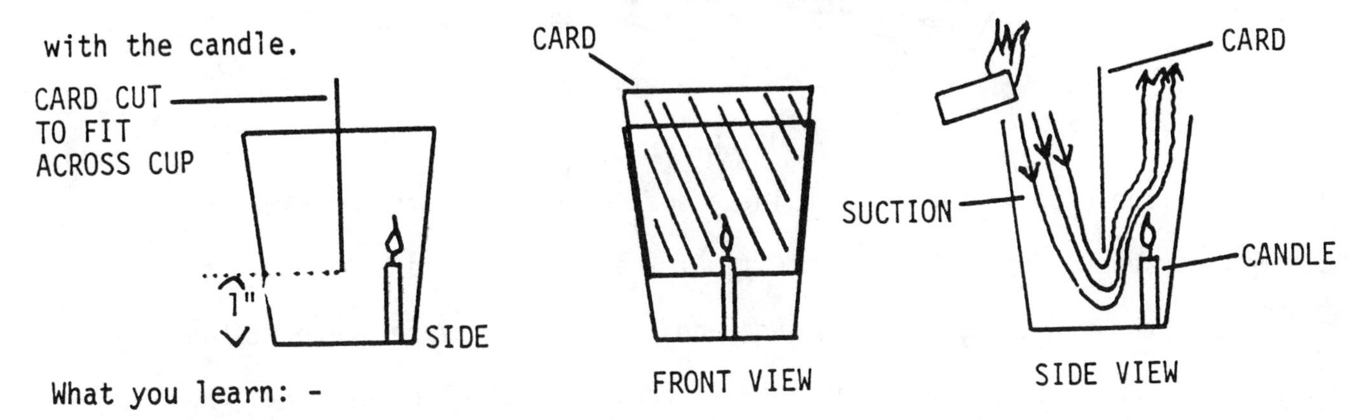

What you learn: -

Qualitative: - The hot air from the candle,being less dense,rises and draws the denser cold air from the other side of the card.

This is the principle by which the hot air rises up a chimney, and draws the smoke with it.

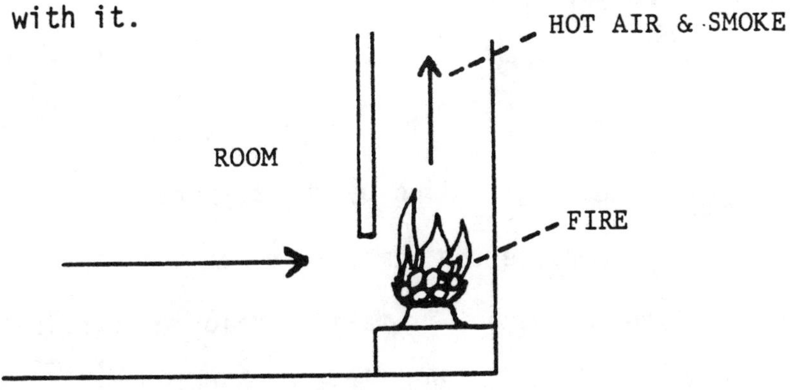

Also, the sea breeze during the day at the beach arises in the same way.

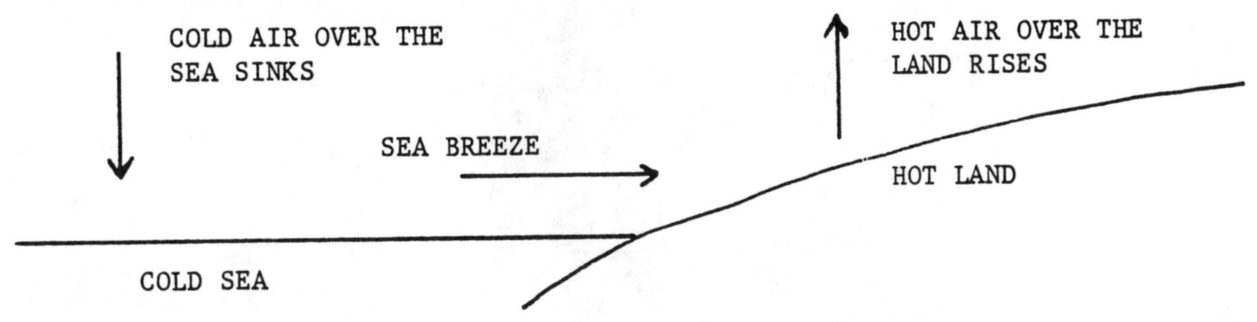

Experiment 4.08: Thermal Capacity

Materials: two styrofoam cups, thermometer.

Procedure: Half fill one cup with the coldest water. Place the soda straw thermometer in the water. Make sure the trapped air is covered. Mark the

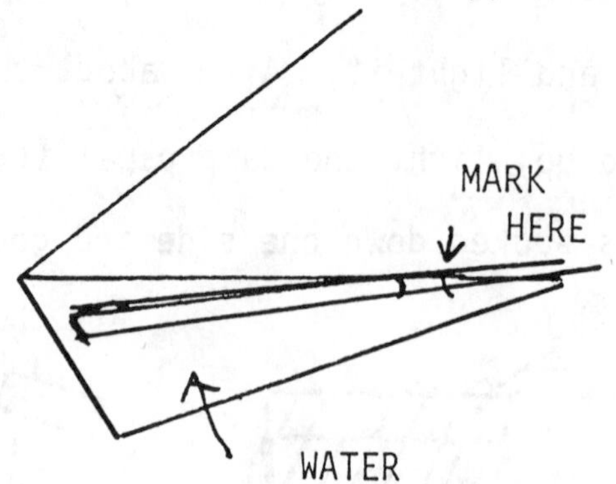

position of the drop in the straw using a soft pencil or a fine marker pen. Half fill a second cup with the hottest water available, and again mark the position of the drop. Now mix the two, and measure the temperature of the mixture. Below is shown the positions marked in an actual experiment. Note

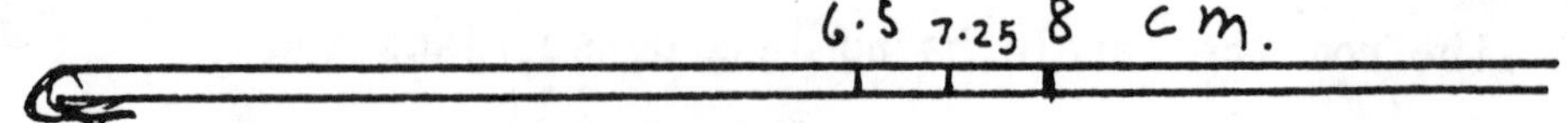

the temperature of the mixture is exactly half way between the temperature of the hot and cold water. Do you find this is so? Why is this?

The quantity of heat contained by water is proportional to the absolute temperature, and the mass of water. It is, in fact, the mass of water multiplied by the specific heat multiplied by the temperature, where the specific heat is the heat to raise the temperature of unit mass of the substance one degree. Since for water, it takes 1 calorie to raise its temperature $1^{o}C$, the specific heat of water is 1. At first we had two masses of water, M, at temperatures T_1 and T_2, a quantity of heat $MT_1 + MT_2$. After mixing, we had $2MT_3$. If no heat is lost,

$$2MT_3 = MT_1 + MT_2$$

or

$$T_3 = \frac{T_1 + T_2}{2}$$

This is what we found, and confirms our hypothesis that the quantity of heat is

1) Proportional to the mass of water.

2) Proportional to the absolute temperature.

Further checks can be made using different masses of water, m_1 and m_2, when

$$T_3 = \frac{m_1 T_1 + m_2 T_2}{m_1 + m_2}$$

The sensitivity may be improved by using the folded straw as thermometer.

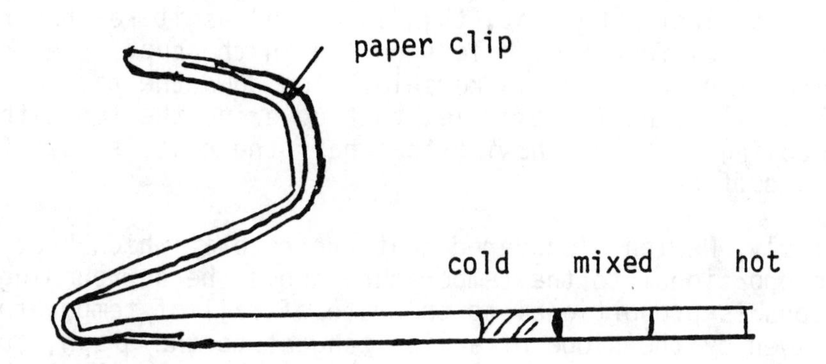

Experiment 4.09. Cooling a cup of coffee. Rate of cooling of a body and thermal conductivity of styrofoam.

Materials: Styrofoam cup, hot water, thermometer.

Procedure: Much about heat can be learned from the way a cup of coffee cools. You will need a Styrofoam cup filled brimful with hot coffee (hot tea or water will do) and the soda straw thermometer (or a regular thermometer)(4.05)made by folding over and taping one end of a straw, and putting a little water in the other end. The straw has an opened up paper clip inside, bent so the drop moves horizontally and does not run out. Make a mark on the straw, using a soft pencil or fine marker pen at the position of the water drop, making sure the coffee covers most of the straw, as shown in Fig. 1. Make readings every five minutes as the coffee cools. Plot the length between the position of the drop at room temperature, and that at a given time as it cools against the time, as shown in Fig. 2. The qualitative question requiring an answer is why the liquid cools faster at first, when hot, than later on, as it reaches room temperature. Does it cool faster with less coffee in the cup? Does covering the top with a piece of paper or a handkershief slow down the rate of cooling, and if so, why? You will find for example, that covering the top with a book cuts the rate of cooling in half, showing that half the heat is lost from the open surface of the coffee.

Quantitatively, Newton discovered that the rate at which heat is lost by conduction is proportional to the temperature above the surroundings T. The rate of heat loss is proportional to the rate of fall of temperature (dT/dt) which is given by the slope of a line tangent to the graph, such as AC. If Newton's law is true, any series of such tangents should strike the axis at a constant distance BC from the point at which a perpendicular AB, dropped from the curve where the tangent touches it, strikes the X axis. Hence, BC = EF, etc.

Now if the time BC = τ, then

$$\frac{dT}{dt} = - \frac{T}{\tau}$$

or

$$- dT/T = dt/\tau$$

Integrating $\ln T = - t/\tau + C$ (where C is a constant)

$$T = C \exp(- t/\tau)$$

τ is approximately 30 minutes with the top open, and 70 minutes with the top covered with a book.

The insulating qualities of a Styrofoam cup ensure that the outside of the cup is almost at room temperature. One can then use cooling to estimate the heat conductivity of the foam. For a six-ounce cup (177 cc) the surface area A is approximately 170 cm^2, and the thickness, d, 0.2 cm. The rate at which heat is conducted across the surface of the cup per second is

$$dQ/dt = kAT/d$$

where k is the conductivity, the heat per unit time crossing the opposite faces of a unit cube of the substance having a one degree temperature drop between those faces.

The specific heat of the water or coffee C is 1 cal/gm, the density S is 1gm./cc., the volume of the cup is V cc., so $(dQ/dt)=-VSC(dT/dt)=-V(dT/dt)$ cal/sec

Therefore $dT/dt = -(kAT)/Vd$

Since $\frac{dT}{dt} = -\frac{T}{\tau}$

$$(dT/dt)/T = -kA/Vd = -1/\tau$$

or $$\tau = Vd/kA$$

so $$k = Vd/A\tau$$

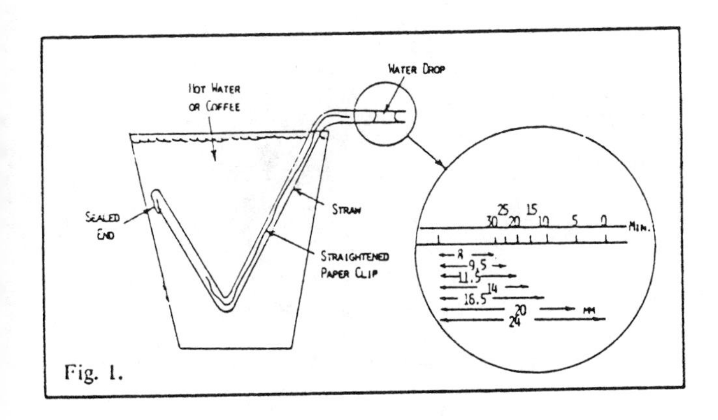

Fig. 1.

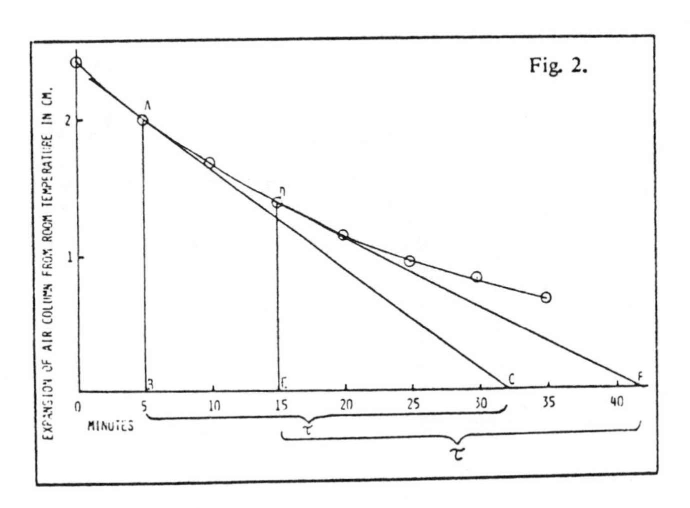

Fig. 2.

The conductivity of foam found by this technique is about 0.00006 cal/(sec, cm, °C) provided heat is prevented from escaping from the top of the cup by putting a book on it. Essentially we are measuring the thermal conductivity of air, which is 0.000057 cal/(sec, cm, °C), because the pores of the Styrofoam are air filled. Each pocket of air is restricted to its cell and cannot remove heat by convection as it does outside the cup.

One can show that the quantity of heat that can be extracted from a body is proportional to its temperature, its mass, and its specific heat using Styrofoam cups, too. Half fill two cups, one with the coldest water, and the other with the hottest water available. Mark the position of the thermometer water drop when it is placed in the two cups (Fig. 3). Make sure the enclosed air is covered. Now mix the two and measure the new temperature. This should lie halfway between the hot and the cold. If we use different masses of water, M_1 and M_2 (use a spoon, or other measuring device, to ladle different known quantities into the cups), the temperature of the mixture is given by

$$T_3 = (M_1T_1 + M_2T_2)/(M_1 + M_2)$$

We can now extend this to different materials. Glycol (antifreeze) makes an interesting example. The quantity of heat transferred from a body is also proportional to its specific heat σ giving us

$$T_3 = (m_1\sigma_1T_1 + M_2\sigma_2T_2)/(M_1\sigma_1 + M_2\sigma_2)$$

where we replace one cup of water, specific heat $\sigma_1 = 1$ with antifreeze, $\sigma_2 = 0.56$. Many common organic liquids (e.g., ethyl and propyl Alcohol) have densities near enough to one, and specific heats of about a half. Avoid liquids giving out heat chemically when mixed with water. The results show immediately that antifreeze cannot absorb as much heat as water for a given temperature rise, which is a disadvantage as a cooling agent.

Fig. 3.

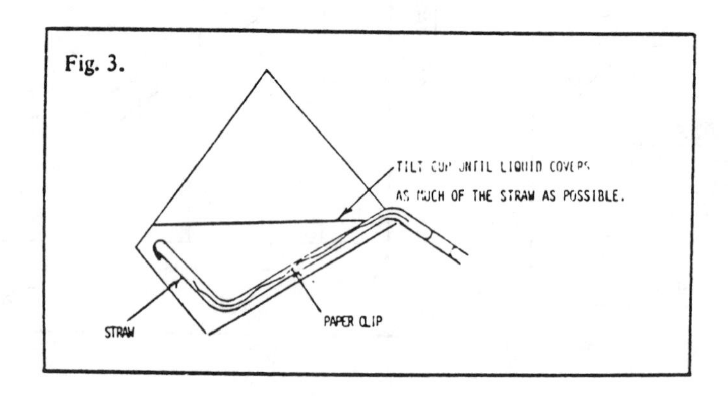

Experiment 4.10 - Relative Humidity with a Drinking Straw Thermometer

Materials: straw, styrofoam cups, tape, paper towels, ice, boiling water

Purpose: to demonstrate how a wet bulb thermometer works and is used to find the relative humidity

Procedure: When asked to measure distances on the straw, always measure to the nearest tenth of a cm.

Place the straw in a cup of water. Allow the straw to fill with water up to about 2 or 3 inches. Put your finger over the top of the straw and remove it from the cup. Turn the straw upside down so that the water is at the top. Fold over the bottom of the straw 2 or 3 times and tape it down. Mark the position of the top end of the water drop.

Wrap a wet paper towel, or wet toilet paper, around the straw until the top end of the water drop is just visible. Wave the straw about gently, and mark the lowest point the water drop reaches, when the straw is held steady.

Measure the distance between the two marks ℓ on the straw and enter it in your notebook. Also measure the distance from the closed end of the straw to the bottom of the drop, L.

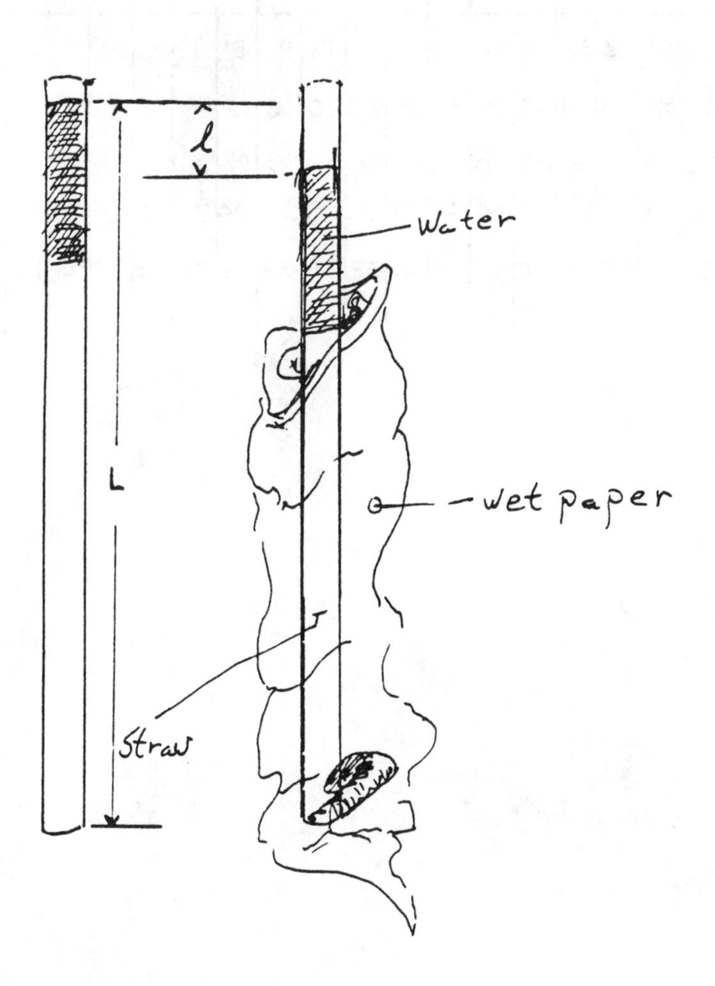

R. Bennett

If we assume room temperature is about 75°F, (24°C) this is 297°K, Charles' gas law tells us the volume of a gas is proportional to its temperature, so in our case, where the volume is proportional to the length, the change in temperature produced by the wet paper is

$$\frac{\ell}{L} \times (297°K)$$

If t is the temperature in °F, the lowering in temperature in °F is given by

$$\frac{\ell}{L} \times (t + 460)$$

The relative humidity may then be found from the table. To give an example, suppose the air column were 18 cm long, and on cooling became 17.6 cm long. If room temperature t were 75°F, the drop in temperature would be 11.9°F corresponding to a relative humidity of 44%.

RELATIVE HUMIDITY

Lowering of wet-bulb thermometer (°F)

Air temp (°F)	1	2	3	4	5	6	7	8	9	10	11	12	13	14	15
60	94	89	83	78	73	68	63	58	53	48	43	39	34	30	26
65	95	90	85	80	75	70	66	61	56	52	48	44	39	35	31
70	95	90	86	81	77	72	68	64	59	55	51	48	44	40	36
75	96	91	86	82	78	74	70	66	62	58	54	51	47	44	40
8	9C	91	87	83	79	75	72	68	64	61	57	54	50	47	44

Air temp (°F)	16	17	18	19	20	21	22	23	24	25	26	27	28	29	30
60	21	17	13	9	5	1									
65	27	24	20	16	12	9	5	2							
70	33	29	25	22	19	15	12	9	6	3					
75	37	34	30	27	24	21	18	15	12	9	7	4	1		
8	41	38	35	32	29	26	23	20	18	15	12	10	7	5	3

Experiment 4.11 The Mechanical Equivalent of Heat

20-30 pennies, 4 styrofoam cups and a thermometer are the only objects required for this experiment.

Make sure everything is at room temperature by leaving it out several hours. Cut a circle of card to fit inside the bottom of the lower cup. Place the coins in one cup, and tape the other cup on top, open ends together as shown in the figure. Push the other two cups over the first two to improve the thermal insulation. Shake the cups so that the coins fall from the top of the top cup to the bottom of the bottom cup 400 times. This takes about three minutes. Since the cups are each 8 cm high the coins fall 2 x 8 x 400 cm or 64 m. If the total mass of the coins is m kg, the work done is mgh which is m x 628 J. h is the height they fall, g the gravitational constant (9.8 m/s^2). Make a small hole in the top cups and poke a thermometer through to measure the rise in temperature of the coins. Let this be t. The heat delivered to the coins is tjmc where j is the mechanical equivalent of heat in Joules/cal and c is the specific heat of the copper, 0.093 cal/gm K. or 93.0 cal/kgK. The specific heat depends on the constitution of the coins. From 1864-1982 American pennies were 95% Cu and 5% Zn. Since then they are 95% Zn, 5% Cu. Luckily, zinc has almost the same specific heat as copper, (.0925 versus .0921) so one can ignore the differences. The mechanical equivalent of heat is given by

$$j = nmgh/mct$$

$$= 628/93t \text{ Joules/cal}$$

If j = 4.18 J/cal., t must be 1.6°C. In fact, we find values of j much smaller than this probably because the tendency is to shake the cups too hard, and generate more energy than given by the pennies falling only 16 cm each shake. The advantage of using coins is that the specific heat being so low, the temperature rise is high. The double walled styrofoam prevents the heat leaking out, and the specific heat of the air in the container is negligible. Nevertheless, it can take several minutes before the maximum temperature is reached in the container. A regular mercury thermometer graduated in tenths of a degree is best, but if a drinking straw thermometer 16 cm long is used (a standard length for straws) the water drop at the top of the drinking straw will move a distance of $16\left(\frac{301.6}{300} - 1\right)$ cm = 0.085 cm which is barely measurable. How can one improve the sensitivity of the thermometer? If you push a large paper clip into the open end of the thermometer, it will spread the circular opening into a flat oval, as shown in the figure, reducing the area by a factor of two or three. The sensitivity of the device is increased by a corresponding amount, but it must now be calibrated against a standard drinking straw thermometer.

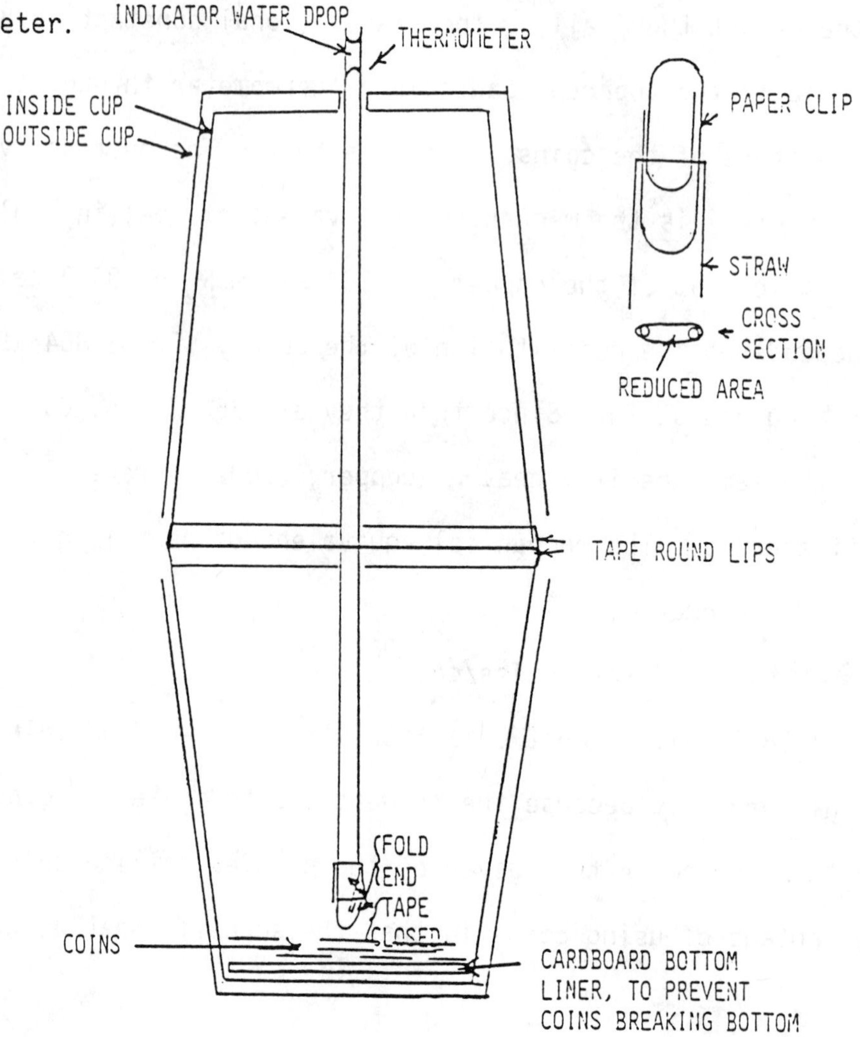

Experiment 4.12 Absorption and Emission of Radiation

We have seen how heat may be transferred by conduction - in a previous experiment the heat is conducted through the styrofoam cup on cooling a cup of coffee, the foam behaving essentially as air unable to convect, which has a low conductivity - a paper cup cools much more rapidly, because conduction through paper is rapid. Convection was demonstrated with the candle in the cup having a vertical division - but what about radiation? This experiment can only be performed on a sunny day. A very simple demonstration consists of a single straw, with a drop of colored water (ink, Coke etc.) in the middle as shown in figure 1. The ends are folded over and sealed with tape, to ensure they are air tight, then one end is blackened, preferably with soot from a burning candle or damp paper, but if not, then with a pen or marker. Then place the device in the sun - the drop moves away from the blackened end, because of the expansion of the hot air. To improve the effect, cover the unblackened end with aluminum foil. For a more quantitative experiment, take two styrofoam cups, cut one side off each one, as shown, and line them with aluminum foil. Blacken one of the foil liners over the aperture, as before. Put the same amount of water in each cup liner - enough to half fill them, and tilt to face the sun - best at noon. The styrofoam cup is merely to prevent heat escaping from the rear of the device. After a few minutes, measure the temperature of each cup, using a thermometer (the drinking straw thermometer from previous experiments works well). The black coated cup heats up much more rapidly. These experiments have the advantage that it is clear the sun is providing the heat through radiation, and not in any other way. One can get a crude value for the solar constant (the energy delivered per unit area and time by the sun at the distance of the earth's orbit) by dividing the mass of water times the temperature by the time multiplied by the absorptive area. The constant is 2 cal/cm^2/sec, but you will get much less than this. Why?

An interesting example of solar radiation is the solar hot air balloon. Take a very thin (the cheapest kind!) black garbage bag, or better till a black garmet bag fill with air, seal the open end with thin thread, which is also the tether. Fill only to about 80% - the air must have room to expand. Put in the sun - morning is best, when the ambient air is cool, on a very still day. The bag will rise up as a hot air balloon - it takes about ten minutes to get hot enough. Rotate the bag to heat uniformly.

Emission of radiation is detected by filling a plastic (not styrofoam) cup with hot water, covering one side with black paper, and the other with aluminum foil. Hold a black coated straw themometer an inch or two from the aluminum face, and the black face, and note the difference. Even more simply, use your hand as the source of heat, put a flat sheet of aluminum foil vertically against your hand, and measure the temperature an inch or so away horizontally. Do the same with black paper. Note the black paper radiates more.

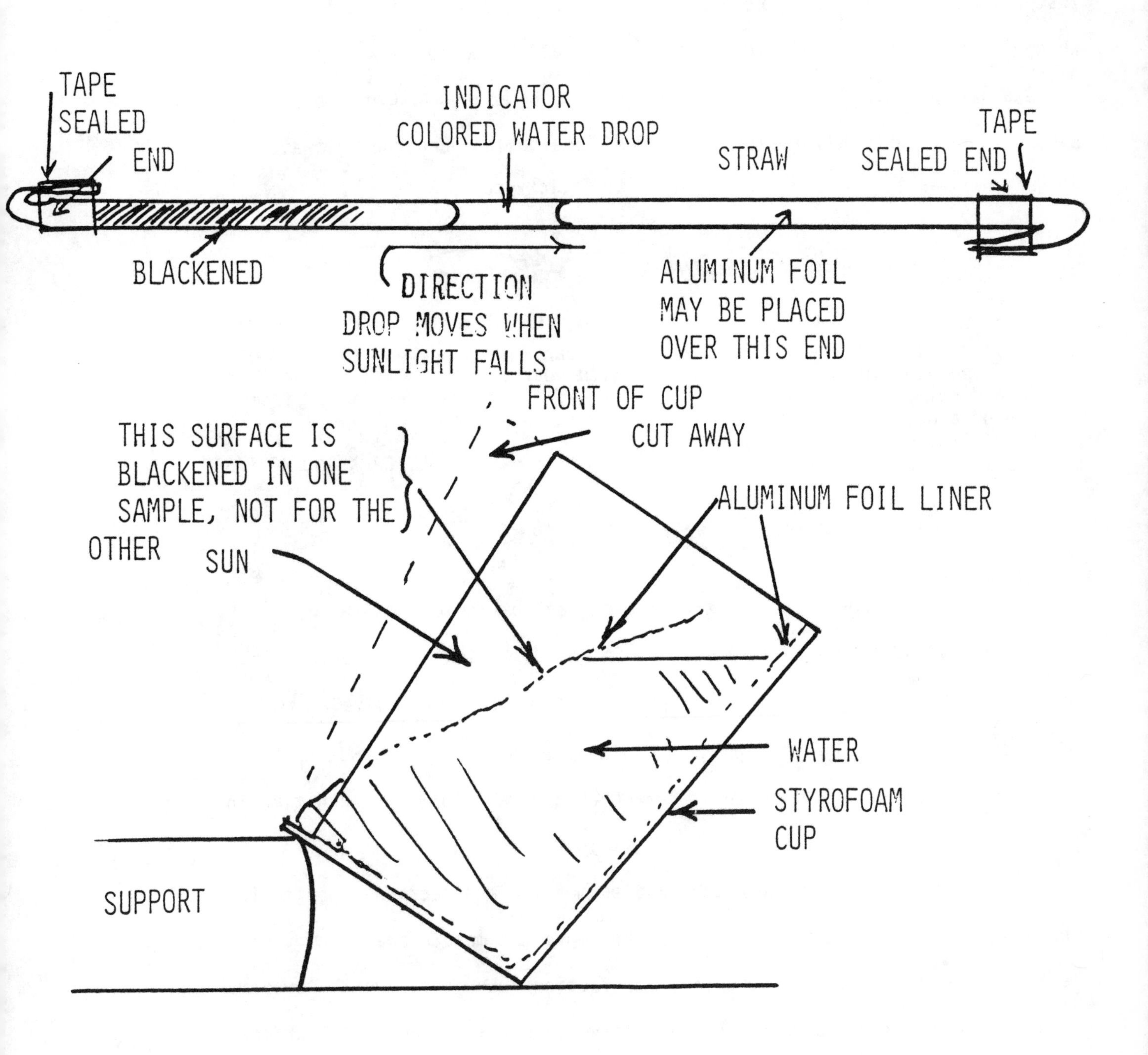
TAPE
SEALED
END
INDICATOR
COLORED WATER DROP
STRAW
TAPE
SEALED END
BLACKENED
DIRECTION
DROP MOVES WHEN
SUNLIGHT FALLS
ALUMINUM FOIL
MAY BE PLACED
OVER THIS END
FRONT OF CUP
CUT AWAY
THIS SURFACE IS
BLACKENED IN ONE
SAMPLE, NOT FOR THE
OTHER
SUN
ALUMINUM FOIL LINER
WATER
STYROFOAM
CUP
SUPPORT

Experiment 5.01 Longitudinal Waves

Materials: - rubber bands, marbles, paper clips tape

Instructions: - The aim of the experiment is to construct a device along which longitudinal waves travel slowly, so that the motion may be followed in detail. Connect sixteen paper clips in a string using sixteen rubber bands, to provide a weak restoring force. To slow the longitudinal wave, attach two marbles to each paper clip with sticky tape.

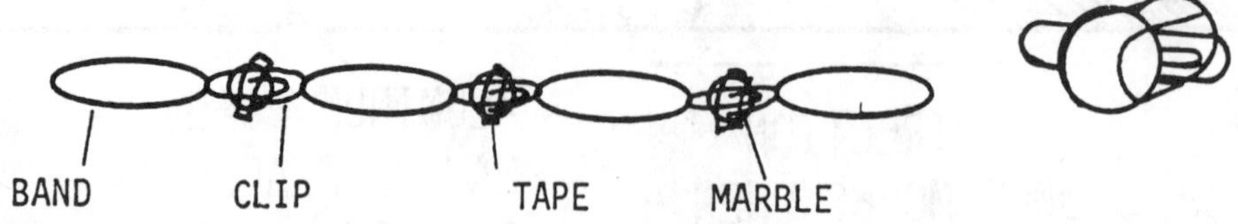

Now, attach both ends to a firm anchor - you can fasten one end to your desk and hold the other with your left hand. With your right, pull back the last marble and release. The device works better if suspended vertically, from the top of a doorway or some other suitable point. A little tension should be provided by hand at the bottom end.

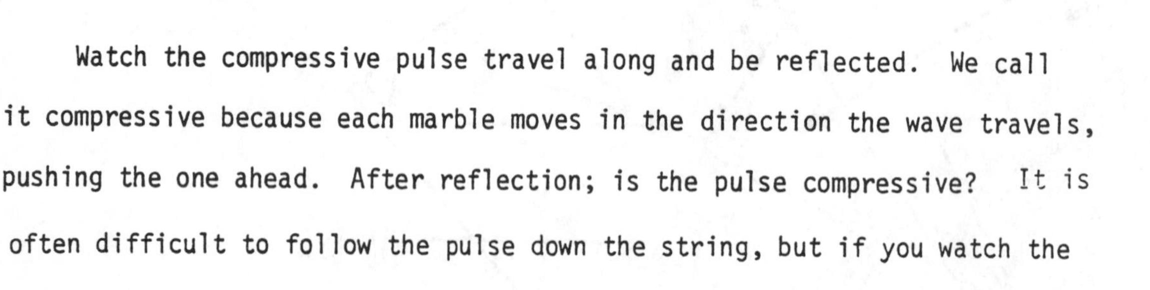

Watch the compressive pulse travel along and be reflected. We call it compressive because each marble moves in the direction the wave travels, pushing the one ahead. After reflection; is the pulse compressive? It is often difficult to follow the pulse down the string, but if you watch the end marble closely, you will see it jerk backwards and forwards each time the pulse passes.

A rarefaction occurs where the marbles move in a direction opposite to that in which the pulse travels - so if you move the marble away from your left hand before releasing, you get a rarefaction traveling down the system. Sound waves in air are composed of successive compressions and rarefactions. You have looked at reflection from a fixed end. Such reflections occur with sound waves in organ pipes closed at one end. Standing waves are built up in such pipes. You can simulate such a standing wave by moving the hand holding the rubber band backwards and forwards until you hit a resonance. The marble near the far end must be stationary. This is called a node. The marble in the middle moves rapidly, which is called an antinode.

To examine what happens if we have an open organ pipe, attach three or four rubber bands without marbles or paper clips between the far end of the string and the table.

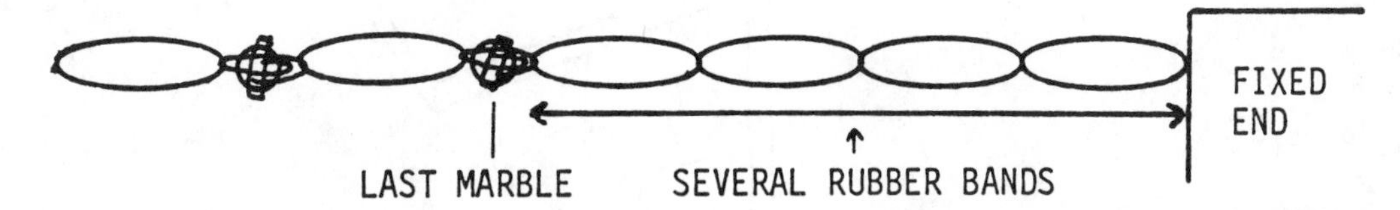

Now, if you feed in a compressive pulse, is it reflected as a compression or a rarefaction? Try producing standing waves You will find the end marble, which was stationary, now moves more than all the rest - so what was a node for a closed pipe, is an antinode for an open one. This arrangement can also be used for transverse waves, but the soda straw device, mentioned elsewhere, works better.

Torsional Waves

If the device is hung vertically from a support, torsional waves may be generated by rapidly twisting the lowest rubber band by rubbing it between the thumb and forefinger. The bottom clip and marbles spin rapidly, and pass this motion very slowly to the top of the chain. Here the pulse reverses and the marbles spin in the opposite sense. Reaching the bottom, which reflects like an open ended pipe, the marbles continue to spin, and wind up in the same sense, the pulse again traveling up the chain and reversing at the top. This proves a dramatic demonstration of the difference between reflection at an open and closed end.

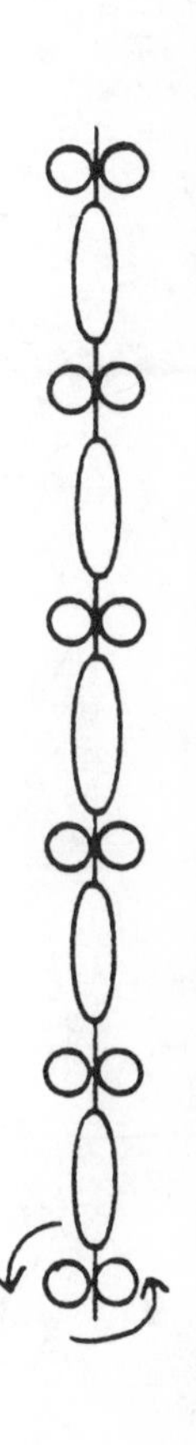

Experiment 5.02 Transverse Waves

Materials: Sticky-tape, about two dozen drinking straws, paper clips.

Procedure - Attach one end of the tape to the table top, pull about two feet off and let it hang down

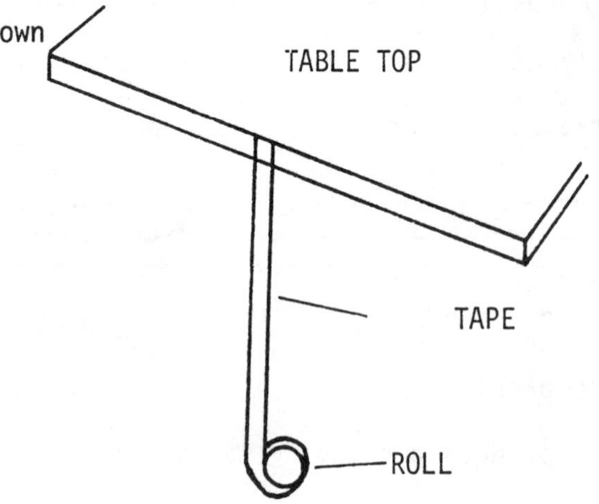

Place one paper clip in each end of each drinking straw

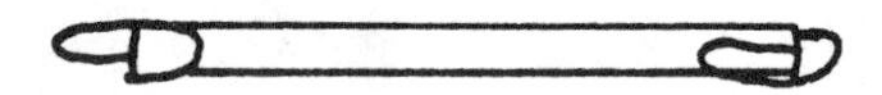

Stick the center of the straws at one inch intervals along the sticky tape, until you

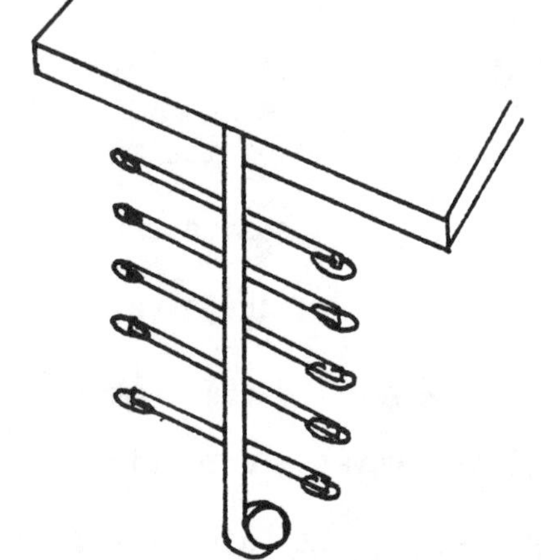

have about 24 of them attached. Now, looking end on at the straws, pull the tape reel, to make the strip taut, and give the bottom straw a tap. You will see a transverse wave pulse travel up the strip, and be reflected at the top

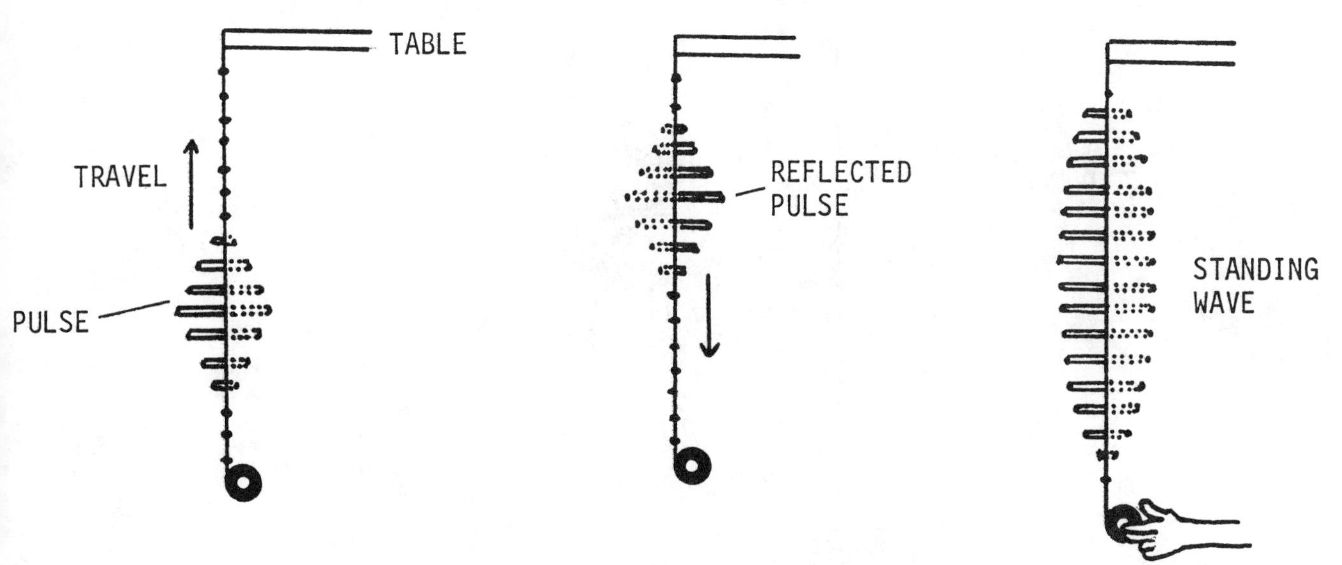

You may induce standing waves by rotating the bottom straw too and fro with the right period. If you unreel a length of tape, you may study reflection from a free end, just as you did reflection from a fixed end

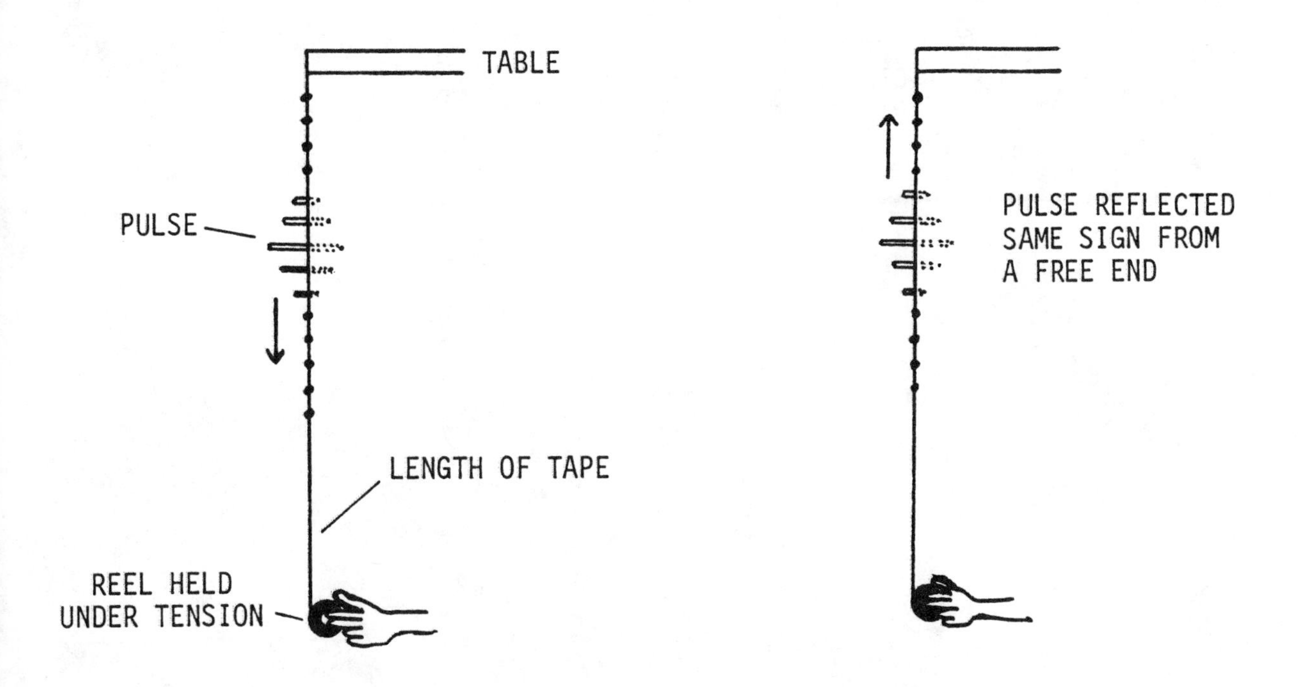

For the last foot or so of the tape, put two paper clips at each end. Now you can study the reflection of a wave traveling from a less dense to a denser medium (top to bottom) or vice versa (bottom to top). Note how, in each case, part of the wave is reflected at the intersection; but in one case it changes sign (phase)

and in the other case it does not.

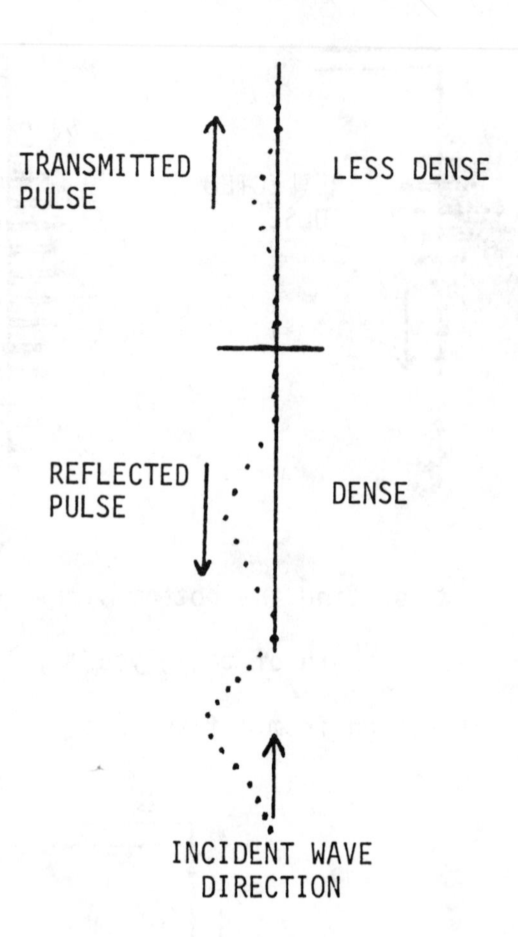

Experiment 5.03 Twin Slit Diffraction

Materials: ruler, pencil, protractor, scissors

Procedure: Cut out the four sheets having the white semicircles on them. The white dot on the bottom of each sheet represents a source of circular waves, the rest of the pattern being similar to that produced by a point source in a ripple tank, (or a monochromatic line source of sound or light). The long vertical line below this point is to help identify the source, and several additional lines have been placed to the right and left as reference.

The separation between two white circles is one wavelength. Take the two sheets with the smaller wavelength, and place them face to face, with the two white dots touching one another. Look through the two sheets, by holding them up to a window pane, or to some other light source. You will see figure 1. Slowly slide one sheet over the other to separate the sources horizontally, and obtain the rest of the patterns shown. The bright central line arises because there is a maximum of intensity here. Each point on the central line is the same distance from each source, so the waves always arrive here in phase, each crest or trough constructively adding to the one from the other source. Away from the center, the waves arrive out of phase--the crest from one wave arrives at the same time as the trough from the other. However, if we separate the sources by a distance greater than λ, we see two further bright lines, which arise because the distance from each point on this line to one source is exactly one wavelength more or less than it is to the other, so that again, the crests and troughs from both sources arrive at the same time. Further out are other bright lines where the difference of path length to one source is exactly two wavelengths more or less than to the other. Draw lines on the sheet of paper where the bright line is with the sources separated by 2λ, and answer the following questions:

1) As you separate the sources, do the bright lines move inward (i.e. closer together) or outward?
2) Separate the sources to 4λ. What is the relationship of the new bright lines to those you drew separated by 2λ? Do you find that doubling the separation of the sources halves the angle between the lines? As you separate the slits how many sets of bright lines do you see? This would show an inverse or reciprocal relationship between the diffraction pattern and the separation of the source.
3) Now take the second set of semicircles, and note that the wavelength of these is twice that used previously. Again, superpose them face to face, and compare the pattern produced when the sources are separated by a fixed number of markers, say two, with that using the shorter wavelength sheets having the sources separated by the same amount.
Do you find the bright lines spread out as the wavelength increases? If so, does the pattern double in width if the wavelength doubles? This would show that the pattern width is roughly proportional to the wavelength.
4) Measure the angle between the center bright line, and one of those you drew with the sources 2λ apart. If one source is λ further from the line than the other, we can make the following calculation.

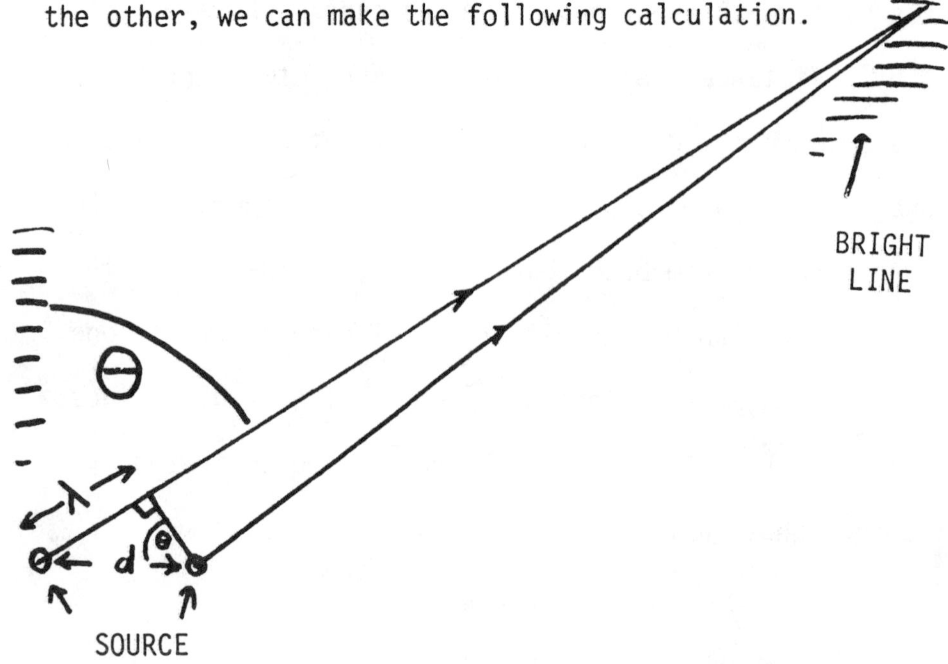

From the diagram, approximately $\sin\theta = \frac{\lambda}{d}$ (if the first line is sufficiently far away).

Then $d = 2\lambda$

so $\sin\theta = \frac{1}{2}$ or $\theta = 30°$

How close is your measurement to this? Why does it not agree exactly?

You can regard the shorter wavelength as blue light, and the larger as red, since these are roughly a factor of two apart, or you could think of the larger wavelength as middle C, and the shorter as the C above this, since these are a factor of two, an octave apart.

Metrologic, the company who first produced this type of diagram, suggested using twin slits in front of a laser to demonstrate the effect using light, but for sound you could use twin speakers fed by a constant tone, and note the increase and decrease in sound volume as you walk across the room because of constructive and destructive sound interference.

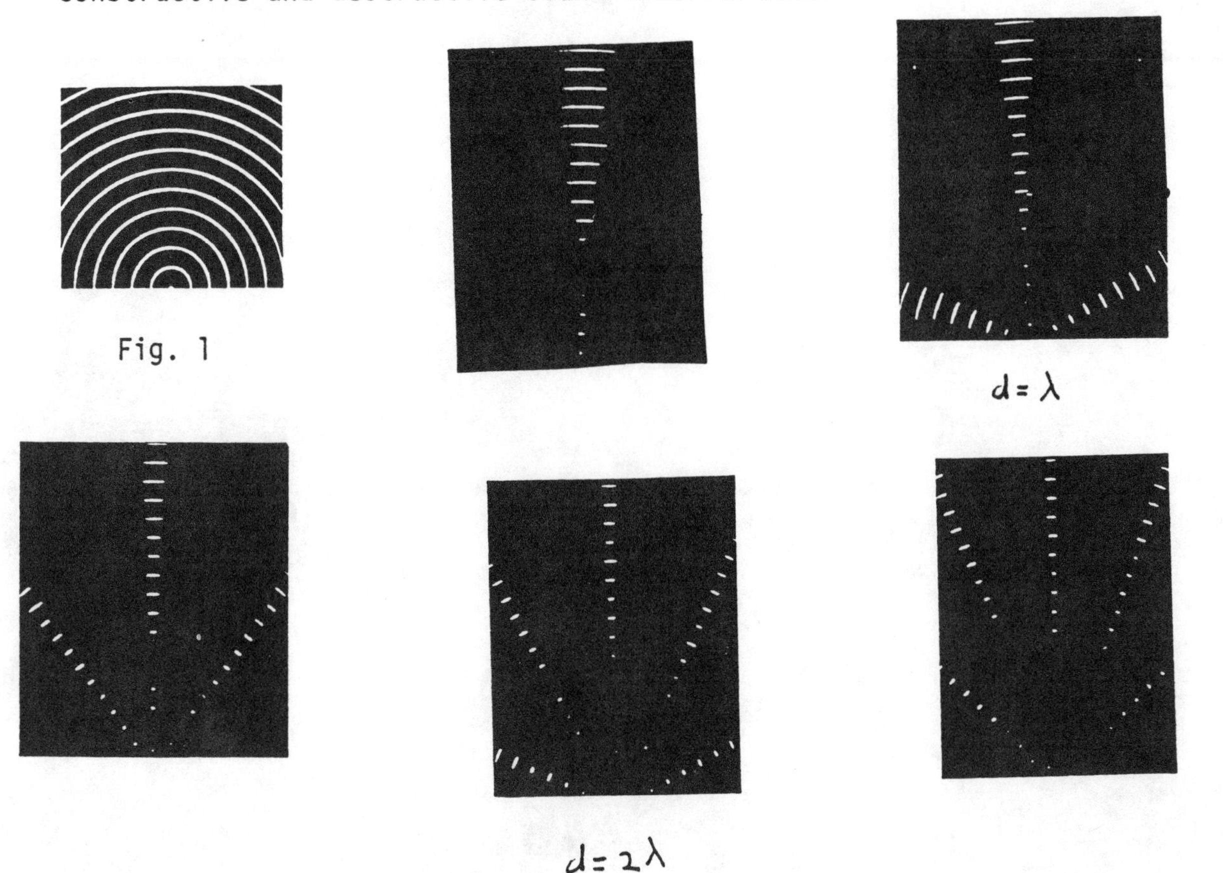

λ

Experiment 5.04 Longitudinal Waves - Crova's Disc

Materials - Scissors, pin.

Procedure: It is sometimes difficult to visualize the process by which longitudinal waves, both travelling and standing, progress. Crova's discs have the advantage that the motion can be made as slow as one wishes. Cut out the disc labelled A, which represents travelling waves. The other disc, stationary waves, will be on the reverse side. Cut out the slot, and push a pin through the point labelled B, through the center of the disc, and into some suitable object, such as a desk or table. The disc, as seen through the slot, should appear as below:

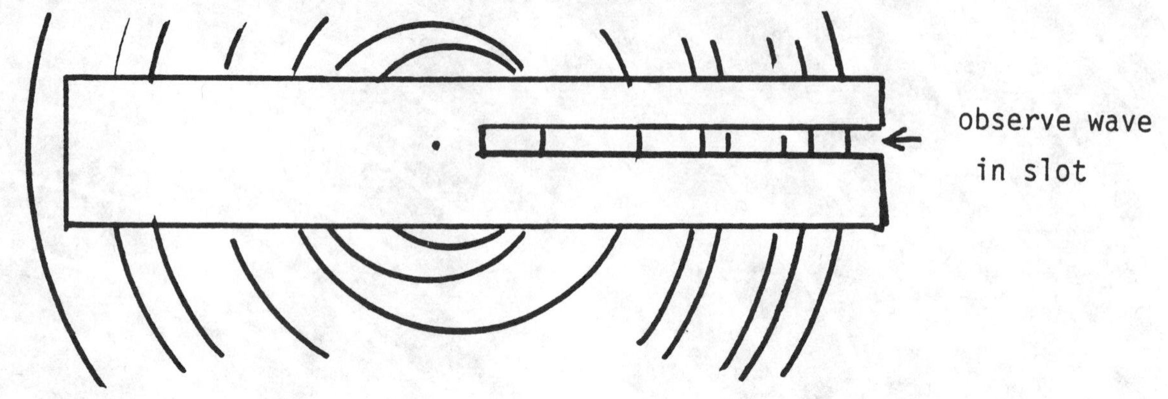

The wave is supposed to take place longitudinally in this slot, as would air in an organ pipe.
Now rotate the disc under the slot. Each pair of lines separates a layer of gas, say, and the motion of this layer, its compression and expansion, can easily be seen, the compressed layer pushing the next layer, and so on. When the standing wave is placed in the slot, it has a node at each end, and two nodes in the middle. The positions of these are marked on the slit.

What is the wavelength of the two waves, in centimeters, and what is the amplitude of the excursion?

The wavelength of the travelling wave is the distance between successive compressions, (where the lines in the slot come closest to one another). It is also the distance between successive rarefactions. This corresponds to the distance between crests , or the distance between troughs in a transverse wave.

The amplitude of the excursion can be found in the travelling wave by noticing that each line in the slot moves backward and forward. There is no net motion of the lines, as there is of the wave. Make a mark at the edge of the slot where any one line is furthest to the right. Now rotate the disc until it is furthest to the left, and make another mark. The distance between these marks represents the maximum excursion of this line, and half this (i.e. from the center to the maximum, or minimum) is the amplitude of the wave.

The wavelength for a standing wave is twice the distance between nodes (where there is no motion of the air or of the line in the simulation) or twice the distance between antinodes (where maximum movement of the line in the slot occurs). The amplitude at the antinodes can be found by marking the slot when the line at the antinode is farthest right, and doing the same with it farthest left. The distance between the two marks is twice the amplitude of the wave.

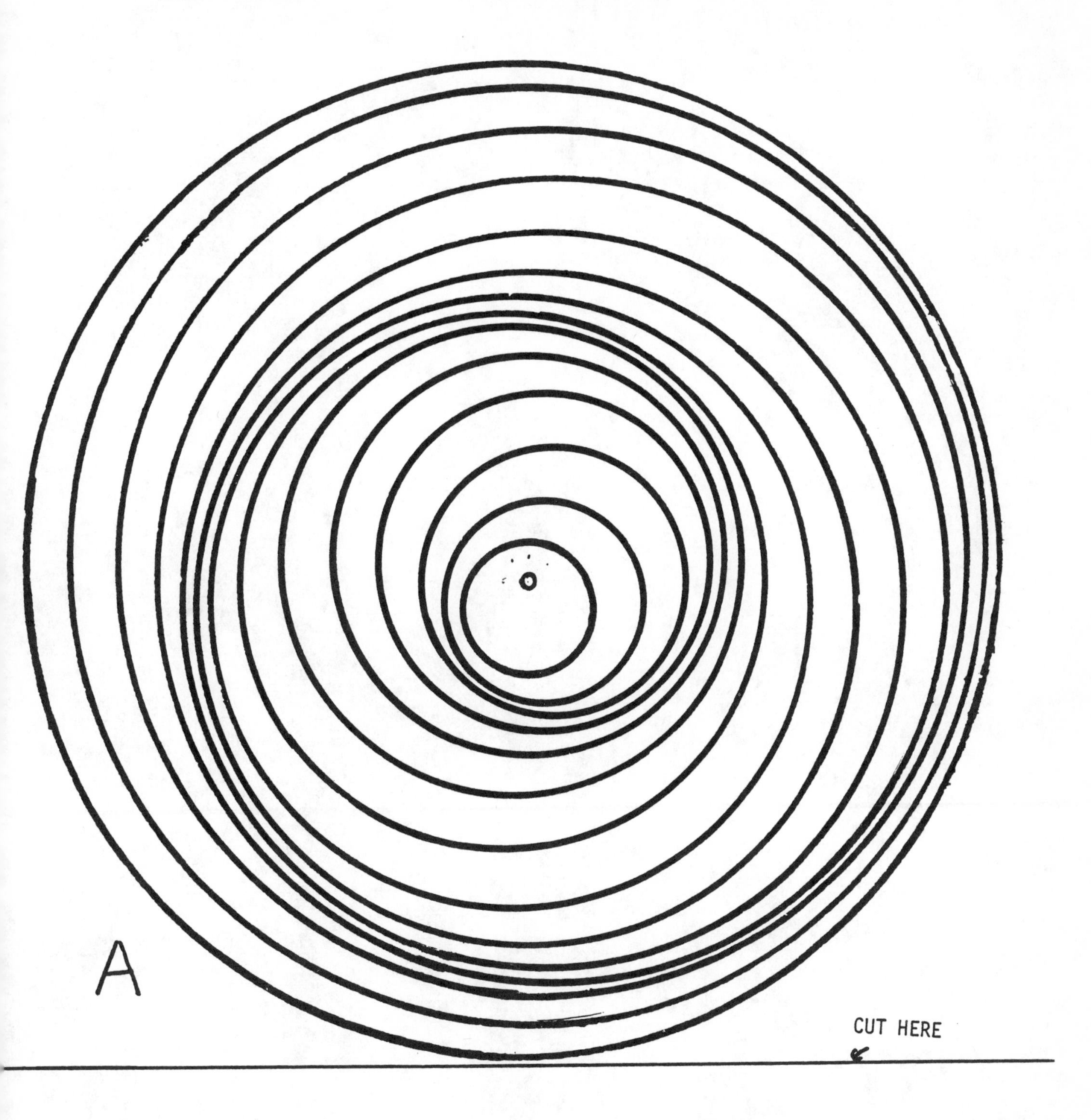
A
CUT HERE

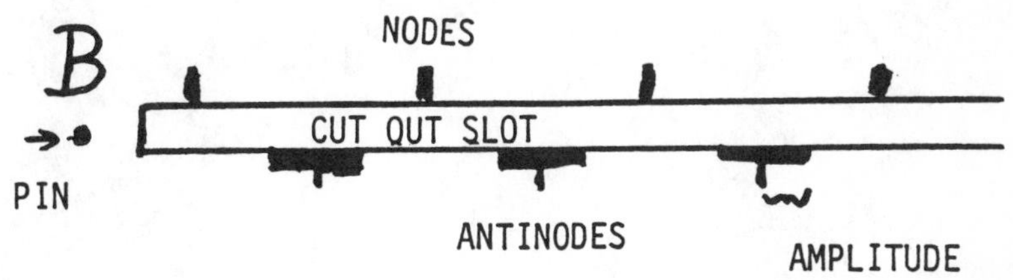
B
NODES
CUT OUT SLOT
PIN
ANTINODES
AMPLITUDE

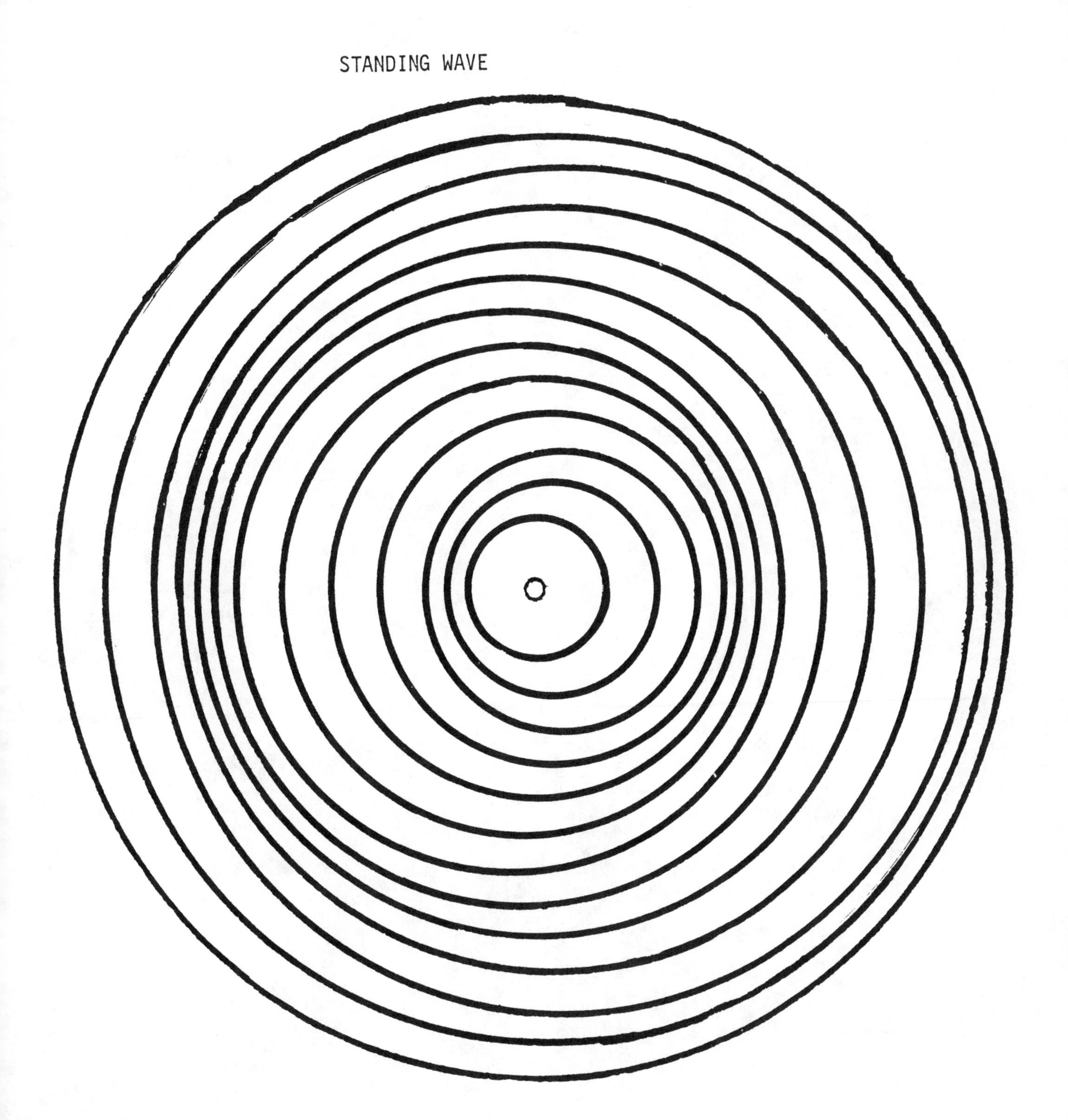
STANDING WAVE

5.05 YOUNG'S SLITS WITH PAPER WAVES

The Moire pattern method of demenstrating Young's slits is expressive, but an even simpler way uses long strips of paper.[1] Cut two long strips of paper from a newspaper, or a brown paper wrapper. Fold them too and fro and cut a waveform out of them much as one does in making paper dolls. Unfold the strips and tape the ends to two "slits" drawn on the wall, or blackboard as shown. Hold the opposite ends of the strips. The paper waves represent an instantaneous snapshot of the waves coming from the slits. Equidistant from both slits, the waves interfere constructively-but the wave pattern oscillates up and down with time, of course. Keeping the strips taut, slide them over one another to the point where destructive interference occurs - the peak of one and the trough of the other coincide - the path lengths here differ by half a wavelength. Quantitatively - relate the horizontal distance the strips must be moved to go from one region of constructive interference, to the next, X say, with the separation of the slits d, the distance to the slits D, and the wavelength λ (two sets of waves can be made of differing wavelength).

$$X = \frac{\lambda D}{d}$$

1) R. A. Lohsen, Physics Teacher 21, 532 (1983).

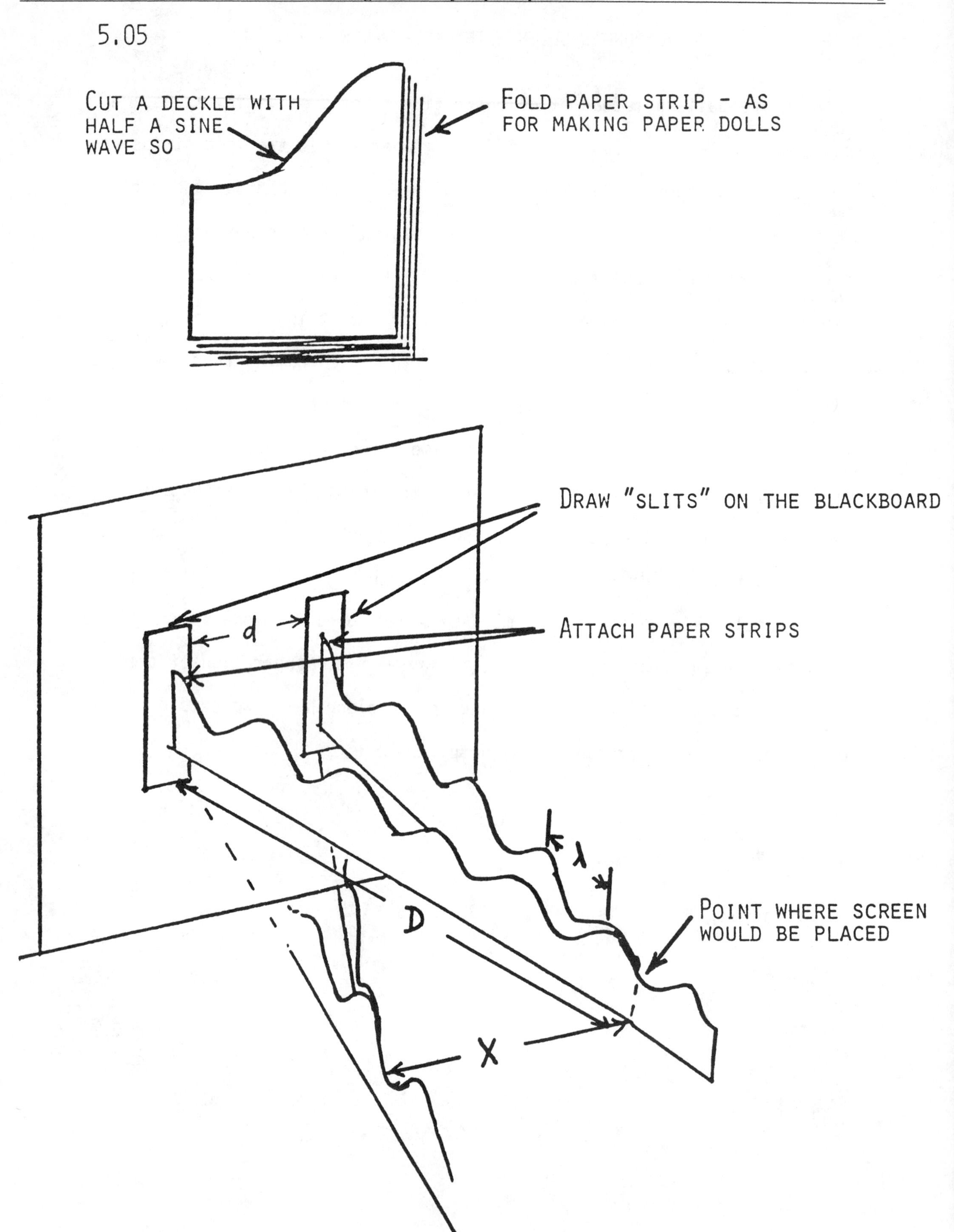
5.05
CUT A DECKLE WITH HALF A SINE WAVE SO
FOLD PAPER STRIP - AS FOR MAKING PAPER DOLLS
DRAW "SLITS" ON THE BLACKBOARD
ATTACH PAPER STRIPS
d
λ
D
X
POINT WHERE SCREEN WOULD BE PLACED

Experiment 5.06 Adding Oscillations

Music students forced (generally unwillingly) to take their first course in physics (often musical acoustics) have great difficulty understanding what is meant by the superposition - "adding" - of two simple harmonic motions - vibrations - at a point (as the effect on the ear drum of two tuning forks of different frequency). To help them picture what occurs, a simple device employing drinking straws, cardboard and tape can be used.

Figure 1 shows three time-displacement curves. Glue or tape this to a piece of card. The middle curve is the sum of the other two. To see, graphically, how this summation occurs, translucent drinking straws are lined up parallel as shown. The two end straws are taped to the back card, on either side of the bunch of straws, which may easily slide up and down between them. The bottom of the straws are aligned along AB using a ruler or you can tape a strip of card along the top, so that when the straws push against it, they are aligned along AB. A strip of stiff card, CD, is taped at the ends over the straws to prevent them crossing over one another. It is a good idea to fold the card as shown, and also the top of the back card, to provide the necessary rigidity so the straws do not cross over. A line is drawn around each straw with a suitable pen (felt tip, fiber etc.) where that straw passes over the top curve. The straws may be numbered at the top end in case they get shifted. The straws can then be slid down so their lower ends touch the bottom curve, and the marks around each straw will now follow the center curve, since we have added the length of straw between the bottom of the straw and the mark, to the bottom curve. Students unfamiliar with mathematics can follow this graphical act of addition much more clearly this way then by a written equation.

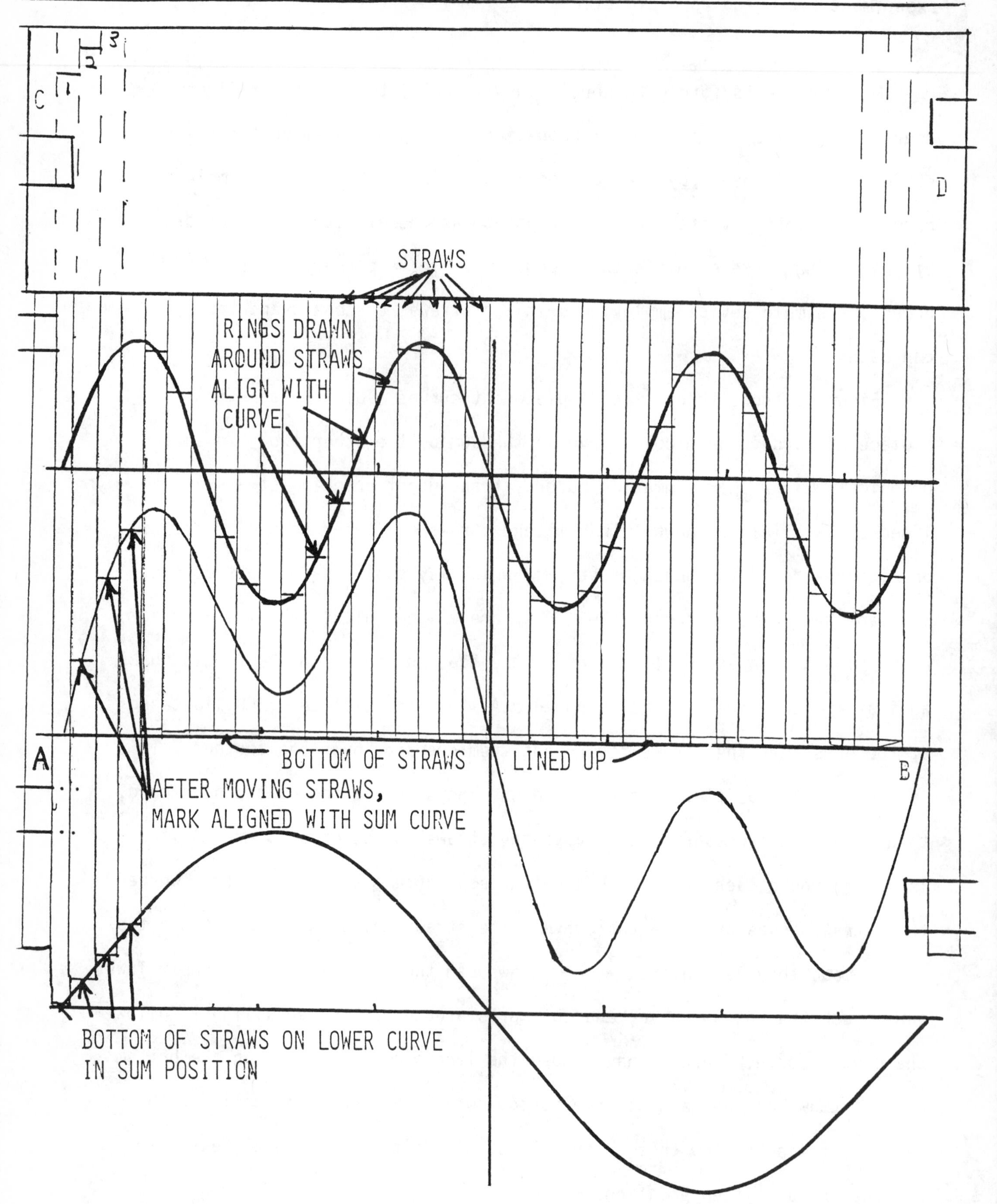
FOLD DOWN HERE
C
1
2
3
D
STRAWS
RINGS DRAWN
AROUND STRAWS
ALIGN WITH
CURVE
A
BOTTOM OF STRAWS LINED UP
B
AFTER MOVING STRAWS,
MARK ALIGNED WITH SUM CURVE
BOTTOM OF STRAWS ON LOWER CURVE
IN SUM POSITION

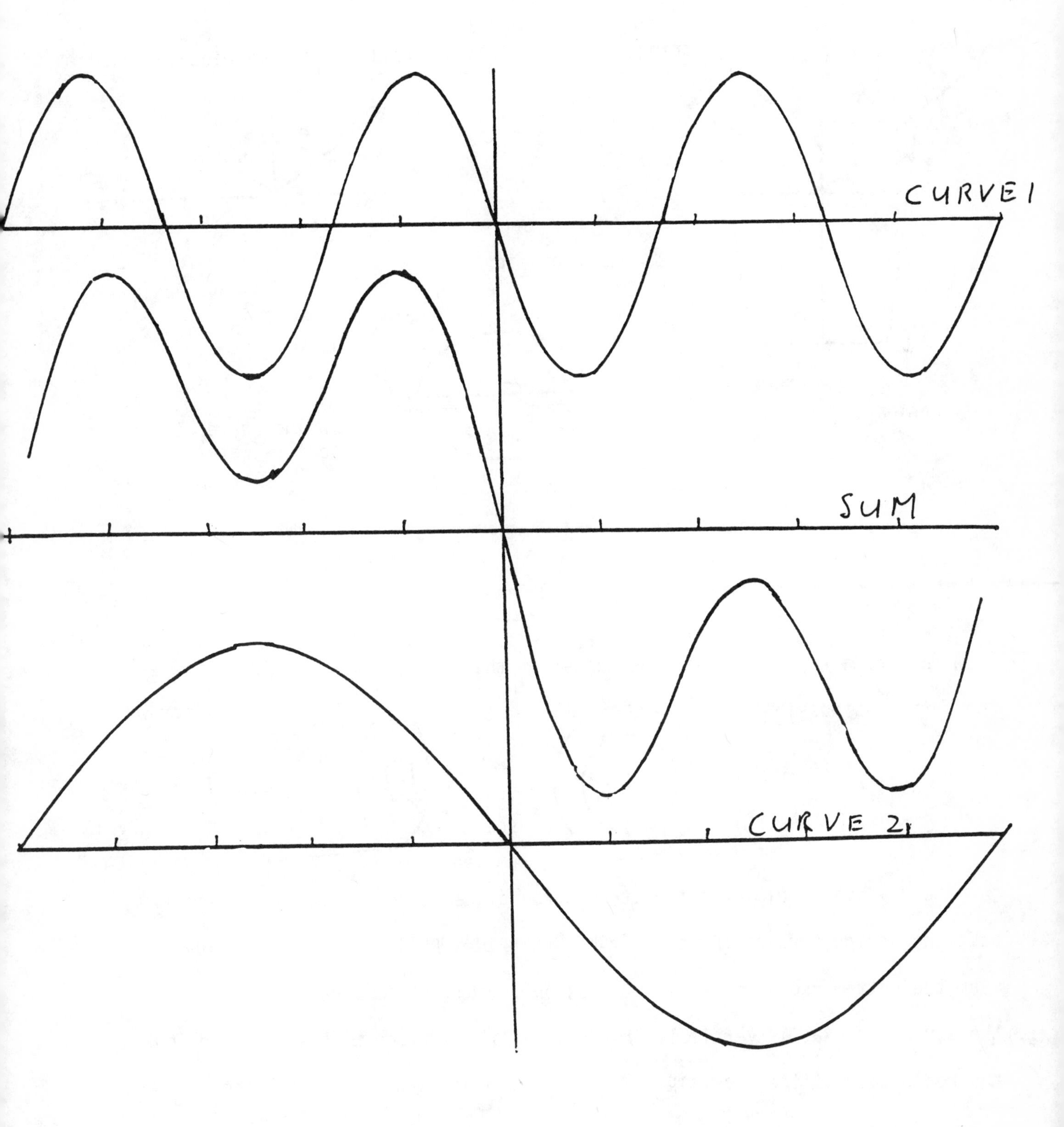
CURVE 1
SUM
CURVE 2

Experiment 6.01 Reflection of Light

Materials: Foam or paper cups, string, sticky tape, soda straws

Instructions: -

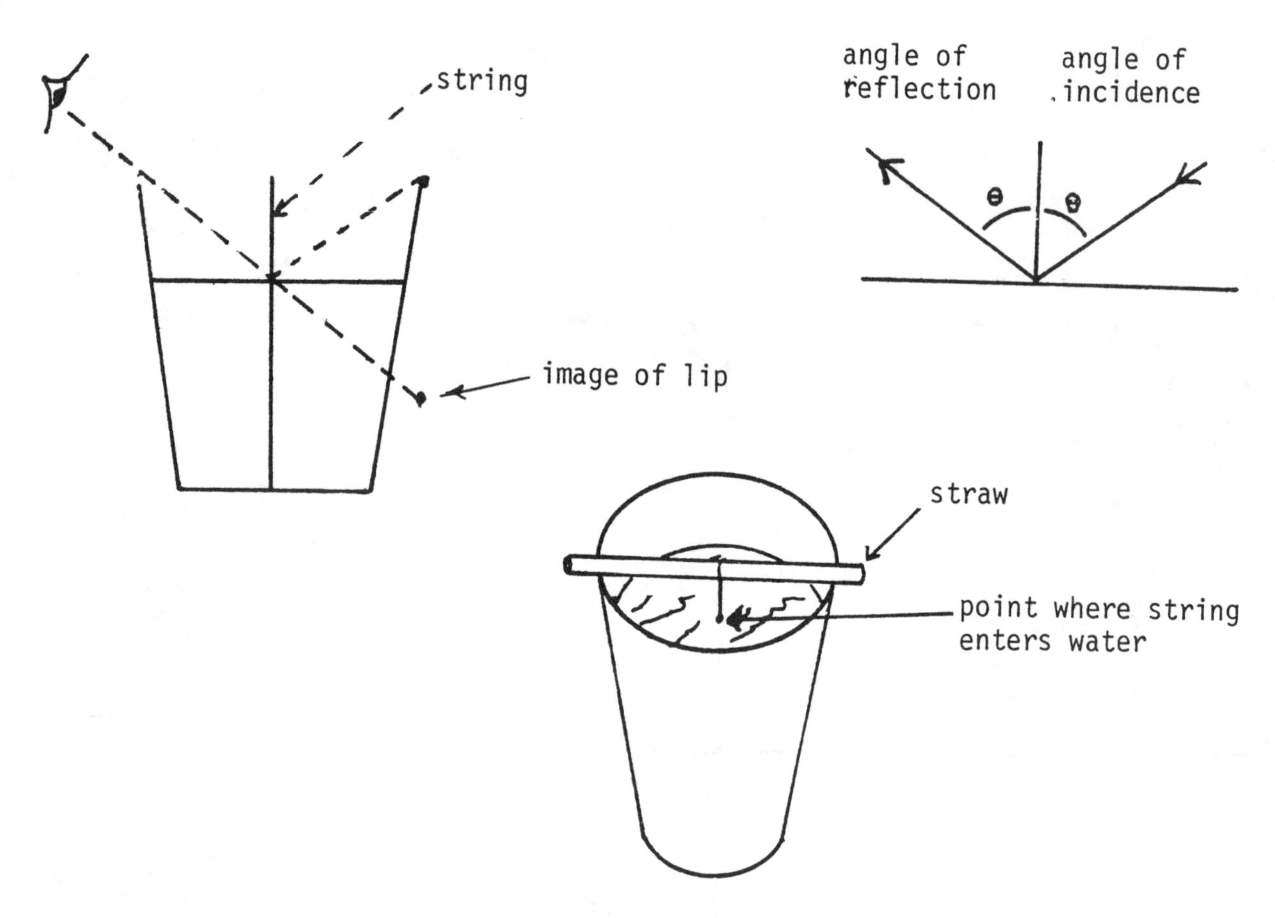

Make a mark, as close to the center of the cup on the inside of the bottom as possible.

Mark
tape seal
string
straw

Make a small hole, tie the string around the center of the straw and pass the string through the hole from the inside, pulling taut and taping over the bottom, so water will not leak out. Adjust the straw so that the string is exactly vertical. Now pour a little water in the cup, and look at the reflection of the far lip. Lower your eye until the reflection of the far lip is exactly on the rear lip. See if the string enters the water at this point. Examining the figure, you can see that if this is so, by symmetry, the angle of incidence on the water is equal to the angle of reflection. You can tilt the cup a little to adjust until where the string enters the water does lie exactly on the near lip, and the reflection the far lip. Now, pour in some water, and repeat the experiment. This increases the angle of incidence. Repeat for several angles of incidence.

Qualitative Questions: - What does this show about the reflection of light?

Quantitative Questions: - With what accuracy have you shown the angle of incidence is equal to the angle of reflection? (estimate this from the tilt you gave the cup) 10%? 1%? or 1/10%?

Note:- it is a good idea to illuminate the string, and employ a dark background

Experiment 6.02 Refractive Index of Water

Materials - cup (as deep as possible), pencil, ruler

Procedure -

Qualitative - Place a pencil in water in the cup. Notice it appears bent as it enters the water. This is because the light travels more slowly in the water, and rays of light are bent as they leave the water surface.

Quantitative : Draw a line straight across the middle of the dixie cup inside. Fill the cup to the brim with water.

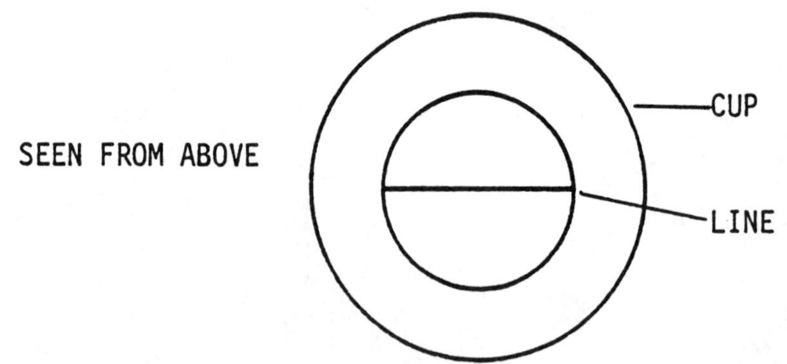

Place a pencil point against the outside of the cup where the line appears to be and move the head up and down. Adjust the pencil until it is at the apparent depth of the line.

Make a mark on the side of the cup. Measure the distance from the lip to the mark, and to the bottom where the line was drawn.

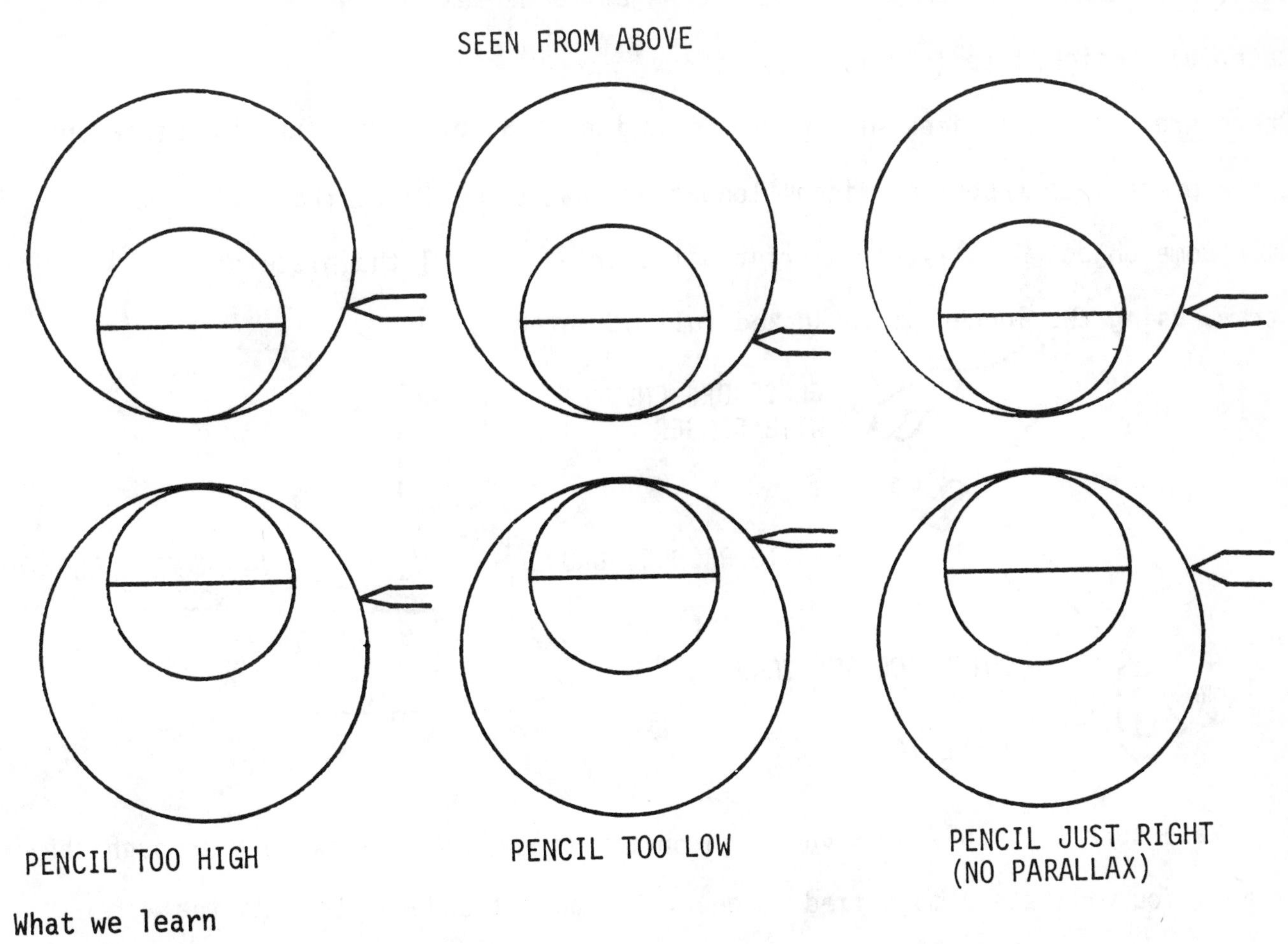

What we learn

The refractive index is the ratio of the sine of the angle of incidence to the angle of refraction. For small angles, this becomes the ratio of the angles, as shown.

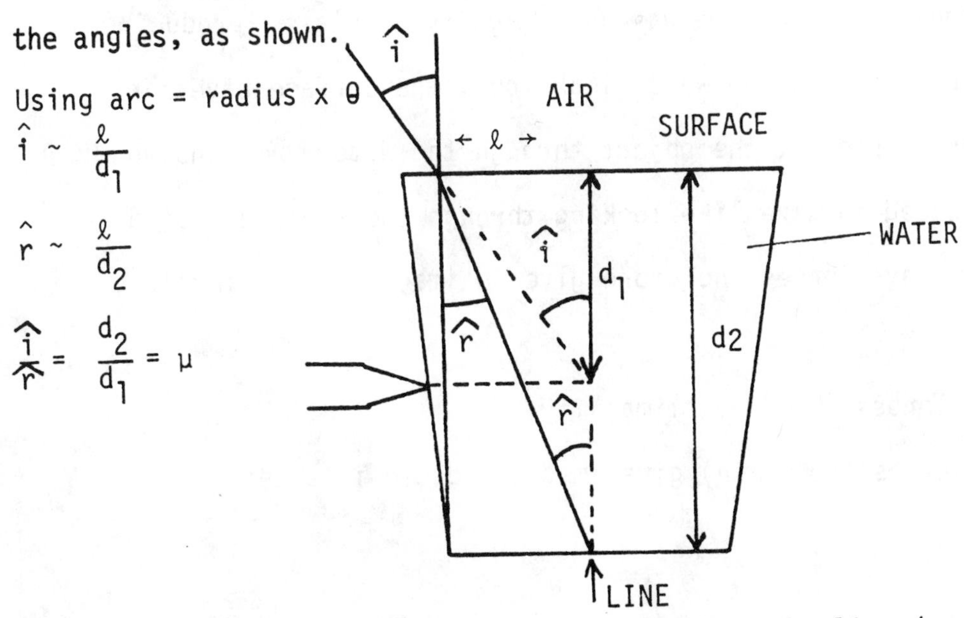

Hence, the ratio of the depth from the brim to the line drawn in the cup, to the depth of the pencil, is equal to the refractive index, which is 4/3 for water.

Experiment 6.03 Optics - Positive and Negative Lenses

Materials sticky tape (clear), soda straw, water

Procedure - Stick a piece of transparent tape flat over the end of a straw and cut a piece 1 cm (about 1/2 inch) long from that end. Place the tape over some object - a fly, or a printed letter - and fill the piece of soda straw, using the longer piece sucked full of water.

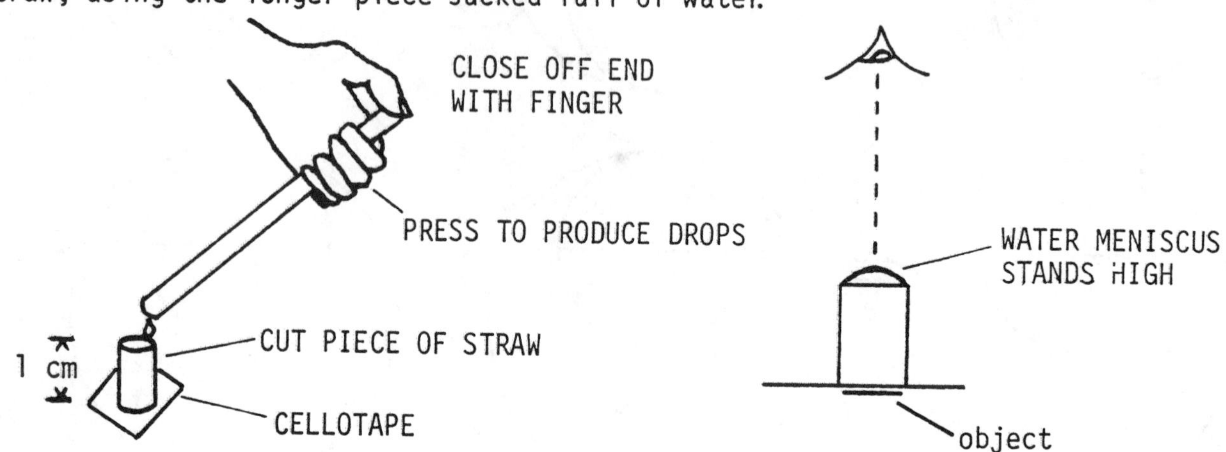

Make sure the meniscus stands high on the straw. Now, look down through the straw. You will see a magnified image of the object underneath. As the water leaks out of the bottom (or, you can soak up a little with toilet paper) the meniscus changes shape. When it stands high, we say it is convex. Hence, a convex lens can magnify in the same way as a magnifying glass, producing an image larger than the object. As the water leaks out, the surface caves in, and we say it is concave. Look at the object through the water now, and you see it appears much reduced in size like looking through the wrong end of a pair of binoculars. Concave lenses therefore give an image reduced in size. What do we learn?

Qualitative - Convex lenses (bowed out) magnify

Concave lenses (bowed in) give images reduced in size

<u>Quantitative</u> - The ray trace of the system is shown below

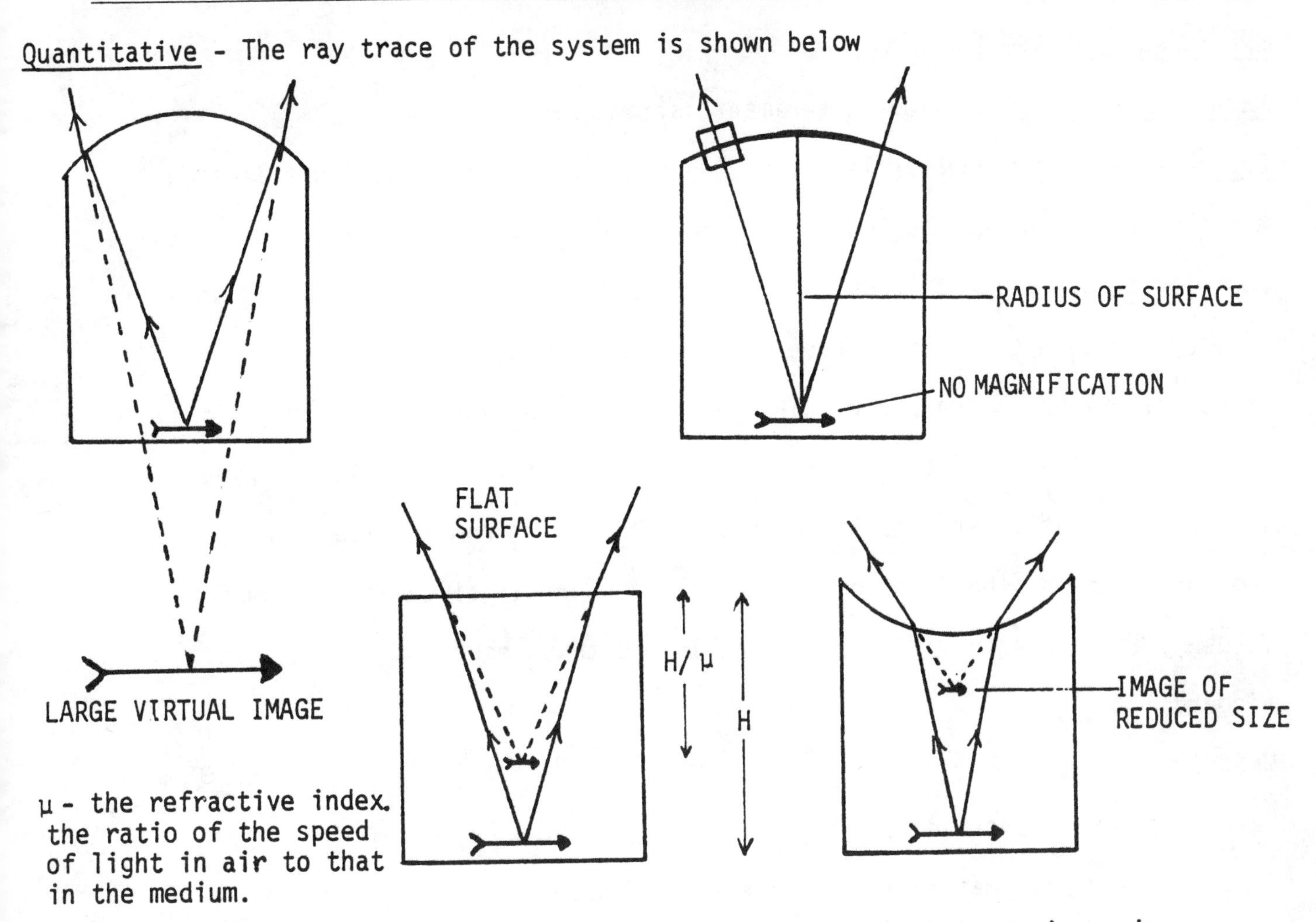

Look at the little square below through the water lens. When the meniscus is

convex

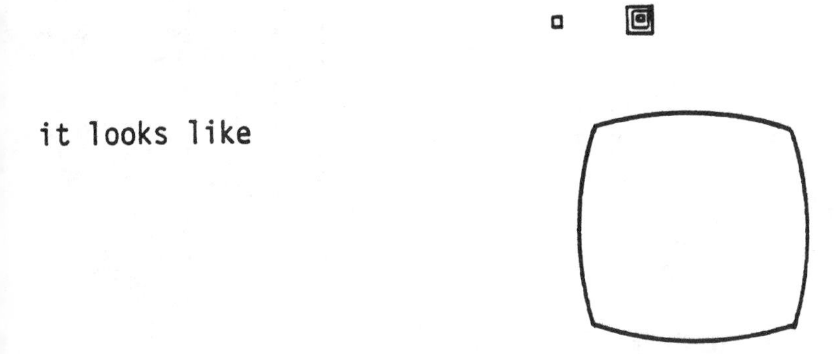

it looks like

This is known as barrel distortion, because the square image is distorted to look like a barrel. Distortion of this kind occurs with all lenses having spherical surfaces, such as this. When the lens becomes concave, the distortion changes and becomes pincushion distortion, so

Experiment - 6.04 Real Images

Materials - As for previous experiment, straw, water

Procedure - Take the water lens from the previous experiment, and fill it until the surface is convex (bulges out). Now, hold the lens vertically under the room light, a few inches above a sheet of paper. Move the lens up and down until you get an image or picture of the light on the sheet of paper. Because it actually lies on the paper, this is called a real image. The image is distinct, but not very clear. Notice as water leaks out, (or if you soak up a little on toilet paper) the image gets farther away from the lens. We say the focal length (the distance from the lens where a point very far away focuses) is increasing. Notice also that, as this occurs, the image gets bigger.

What we learn

A convex lens can form a real image, as shown — In the case of our water lens, where the bottom surface is flat, we can calculate the radius of the top surface from the formula for a thin lens-

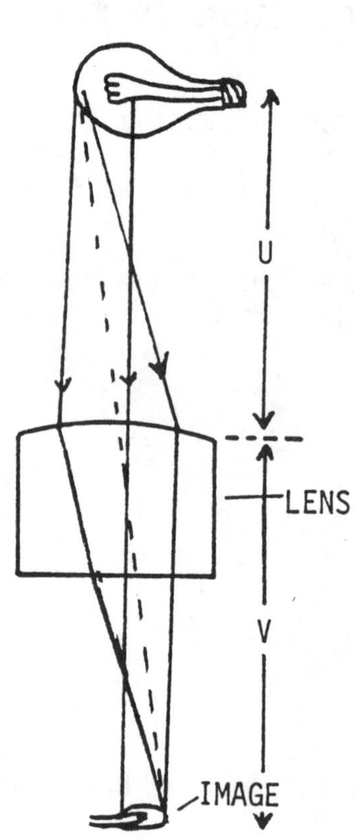

$$\frac{1}{u} + \frac{1}{v} = (\mu - 1)\,\frac{1}{r}$$

where μ = refractive index of water = 1.33

u = distance of object from lens

v = distance of image from lens

r = radius of curvature of the water surface

EXPERIMENT 6.05 Refraction of Particles

Materials: Chalk dust, cardboard, marble, ruler, protractor (from expt. 2)

Method: Fold about 1" down the middle of the sheet of cardboard, as shown, and place the top part on a book. Dust chalk over the surface.

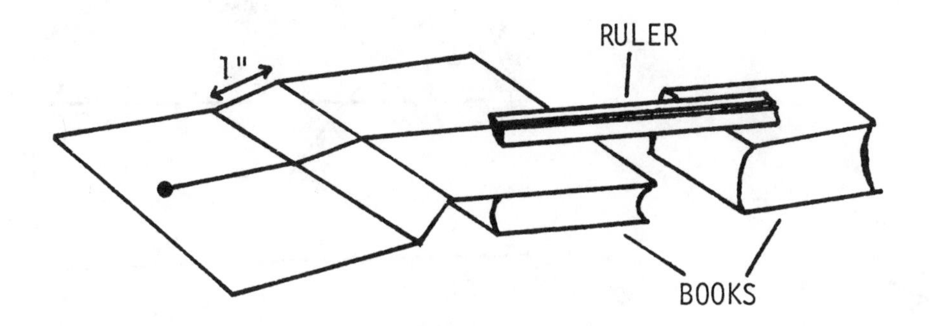

Now roll the marble over the dust, down the ruler, from the same height, but in different directions.

Qualitative: Does the particle bend in the same way light would in going from air to water? Do you think light could be a particle, like the marble?

Quantitative: Snell's law states

$$\frac{\sin i}{\sin r} = \text{const.}$$

Is this true for the trajectory of the marble?

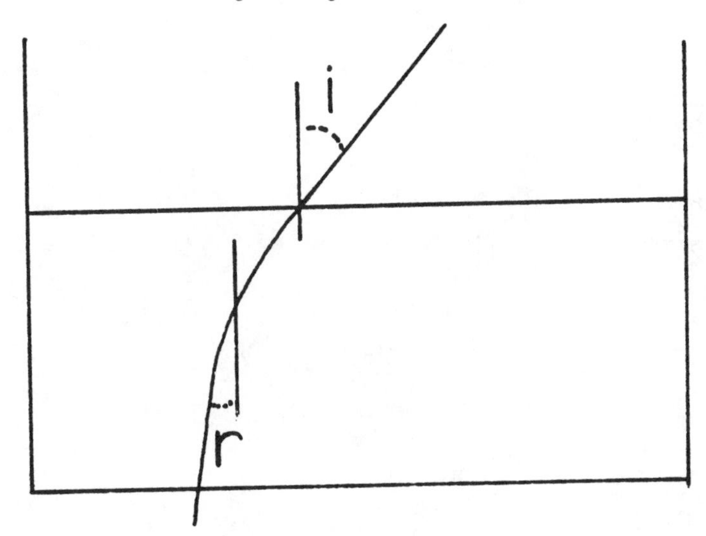

To see how the trajectory of the marble compares with the path of a light ray, let us set the ball free down its shute from a height h, above the upper plane, and let the upper plane be a distance h_2 above the lower plane, as shown.

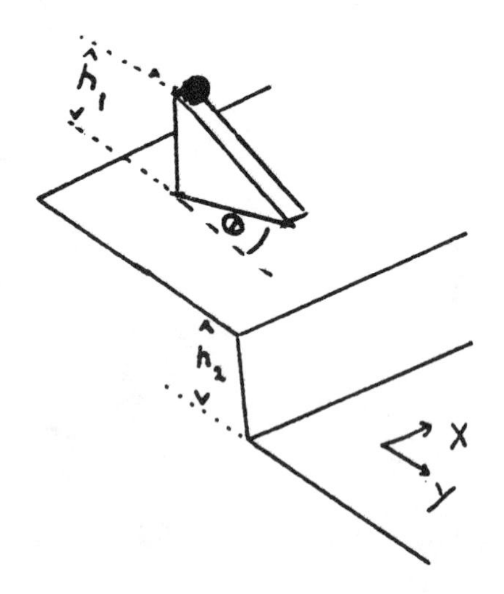

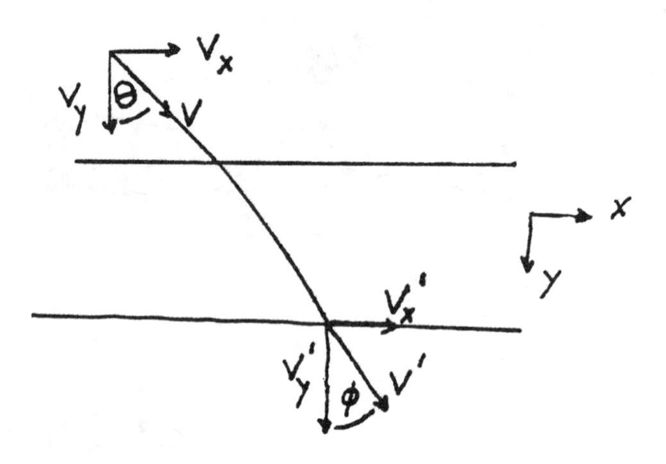

Using conservation of energy, as the ball rolls from one plane to the other

$$\frac{1}{2} mV'^2 = \frac{1}{2} mV^2 + mgh_2$$

$$V'^2 = V^2 + 2\ gh_2$$

We may divide this equation by $V_x^{\ 2}$

$$\frac{V'^2}{V_x^{\ 2}} = \frac{V^2}{V_x^{\ 2}} + \frac{2gh_2}{V_x^{\ 2}}$$

Since there is no acceleration in the x direction, $V_x = V_x{}'$,

$$\frac{V'^2}{V_x{}'^2} = \frac{V'^2}{V_x^{\ 2}} = \frac{V^2}{V_x^{\ 2}} + \frac{2gh}{V_x^{\ 2}}$$

Now, $\sin\theta = \dfrac{V_x}{V}$, $\sin\phi = \dfrac{V_x{}'}{V'}$

so $$\frac{1}{\text{Sin}^2\phi} = \frac{1}{\text{Sin}^2\theta} + \frac{2gh}{V_x^{\,2}}$$

multiplying by $\text{Sin}^2\theta$

$$\frac{\text{Sin}^2\theta}{\text{Sin}^2\phi} = 1 + \frac{2gh}{V_x^{\,2}}\,\text{Sin}^2\theta = 1 + \frac{2gh}{V^2}$$

For light, the refractive index μ is given by Snell's law to be

$$\mu = \frac{\text{Sin }\theta}{\text{Sin }\phi}$$

But $\frac{\text{Sin }\theta}{\text{Sin }\phi} = \sqrt{1 + \frac{2gh}{V^2}}$ for our marble

Also, we have, since the marble rolled down a shute from height h_1, that $mgh_1 = \frac{1}{2} mV^2$

$$\frac{\text{Sin }\theta}{\text{Sin }\phi} = \sqrt{1 + \frac{h_2}{h_1}}$$

If we make $h_2 = 1.25\ h_1$, then

$\frac{\text{Sin }\theta}{\text{Sin }\phi} = 1.5$, the refractive index for glass,

so the trajectory of the marble should follow the path of a light ray entering a glass surface.

EXPERIMENT 6.06 The Phase amplitude diagram for a 12 slit interferometer.

Materials: Soda straws, string.

Procedure: It is frequently difficult to visualize the "curling up" of the phase amplitude diagram in diffraction. This provides a model which helps to demonstrate it.

We may represent the light passing through a narrow slit by a vector which is a straight line, and the phase of this vector relative to that of light passing through a second slit by the angle these two lines make. Thus, for two slits in a Young's interference experiment, if the light at a distant screen from one slit is in phase with that from the other, we represent it

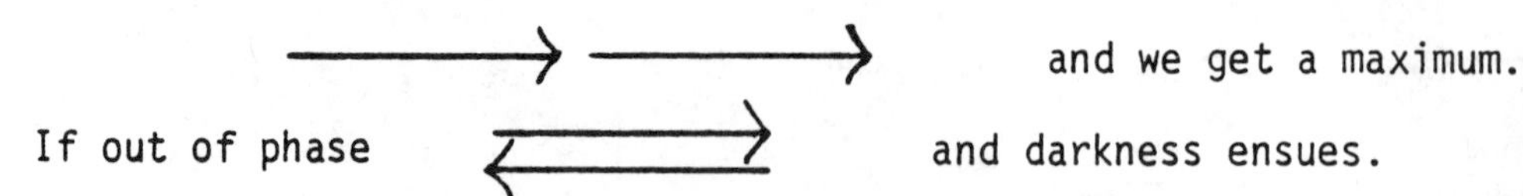

and we get a maximum.

If out of phase and darkness ensues.

Extending this to twelve slits, equally spaced, we can represent them by twelve straight lines of equal length, which, when in phase appears

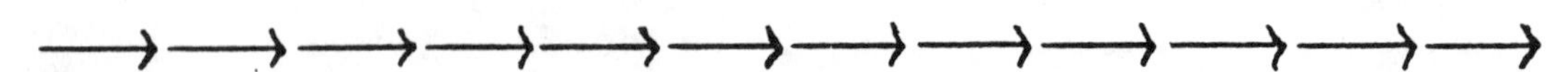

String 12 straws together as shown, attaching half a straw at each end as a handle.

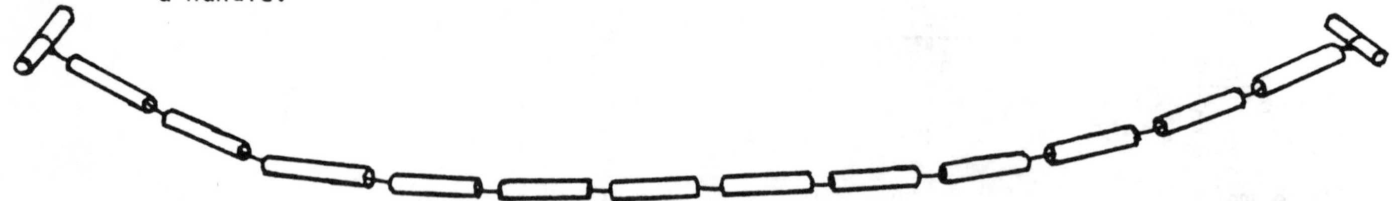

We use each straw to represent the electromagnetic vector coming from one slit. When laid out in a line, the distance between handles represents the sum of the vectors with all the slits in phase. Pick up one handle; the vectors make roughly constant angles, and the closing vector gets shorter. Ultimately it will close on itself,

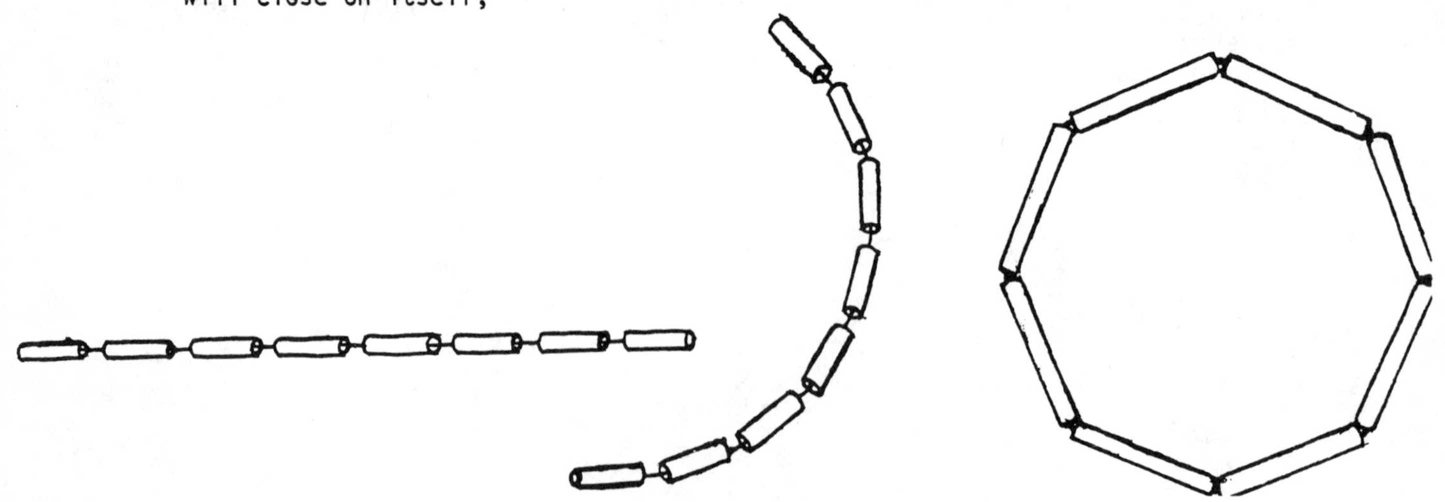

and we can reproduce this with the straws. Then, the diagram curls up some more, and we get another maximum.

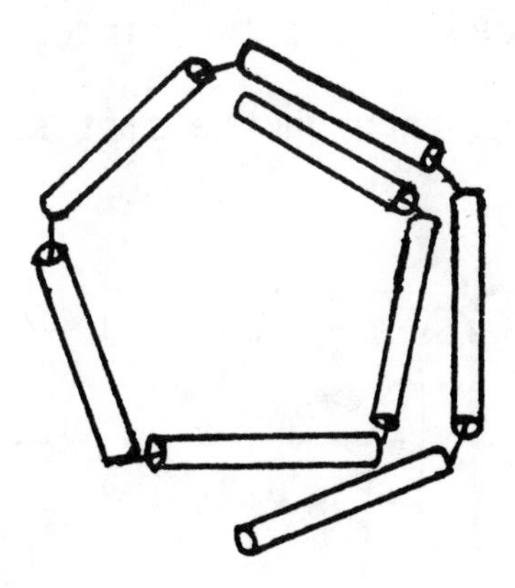

The straws greatly help visualize the process.

Experiment 6.07 Diffraction

Diffraction is the bending of light around objects and is a feature of all multi-dimensional wave motion. Place two fingers very close together, and examine a light source through the slit so formed, holding your hand 10 to 20 cm from the eye.

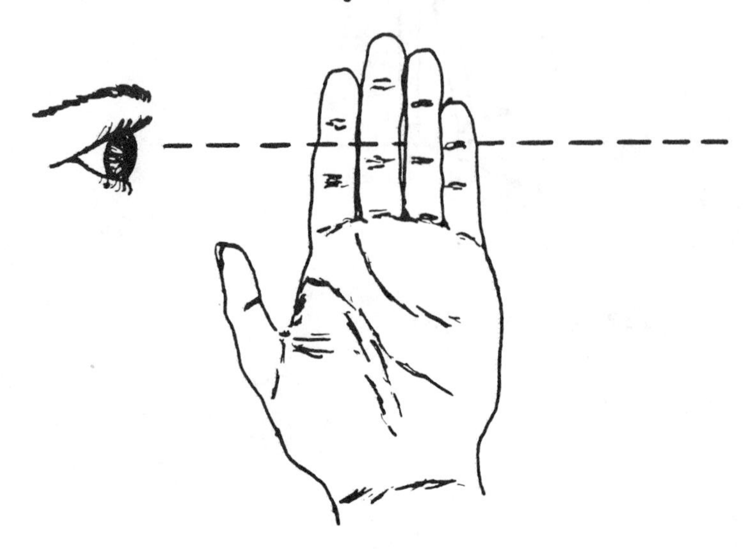

You will observe lines of light and dark parallel to the slit in the gap. A clear incandescent bulb gives the best effect, with well defined images produced by diffraction, but it is possible to observe dark and

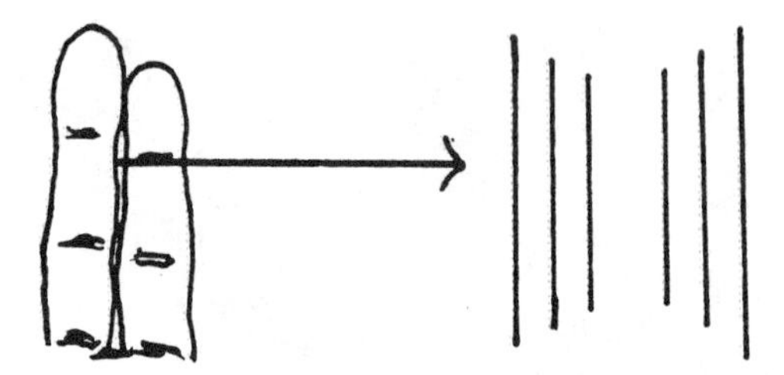

Appearance of lines looking through gap

light lines with any light source.

These dark and light lines arise because we must consider what happens to the light wave over the whole gap in order to work out the light intensity in any particular direction. As shown in the diagram below, one part of the wave front in the gap may cancel another, leading to darkness in one direction, whereas they may reinforce in another direction, giving rise to a bright line.

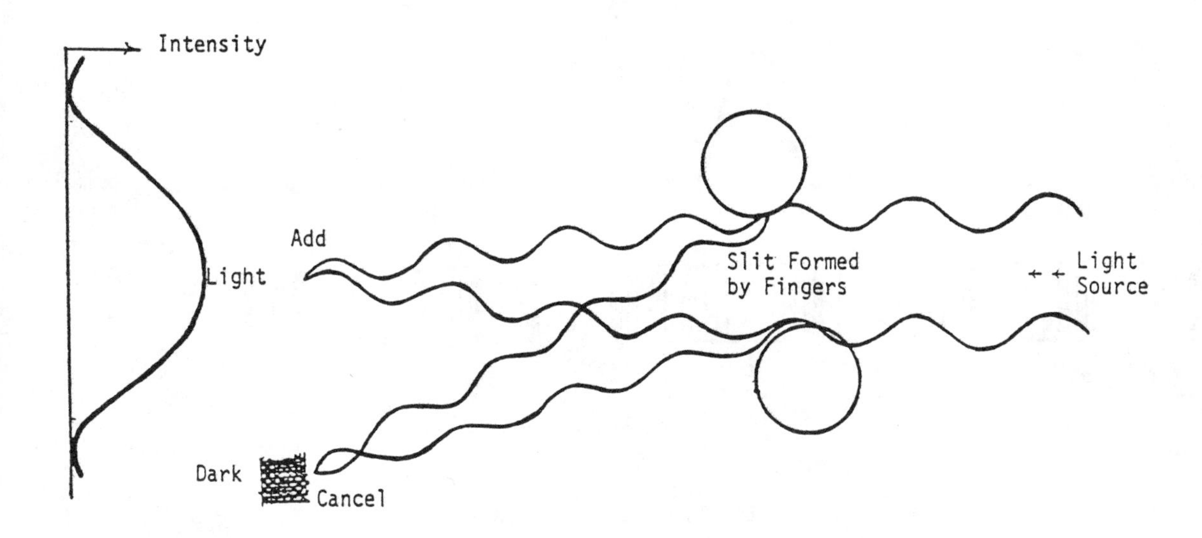

The diffraction pattern for a narrow light or slit, such as a filament lamp, is shown below. The filament appears blurred and spread out, and there are two subsidiary bright spots, one on each side. Now, try widening and narrowing the gap between your fingers. As the gap gets wider, the blurring gets less because the diffraction pattern gets narrower. If the slit gets very narrow, however, the bending of the light causes the image of the filament to spread out, so that it gets very blurry. This is a general principle - the narrower the slit, or the object causing the diffraction pattern, the more spread out is the pattern - width of pattern $\alpha \frac{1}{\text{width of slit}}$.

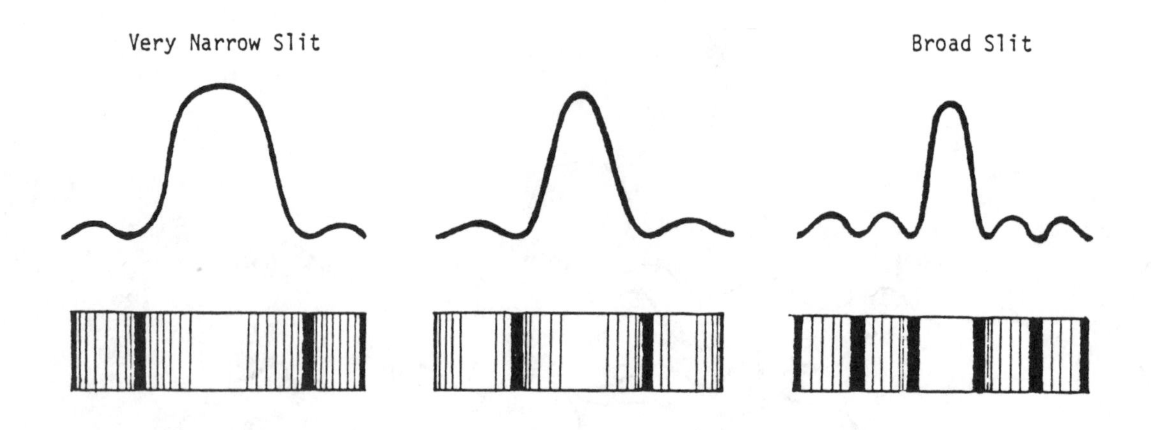

With a narrow slit, and a narrow light, the edges of the pattern appear reddish on the outside, and bluish on the inside.

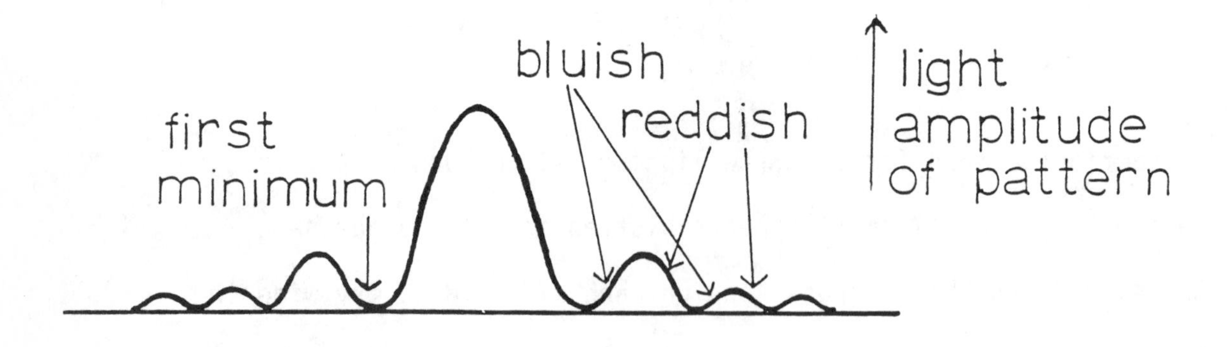

This is because the red light is bent or diffracted more than the blue light, because it has a longer wavelength, and, in fact

$$\text{angle of first minimum} = \frac{\text{wavelength of light}}{\text{diameter of slit}}$$

This type of diffraction occurs also in sound. Long waves bend more readily than short waves. This may be noticed by listening to music outside an open door not in direct line with the performer. The bass notes sound louder, since they bend more.

Experiment 6.08 Pinhole Camera

Materials: Two cups, tape, thin paper, aluminum foil

Procedure: Punch a hole with a pencil in the bottom of one cup, and cut a hole about an inch in diameter in the bottom of the other as the eyehole. Tape a sheet of paper over one cup, and attach the other, as shown.

Now, cover the whole of the two cups apart from the eyehole with one sheet of aluminum foil (such as Reynold's Wrap). The pinhole is made with a paper clip in the foil over the front cup. If aluminum foil is not available, make the pinhole in the front cup, then render the two cups opaque with black ink, or a dark cloth.

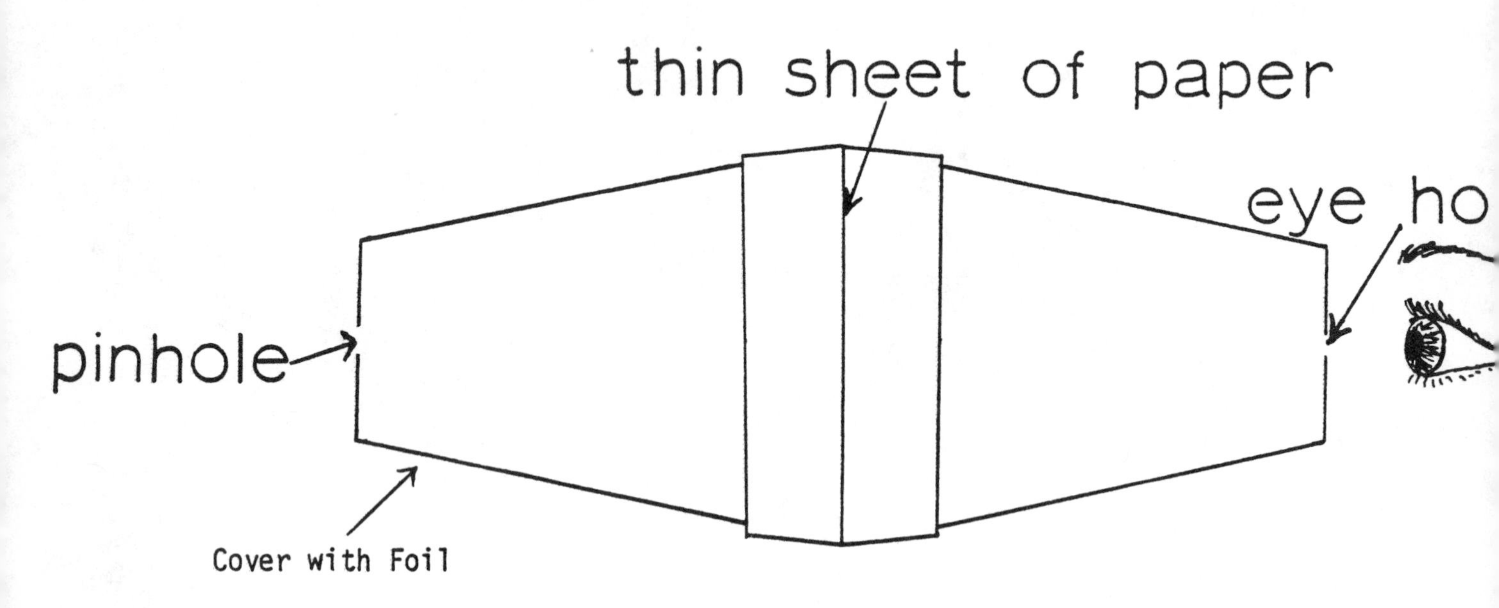

Pinhole viewer

Holding the cups up, an image is projected on the paper by the pinhole, and may be observed through the peep hole.

Widen the pin hole. What happens to the definition (sharpness) of the image? Is the image brighter? The image of an incandescent lamp, or other light is easiest to observe. What is the relationship of the size of the image to the size of the object? Is the image erect or inverted?

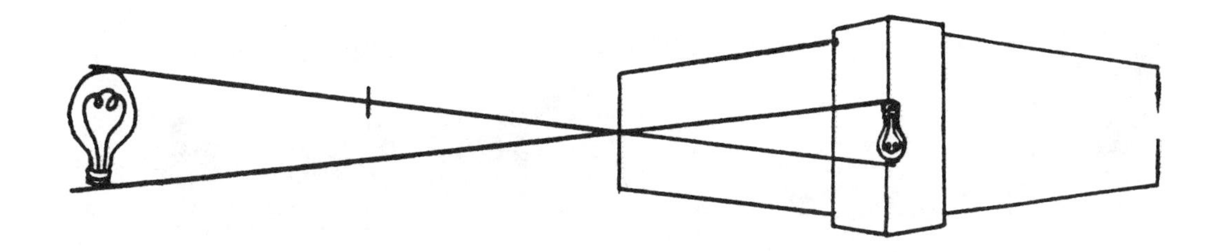

Experiment 6.09 Triboluminescence

Materials: Sticky, cellulose tape.

Take a roll of the tape into a very dark room. Rapidly pull a little off the roll. Where the tape pulls off the reel, light is given out. However, it is a weak source, and for such sources, the outer regions of the eye are more sensitive, so do not look directly at the roll, but about a foot or two away from it. What color does it appear?

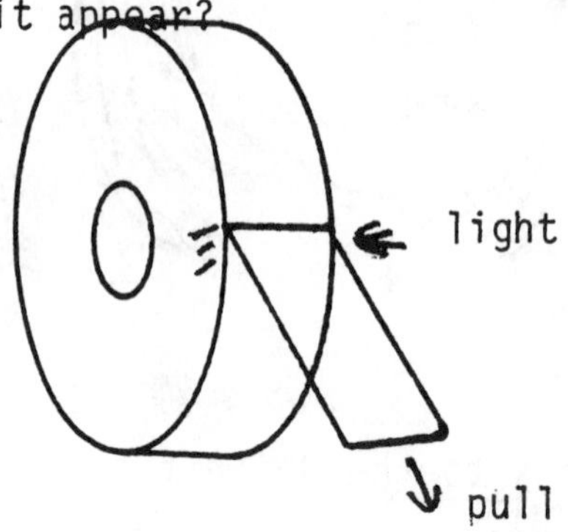

Qualitative: This effect is known as "triboluminescence" see "The Flying Circus of Physics" 6.11 *

The bluish color arises because the rods of the eye are more sensitive to blue light, so all dim illumination appears bluish, since the rods are very sensitive to weak light.

* The Flying Circus of Physics. - Jearl Walker (Wiley 1977-)

Experiment 7.01 The Rubber Band Guitar

Materials: - soda straw, rubber band, cardboard

Instructions: - stretch the rubber band over the straw as shown (the ends of the straw may be notched). A small piece of straw acts as a bridge. A piece of cardboard may be inserted as shown

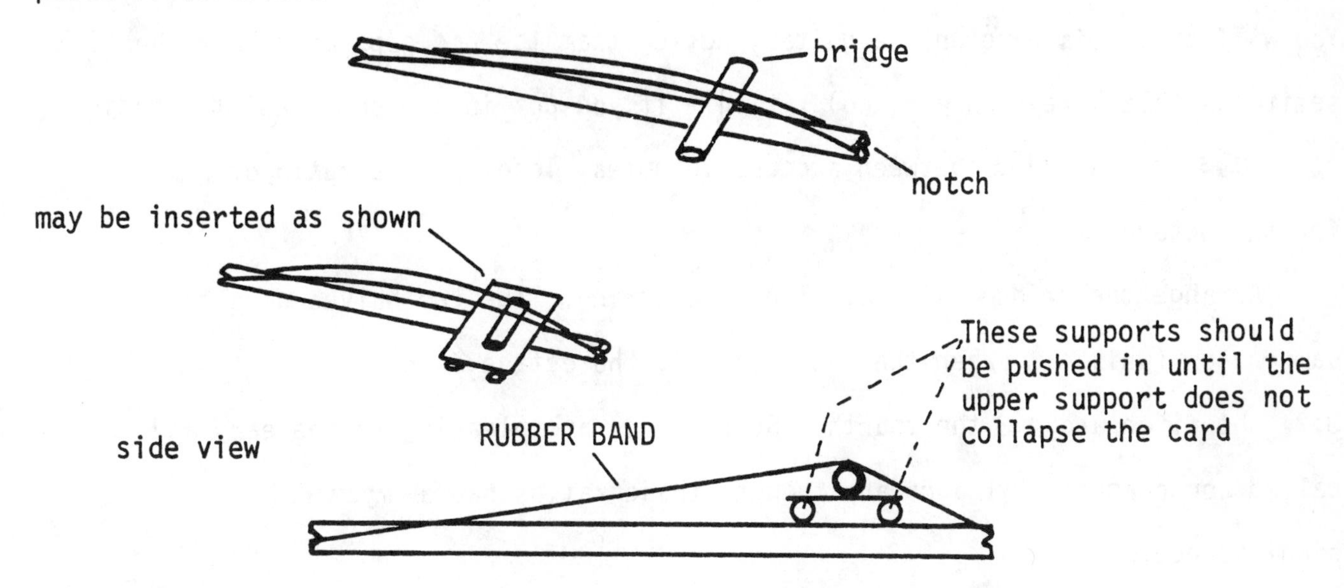

Does the guitar sound louder with or without the card in place?

Starting with the open band as "doh", place your finger on the band to shorten it, and mark on the straw the position of the notes of the octave. Are they equally spaced? Measure the distance of the marks from the bridge. Now divide the doh length by

length

the ra length, the ra by the mi, and so on. Tighten the rubber band a little. Does the pitch go up or down?

Note: If you stretch a rubber band between your fingers and pluck it, it may go up or down in frequency, or even stay the same, as it is stretched. Think how the frequency depends on length, tension, and mass per unit length of the string to explain why this is.

Qualitative: - You have seen how the pitch of a plucked string depends on its length. How much shorter must a string be to give the octave? The pitch also depends on the tension on the string.

Quantitative: - The ratio of lengths, for successive notes, starting with do, should be ·89, ·89, ·95, ·89, ·89, ·95.

You will find this also on the guitar, notice that the big gaps are where the semitones (black keys on a piano) fit in. If you put in the semitone, the ratio is always ·95 (17/18) between successive notes, leading to a ratio of 2 for the octave.

Arrange the bridge half way along the string. The two halves give the same notes (unison). When the ratio is 2:1, the octave is heard 3:2, the fifth and 4:3 the fourth. Such ratios are pleasing to the ear, and called consonants. Pythagoreans thought these ratios had a mystical significance.

Experiment 7.02 The Way the Tension and Length of a Plucked String Affect its Pitch.

Materials: Two styrofoam (or paper) cups, string, marbles and a ruler.

Procedure: Hang one cup from the string, as shown, passing the string through holes poked in the top on opposite sides, and pass the other end of the string through a small hole poked in the bottom of the other cup. Tie small knots in the string 15 cm. , 30 cm. , and 60 cm. from A. Put ten marbles in the cup and pluck the string. Hold your ear over the cup, and

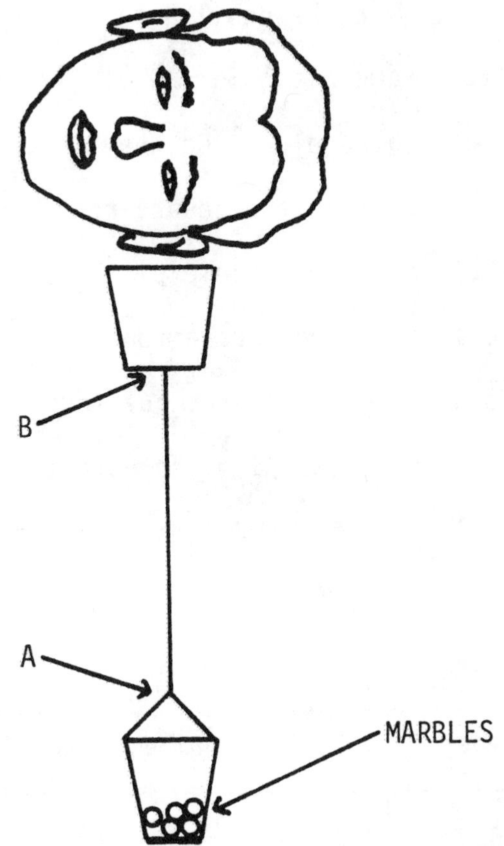

you will hear a clear tone, the string vibrating between A and B. Now, pull the string up until the 15 cm.knot is at B, and pluck again, noting the pitch. Drop to the 30 cm. knot and pluck again, and the 60 cm. knot. Does the pitch drop an octave in each case? If it does, the frequency also drops by a factor of two each time, since it is known, and we must assume that a note an octave higher than a second note, has twice its frequency. Depending on the length of the string you use, a clearer tone may be heard using more marbles.

Now set the string at the 30 cm. knot and pluck it. Add ten more marbles, and drop the string until it gives the same pitch as before. Measure the new length of string from A to B, using the ruler. You can go up and down, adding and taking away ten marbles until you get it right. What is the new length? Is it twice the old length? No, it is much less, showing that if you double the tension, the frequency goes up less than twice. In fact, the frequency should be proportional to the square root of the tension. Now, we doubled the tension by putting in twice as many marbles, so the length of string should be $\sqrt{2}$ larger. $\sqrt{2}$ is 1·412, so the new length should be 42.36 cm.

Find the length having the same pitch for 10, 15, 20 and 25 marbles in the cup. Put the numbers in the table below. Do they agree with the calculated values? If so, you have shown that the square root of the tension is proportional to the length of string for the same pitch, and we deduce that, if we keep the string the same length, the square root of the tension is proportional to the pitch. So we need not twice, but four times the tension to make the pitch increase a factor of two, i.e. an octave. Try this out by putting as many marbles as possible in the cup, and then reducing to one quarter. Does the pitch drop an octave?

Number of marbles	$\sqrt{\text{number}}$	3.16 x $\sqrt{\text{number}}$	Experimental length, cm.
10	3.1622	10	
15	3.8729	12.24	
20	4.472	14.142	
25	5	15.8	

What happens to the pitch with a thick string? Attach three strings to the top of the lower cup, with, say twenty marbles in. Pluck the string. Now, spin the lower cup. The three loose strings will wind around the one taut one, so

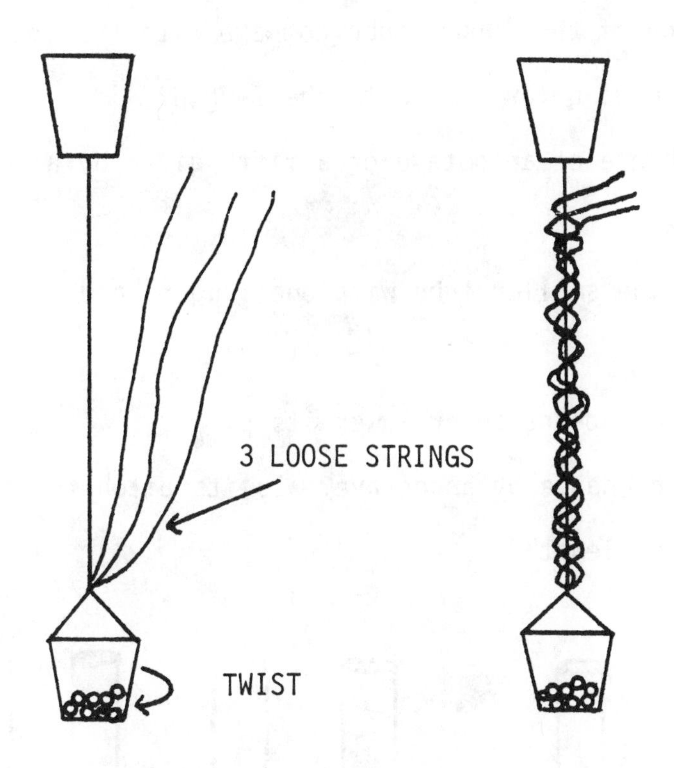

it is four times as heavy--the mass per unit length is four times as great. Again pluck the string. What happens to the pitch? Take out marbles until you have the same pitch as before.

Question: How many marbles must be removed? Is the pitch proportional to 1/(mass /unit length) or $\frac{1}{\sqrt{\text{mass/unit length}}}$

Theory shows that the frequency of the fundamental mode $f \propto \frac{1}{L}\sqrt{\frac{T}{m}}$ where m is the mass per unit length of the string, T is the tension in the string, and L is its length.

Experiment 7.03 Waves in Pipes

Materials: Soda Straws

Procedure: Take two straws. Cut one in half. Blow across the end of each tube. How does the pitch of the longer tube compare with the shorter.

1. Does the pitch go up or down with the length?

2. Is a pitch change of an octave or a fifth given by a length of a factor of two?

Close the bottom end of the smaller tube with one finger, and again blow across the top end.

3. Does closing a pipe raise or lower its pitch?

4. Does its pitch change by an octave? Is its pitch the same as an open pipe twice the length?

Why is this?

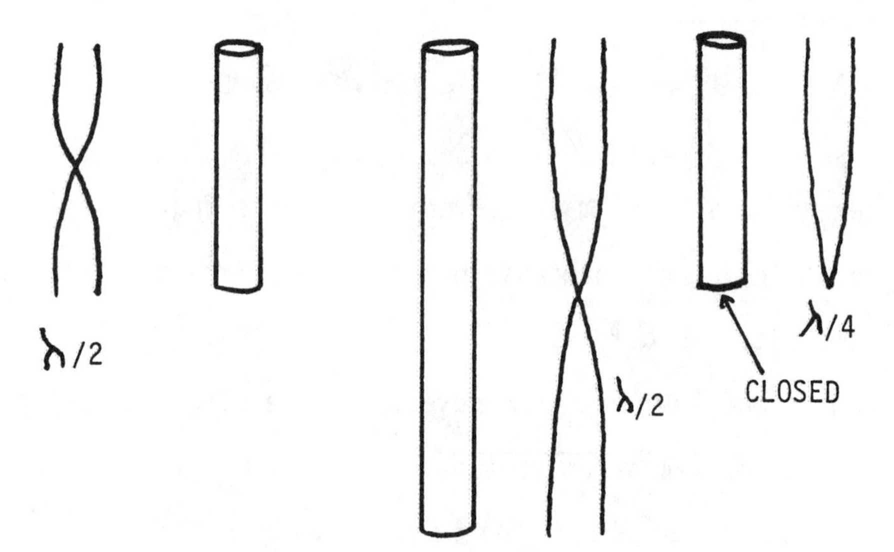

The closed tube has a node at one end, and its fundamental mode is one quarter of a wavelength long. In this mode, air rushes in and out of the open end, the

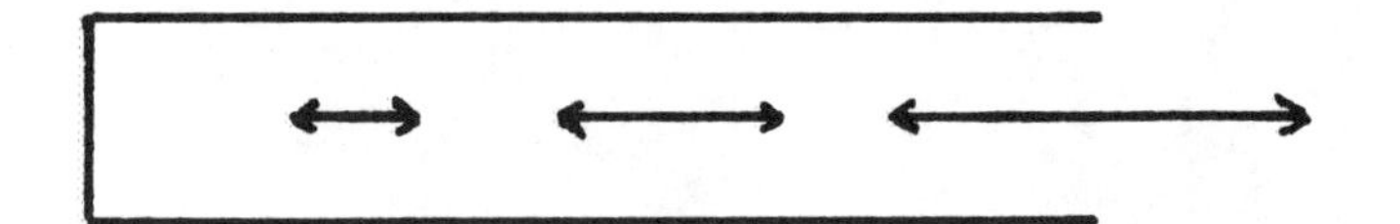

motion diminishing until at the closed end it is zero. The open tube has a node at the center, and is half a wavelength long.

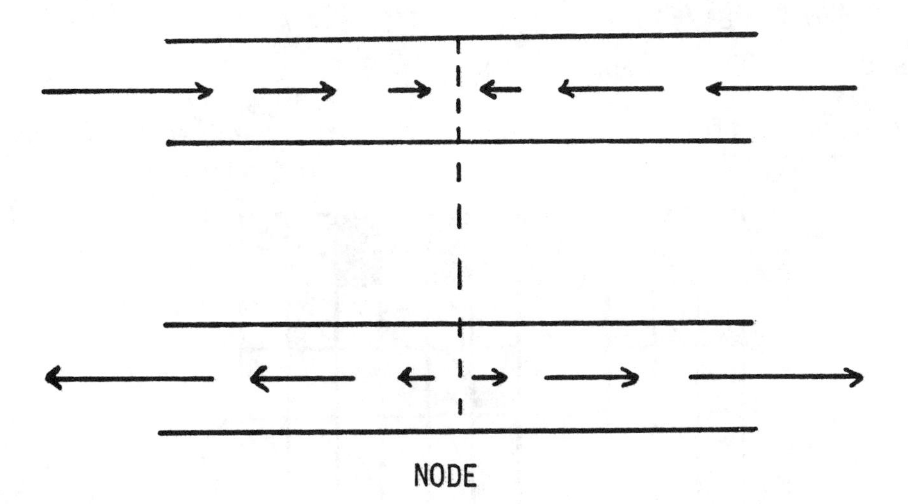

Listen to the timbre of the closed pipe, half a straw long, and the open pipe, a whole straw long. They have the same pitch, but sound quite different. The open tube has all the harmonics, the closed tube only the odd harmonics.

To determine frequency threshold of hearing, cut a straw into shorter and shorter lengths until the frequency can no longer be heard on blowing across the top. The upper frequency is about 16,000 Hz, which is produced by a straw 1 cm. long, roughly half an inch.

Equal Temperament

The ratio of frequency for successive notes on the scale of equal temperament is a constant - 1.059, which is approximately $\frac{17}{18}$. To make a scale based on equal temperament using straws, the ratio of successive lengths should be $\frac{17}{18}$. Cut the straws:

8"*, 7.55", 7.13"*, 6.73", 6.35"*, 6"*, 5.66", 5.34"*, 5.04", 4.76"*, 4.49", 4.24"*, 4"

These will form a scale of equal temperament, those with asterisks forming the natural octave. Join them together with tape, to form pipes of Pan, as shown. The lengths are correct, as drawn and can be used to measure the straws.

Equal temperament never gives a chord having exact fifths, i.e. the ratio of two notes is never exactly 3/2. However, you can construct an octave of pipes according to the Greek Pythagorian scale, which does give exact fifths.

The straws should be cut to lengths in inches:

	8	7.111	6.32	6"	5.333	4.74	4.21	4
frequency ratio	1	9/8	81/64	4/3	3/2	27/16	243/128	2

interval: 9/8 9/8 256/243 9/8 9/8 9/8 256/243

The perceptual differences between Pythagorean and equal temperament are small, but noticeable.

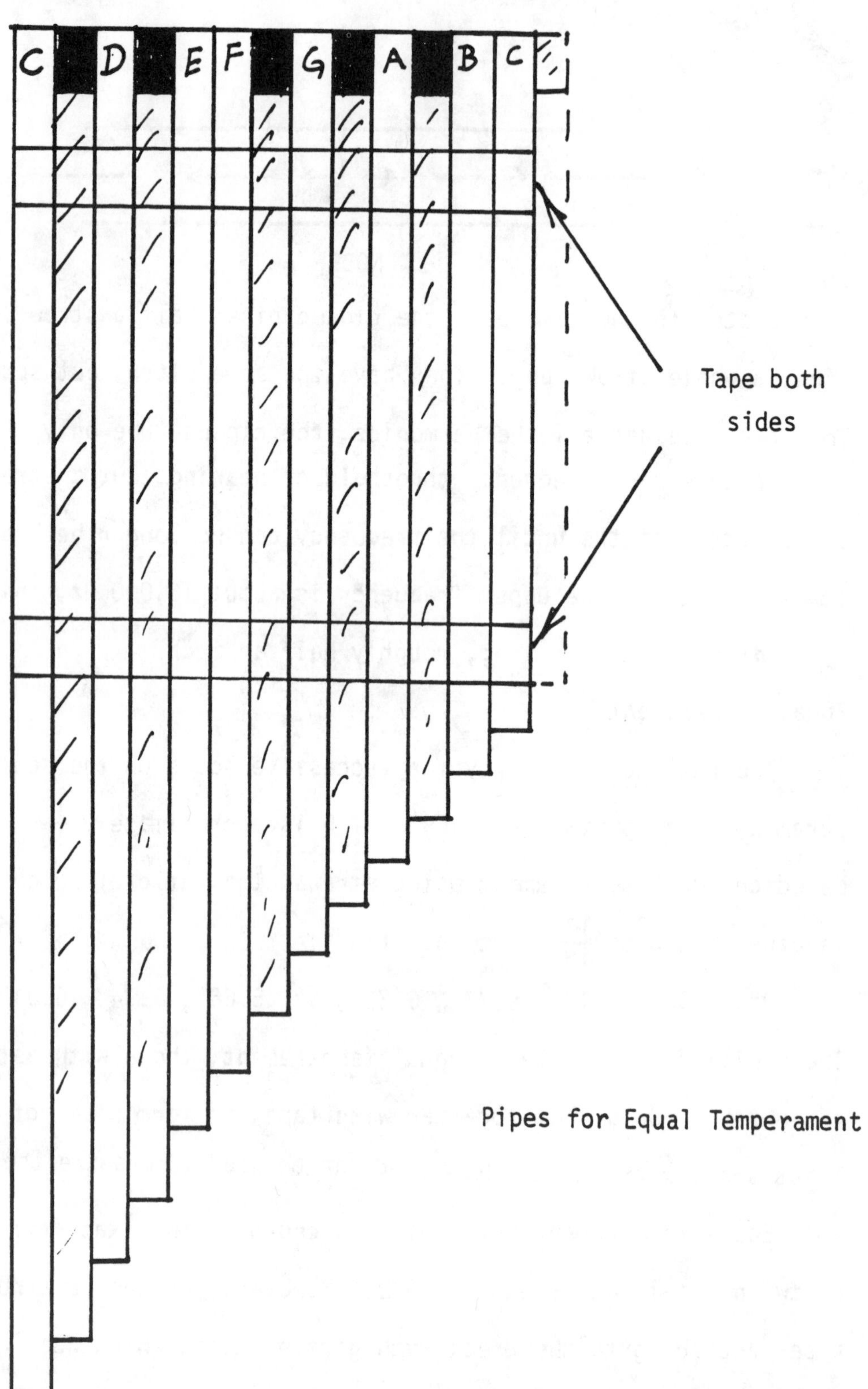

Pipes for Equal Temperament

Experiment 7.04 The Reed

Materials: paper scissors and tape

Procedure: Take a six to 8½ inch square of paper and fold along one diagonal. Then open out and proceed to roll the paper tightly around a pencil from one end of the diagonal crease to the other end, so that the diagonal rolls along itself as shown in the figure. If a six sided pencil is used, do not wrap too tightly, or the pencil will not come out. When completely rolled it should look like the lower figure. Push the pencil out and paste the last fold at A., or hold it in place with a rubber band or strip of tape The ends of the roll will look like the figure below.

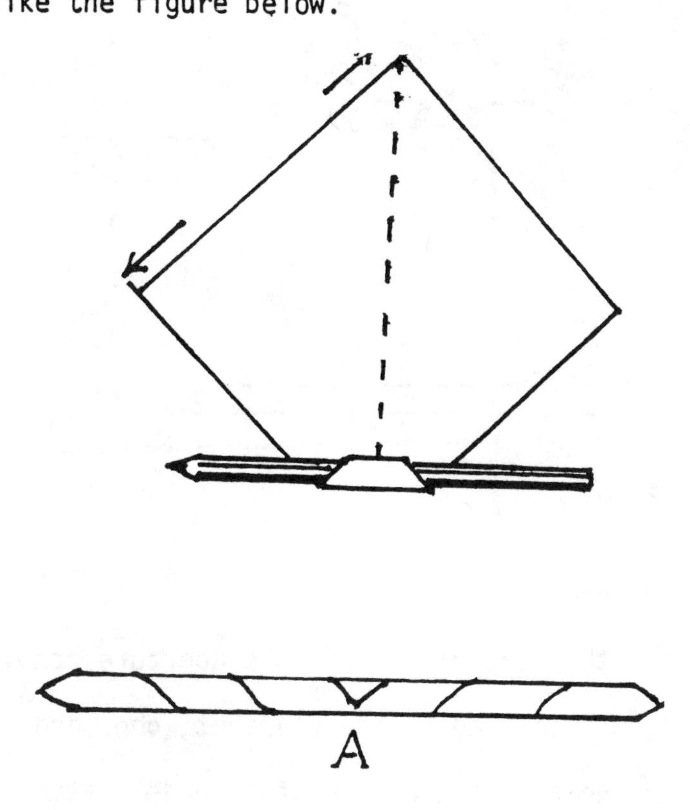

Now from point marked B. at one end cut away on each side, in direction indicated by the small arrows, until the end piece may be opened out into a triangle shape C. The cuts must be at right angles to the main roll and are each a trifle over one third of the circumference of the rolled tube. Now fold the triangle piece at right angles to the tube so that it forms a little cover over the end (see figure below). Trim away a small part of the triangle on each side along the dotted lines indicated in sketch, but do not trim too close. Now place the other end of the tube in your mouth, and, instead of blowing, draw in your breath. This action will cause the little triangular paper lid to vibrate and the instrument will give a bleating sound. The noise can be made louder by rolling a cornucopia or horn and putting on the tube as shown in the last figure, or the tube can be poked through a hole in the bottom of a styrofoam cup, to act in the same way.

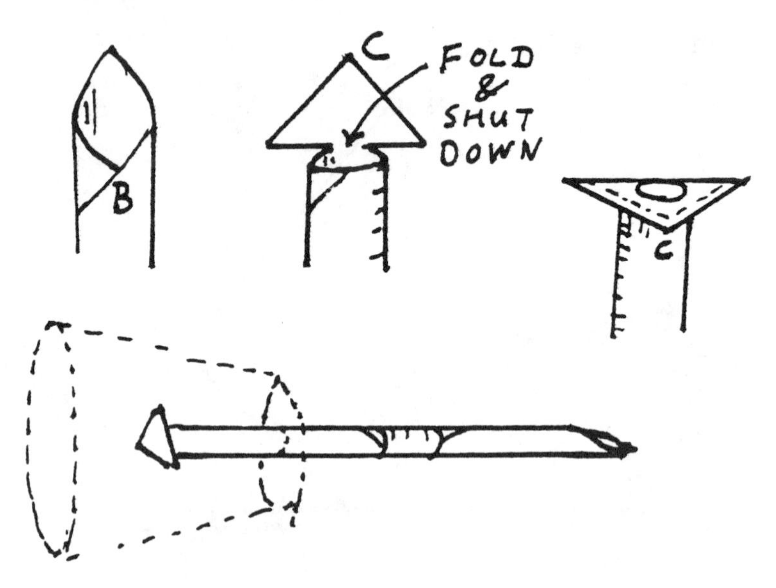

Qualitative: How does the reed work?

This type of reed, in which the flap closes off the aperture completely, is known as a "beating reed", and is used in the clarinet, oboe and bassoon. The clarinet has one reed, and the oboe and bassoon two reeds beating together. The vocal cords are also similar.

A puff of air is allowed into the pipe each time the flap opens and closes, and such sharp puffs have a lot of high frequencies in them, in addition to the lowest frequency, or fundamental. Now, cut the corners off the reed, as shown

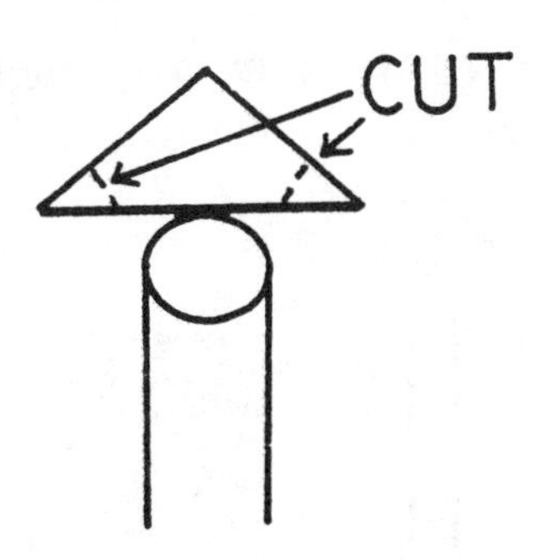

(Take care that the reed covers the tube--otherwise it will not work). Does the pitch go up or down? The natural frequency of vibration of the reed, (as with all simple harmonic motion) is higher the smaller the mass oscillating.

Why does putting the horn over the reed make it sound louder?

In a wind instrument, the reed is not free to vibrate at its natural frequency, as here, but is forced to oscillate at the resonant frequency of the tube of the instrument. The paper reed can only vibrate at a low frequency, so you would need a tube about three feet long to be able to bring the reed into resonance. If you have such a tube, you could try it out.

We can examine the way the voice works using this device. The foam cup over the end of the tube has its own resonant frequency. The vocal tract (larynx, mouth) behaves similarly in the case of the voice. The resonance is at a high frequency, and tends to emphasize frequencies produced by the reed in this vicinity--these resonant frequencies are called formants in the case of the voice, and determine whether you are saying "oo" or "ah", even if our voice holds the same basic fundamental pitch.

While sucking on the reed, close the cup partially with one hand, then open it again. Doing so alters the formants, and it is quite easy to get the device to say "ma ma" or even more difficult vowel sounds with a little practice.

Try cutting the tube shorter, and shorter. Does the pitch of the reed change? At what length does the reed stop functioning? Clearly, the air in the tube is necessary to the functioning of the reed, even though the length does not determine the pitch in the same way as it does for an open pipe.

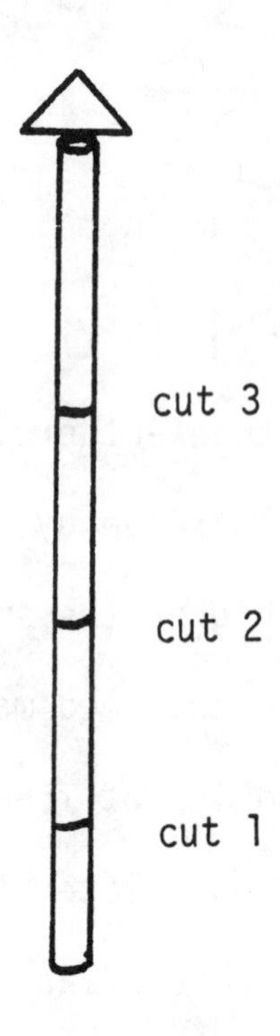

After cutting the reed shorter, around the outside roll a tube of paper and tape it as shown. Slide the outer tube up and down. Does the pitch go up or down as you slide the tube in and out? Why?

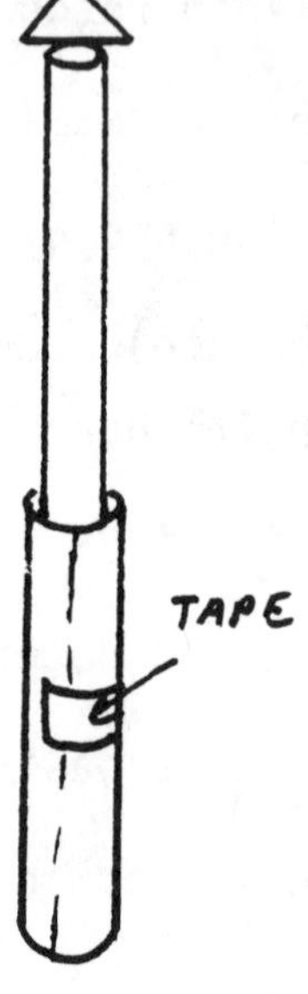

A reed can also be made from a plastic soda straw by cutting slits on either side at the end, as shown

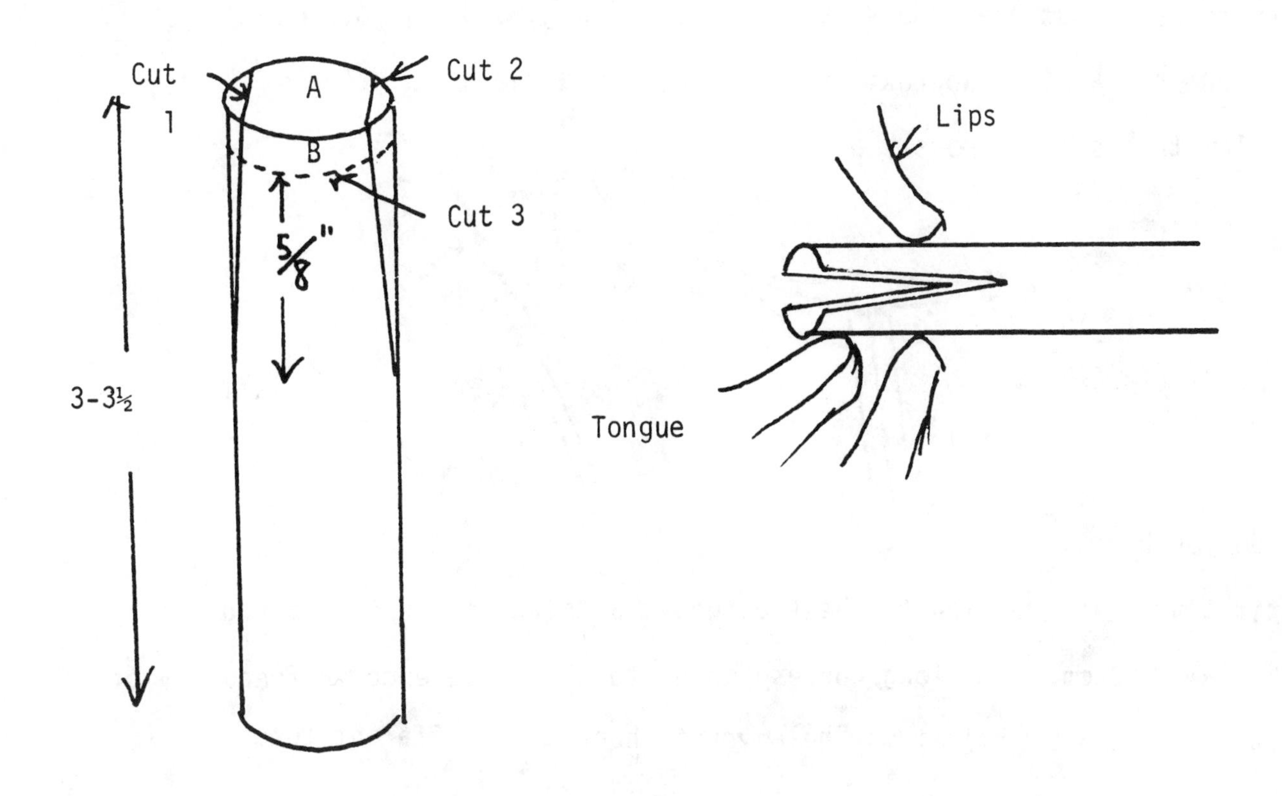

Place the reed in your mouth, and chew the flat ends A and B between your rear molars until they are quite flat and parallel.

Some practice is required in blowing this. Roll the reed between the lips while blowing--the range of pressure under which you can make the reed sound is rather restricted. The top and bottom reeds should be closely the same size. This is similar to the double reed in an oboe, so you can insert it into a paper cone and see how it sounds.

Experiment 7.05 Combination Tones

Materials: - Soda straws

Instructions: - Cut one soda straw 5.0 cm. long, and one 6.7 cm. long. One has a pitch approximately 3300 and the other 2460 Hz. Now, blow them simultaneously as shown

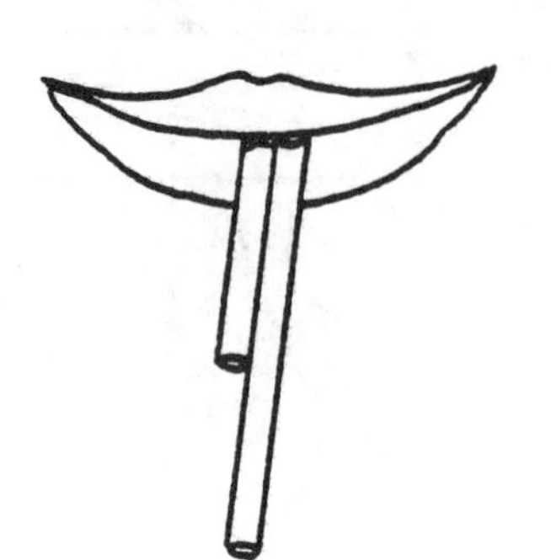

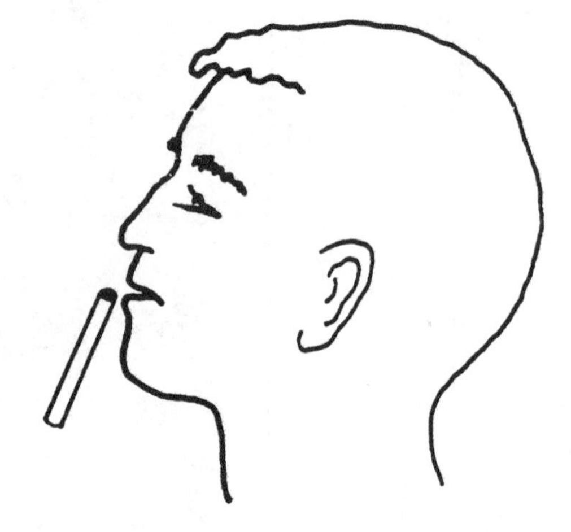

What do you hear?

Quantitative - in addition to the two tones, a third is heard. If you cut a soda straw 20 cm. long corresponding to the difference in frequency of the two straws, (840 Hz) it may help you to hear this different tone.

EXPERIMENT 7.06 Resonance in Sound

Materials: Two Styrofoam or paper cups, straw, sheet of card, sticky tape

Procedure: Poke a hole with a pencil in the bottom of the cup, and insert about two inches of straw, as shown

Attach it firmly by putting strips of tape around the junction. Now carefully insert the straw in your ear, and cover the mouth of the cup with a sheet of card.

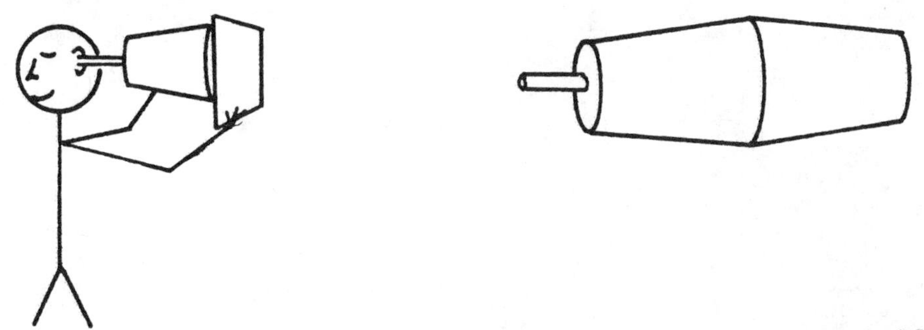

Leave an opening for sound to get in as shown. You have probably, at some time or another, "listened to the sea" by holding a conch shell to your ear, and you will find you can hear the sea just as clearly with the cup as the shell. What is happening is that the cup resonates to a certain unique pitch or frequency. Just as the pendulum responds to oscillations of only one frequency, so the cup responds to only one pitch. So, of all the sounds in the room, you hear this pitch greatly exaggerated. To find which pitch this is, ask someone to sing or hum continuously from low to high. You will hear the resonant pitch stand out, sounding loud compared with the others. Now, the size of the cup determines the pitch, so take a second cup, and put it over the first instead of the card with enough gap between to let sound in (or you can poke a hole through the base). You will find it resonates at a lower frequency and, in fact, it will be about an octave lower, because you have twice the volume of container. You can also vary the resonant pitch by moving the card across the top. When the card covers most of the cup, the resonance is sharp, (i.e., that one pitch is much exaggerated, or amplified) as the card is moved back, the resonance pitch becomes less sharp, until, when completely removed, there is practically no resonance at all.

Qualitative questions: Why should a hollow container resonate in this way? Think about the way sound bounces around inside the cup, and the way, if you blow across the top of a bottle (or across the opening left when the card all but covers the cup) you get a note, which is the resonant frequency of the device--this can be used to find the resonant frequency of your cup-resonator instead of humming.

Quantitative questions: The linear size of the cup is a fraction of the wavelength to which it resonates. Measure the size of the cup. The frequency times the wavelength is the velocity of sound, so to what frequency does the size correspond? The velocity of sound in air is 330 m. or 1100 ft. per second. Is the real frequency higher or lower than this? (Remember, middle C on a piano has a frequency of about 260 cycles per second, and when you blow across a straw 15 cm. long, open at both ends, it vibrates at 1100 cycles per second).

Experiment 7.07 The String Telephone

Materials: two styrofoam cups, string

Procedure: Tie a knot in a piece of string, and draw the string through a small hole in the bottom of a paper or styrofoam cup. Draw the string between the thumb and forefinger.

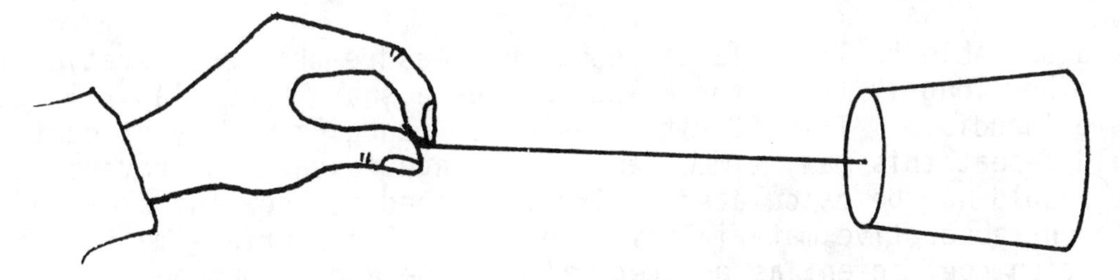

Qualitative Questions: Are the vibrations generated longitudinal or transverse? Why does the cup appear to amplify the sound? Does the pitch go up and down as you hold the string tighter, and put more tension on the string?

Now, push the end of the string through a small hole in the bottom of the second cup. When you talk into it, the speech is clearly audible in the first cup, even with a string ten to thirty feet long.

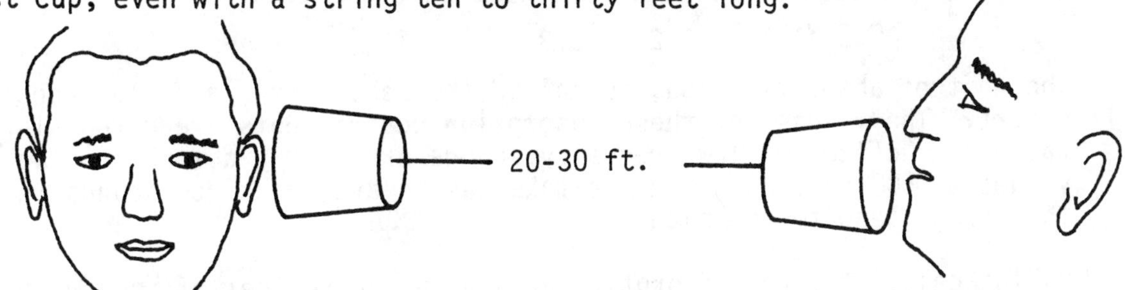

Qualitative Questions:1)Why does its sound travel better through the string than through the air?2)Are high or low frequencies transmitted better? 3)Does this depend on the tightness and density of the string and the size of the cup? 4) Approximately how much more energy reaches the hearer's ear with the cup than without it?

Flying circus of physics 1.9

EXPERIMENT 7.08 Acoustics of Rooms -- Reverberation Time

Materials: A watch, if available.

Procedure: Musicians often comment that a given hall is "live" or "dead". Physically, this is expressed as the reverberation time--the time taken for a loud sound, such as a pistol shot--to fall to the point where it is no longer audible. More accurately, the reverberation time is defined as the time taken for the energy density of sound in the room to fall by a factor of a million--a range of 60 dB.

Pick a suitable hall, or large room, and measure the reverberation time to see how long it takes for a loud sound--a sharp hand clap will do--to become inaudible. Time it with a watch or pendulum. Many students should repeat this many times, and an average taken. The reverberation time should now be calculated. First, we need to know the areas of different absorptive materials in the room. Each surface must have an absorption coefficient associated with it, depending on whether it is strongly absorptive or not. An open window is completely absorptive--nothing of what goes out the window ever gets back--so we may give this an absorption coefficient of unity. We can then say that other materials have some fraction of this -- for example, if we cover the window with plywood, it will absorb $\frac{1}{10}$ of the sound falling on it, that a window of the same size would let out, so its absorption coefficient is $\frac{1}{10}$ or 0.1. Then, the reverberation time would be ten times as long, and we have a new equation, replacing the opening area A by the sum of all the absorptive surfaces a_1, a_2, etc. multiplied by their absorption coefficients, S_1, S_2 etc. The absorption coefficient is the ratio of the intensities of the sound absorbed to that incident on the material.

$$A = S_1a_1 + S_2a_2 + S_3a_3$$

The unit of absorption, Sa, is called the Sabin after a famous acoustics expert. The values for these absorption coefficients are given in table 1. Notice how low frequency sounds are poorly absorbed. The values at 500 or 1000 Hz may be taken as an average since sounds in this range tend to dominate.

Having calculated the absorptive area A in square feet of open window in this fashion, we proceed to calculate the volume V of the room in cubic feet, by multiplying the height by the width by the length of the room. The reverberation time T is then given (approximately) by

$$T = .049\ V/A$$

We can see how this happens by looking at the curve of decay of sound in a room with an open window. In a time t_1, half the energy would have flowed out. It would take the same time again for half of what was left to flow out, and so on. This is an exponential decrease, the power escaping depends on and is proportional to the amount of energy remaining in the room. Now, the time it takes the sound to escape will clearly depend on the size of our open window--like a hole in a bucket, the bigger the opening, the faster the leakage, so the power escaping is proportional to the area, and the time it takes for a fixed amount of sound energy to escape is inversely proportional to the area, so the time for this sound to escape $T \propto \frac{1}{A}$. Similarly, for a given noise level, the bigger the room, the more sound energy it contains, so the longer it takes to empty, hence

$T \propto V$ -- this is like saying it takes longer for a bigger bucket to empty through the same hole.

$$\text{so} \quad T \propto \frac{V}{A}$$

where V is the volume of the room and A the area of the open window.

Now you can see, a hall with a short reverberation time is "dead" -- one with a longer time is more live -- but too long and you can't hear yourself speak. For example, rooms with completely reflecting walls having but one opening are uncommon. However, squash and handball courts closely approximate this. A squash court is 21 ft. by 32 ft. by 20 ft. high, giving a volume of 13,440 cubic ft. The opening in one wall (for spectators) is 4 ft. by 21 ft., an area of 84 sq. ft. This leads to a reverberation time of 7.84 sec. A handball court is 20 ft. by 40 ft. by 32 ft., with an opening 4 ft. by 20 ft, though the latter may vary.

How suitable is the reverberation time of the room you selected, for its purpose?

On the whole, reverberation times for lecture rooms, where speech is the principle use, should vary from 0.4 second for a small room to 0.8 seconds for large lecture theaters. Concert halls must have a longer reverberation time, from 1 to 1.2 seconds for chamber music, to 1.7 seconds for opera and orchestral concerts. It is clear from this why the words in opera are so difficult to distinguish--the acoustics of the opera house blurr them, because of the relatively long reverberation time. For organ music, written to be played in a vast cathedral, a reverberation time of two seconds or longer proves quite suitable.

The reverberation time should be longer for the deeper notes. This is accomplished by ensuring the absorptive properties of the auditorium are larger for the higher frequencies--which is generally true in any case.

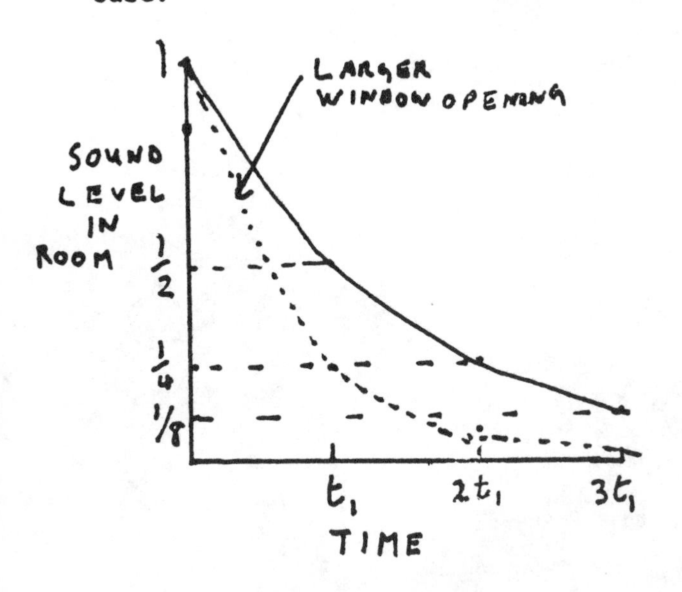

TABLE I

Absorption Coefficients of Some Building Materials

	FREQUENCY--CYCLES PER SECOND					
	125	250	500	1000	2000	4000
Marble or glazed tile	.01	.01	.01	.01	.02	.02
Concrete, unpainted	.01	.01	.01	.02	.02	.03
Asphalt tile on concrete	.02	.03	.03	.03	.03	.02
Heavy carpets on concrete	.02	.06	.14	.37	.60	.65
Heavy carpets on felt	.08	.27	.39	.34	.48	.63
Plate glass	.18	.06	.04	.03	.02	.02
Plaster on lath on studs	.30	.15	.10	.05	.04	.05
Acoustical plaster, 1"	.25	.45	.78	.92	.89	.87
Plywood on studs, 1/4"	.60	.30	.10	.09	.09	.09
Perforated cane fiber tile, cemented to concrete, 1/2" thick	.14	.20	.76	.79	.58	.37
Perforated cane fiber tile, cemented to concrete, 1" thick	.22	.47	.70	.77	.70	.48
Perforated cane fiber tile, 1" thick, in metal frame supports	.48	.67	.61	.68	.75	.50

Experiment 7.09 Velocity of Sound

The velocity of sound is generally measured by resonances in tubes, and other rather complex methods. The simplest method is probably to time an echo, but it is difficult to find a suitably distant cliff or canyon wall. Two people can combine to measure the speed of sound fairly accurately as follows. We can count easily about five hand claps a second. If the speed of sound is 330 m.(1100 ft.)/sec., and we stand 33 m.(110 ft.) from a wall and clap so that the echo coincides with successive handclaps (experience shows this can be done with surprising accuracy - one can easily tell if the clap is a little early or late). The second member of the group then times the claps with a seconds watch. If we find 100 claps (about 20 seconds) takes T seconds, the velocity of sound is 2 x (distance to the wall)/0.01T. This should be repeated ten times, and the average and standard deviation found.

Experiment 7.10 - A paper wave to explain the Doppler Shift

Students with no physics background often have great difficulty understanding the Doppler Shift. It seems incomprehensible to many physics teachers that there is a severe conceptual problem involved here, but discussion with the students shows it is very real.

A simple practical model can be of considerable value. Take a long sheet of paper, and pleat it as shown in figure 1. A simple way to do this for a small class is to fold a 36 cm (14 inch) sheet of lined notepaper alternating up and down at every fourth line (3.5 cm or $1\frac{3}{8}$ inches). This gives about $4\frac{1}{2}$ pleats. Now cut the sheet lengthwise into three equal strips, join the strips together end to end, and you have a transverse paper wave with about twelve oscillations. A larger sheet is necessary for a big class.

The concertina-folded strip is laid on a flat piece of paper whose length represents the distance sound travels in one second. Again, a 22 x 36 cm ($8\frac{1}{2}$ x 14 inch) sheet may be used. For students with little mathematical background, it is convenient to provide two explanations for sound waves, one with a moving source which gives the wavelength shift, the other which gives the frequency shift for a moving observer.
It is a good idea to consider a fixed time such as one second, so to start we take a stationary train emitting a toot one second long. The train may be drawn on a sheet of folded paper. The number of oscillations in one second, f Hertz, (thirteen for the pleated sheet we are using) is contained within the length traveled by sound in one second (V). To demonstrate this, pull the paper wave out from under the paper train, as shown in figure 1. Using the model, it is easy to show $V=f\lambda$.

The second stage in the demonstration is to move the train forward as the wave is slid from beneath it. The train will move forward the distance it travels in one second (v). In our model, we move the train one inch per second, which is marked on the sheet of paper, whose length we have taken to be the velocity of our sound. The waves now occupy a distance equal to V-v, and the wavelength λ' is now $(V - v)/f$, where before λ was V/f. In our model, we moved 14 inches in one second, with 13 oscillations, so $\lambda=14/13$ inches. If the train moved one inch in the second, $\lambda' = (14-1)/13 = 1$ inch. The frequency was originally 13 Hz, and is now $V/\lambda = 14$ Hz. - an 8% increase.

In practice, sound travels at 330 m/sec, so a suitable train speed is 30 m/sec, which is 108 km/hr (67.1 mph). If we take the musical note A for the train whistle, 440 Hz, $\lambda=75$ cm, $\lambda'=68$ cm and $f'=484$ Hz, 10% higher, almost two semitones. If the observer moves instead of the source, we treat the situation differently.

I use a big paper ear, as shown in figure 2, to represent the observer, and the waves now always occupy a length V since the source does not move. With the observer at rest, it takes one second for the f waves to enter the ear. However, if we move the observer away from the source, in one second the sound will travel a distance V, but the ear will have moved a distance v in this time - so those oscillations in the distance v will not have entered the ear. Again, the distance v can be marked on our sheet of paper. Since there were f oscillations in a distance V, there will be $f(V-v)/V$ in the shorter distance, and the frequency heard by the observer is $f'/f = (V-v)/V=1-v/V$.

So, in our model, V = 36 cm (14 inches), v = 2.54 cm (1 inch) f = 13(14-13)/14 = 12.02 Hz, a lower frequency. If we are on a train approaching a railroad crossing at 30 m/s where a 660 Hz whistle is blowing, we get $f'=f(v+v)/V = 660\ (300)/330 = 480$ Hz which is not the same as the previous case when the train approached, at the same speed, showing the Doppler effect differs considerably when the observer rather than the source is moving in the case of sound.

Now, f'= original frequency f plus the difference in frequency df.

$$\text{So } \frac{f+df}{f} = 1+\frac{v}{V}$$

and df/f=v/V It is easy for student to remember that the change in frequency, divided by the frequency is the ratio of the velocity of the observer to the velocity of sound. This equation is accurately true for light, whether we speak of the observer or the source moving, because one tenet of the theory of relativity is that we cannot tell whether the observer or the source is moving at constant velocity. It is approximately true for sound, in the general case for both observer or source moving, if the velocities of the observer or source are small. This pragmatic approach to the Doppler Shift really helps those students whose ability to master abstract concepts is restricted.

We can follow the lead of the Australian aborigines to demonstrate the Doppler shift experimentally. They have only two prehistoric musical instruments. The bullroarer, and the didgery doo. The didgery doo is a hollow log about eight feet long or more, which, when blown produces a deep and somewhat monotrous moan. The bullroarer is a flat piece of wood on the end of a long string. The one I use is a piece of 1/8 inch plywood, 3 by 10 cm with a hole drilled at one end to attach a 1 m piece of string. A piece of thick cardboard about the same size also works well. The dimensions are not critical. This makes a strong buzzing sound when whirled round the lead, and a higher pitch as the bullroarer approaches the observer, and lower as it moves away. Unfortunately, the person doing the whirling, being at the center, doesn't hear any change.

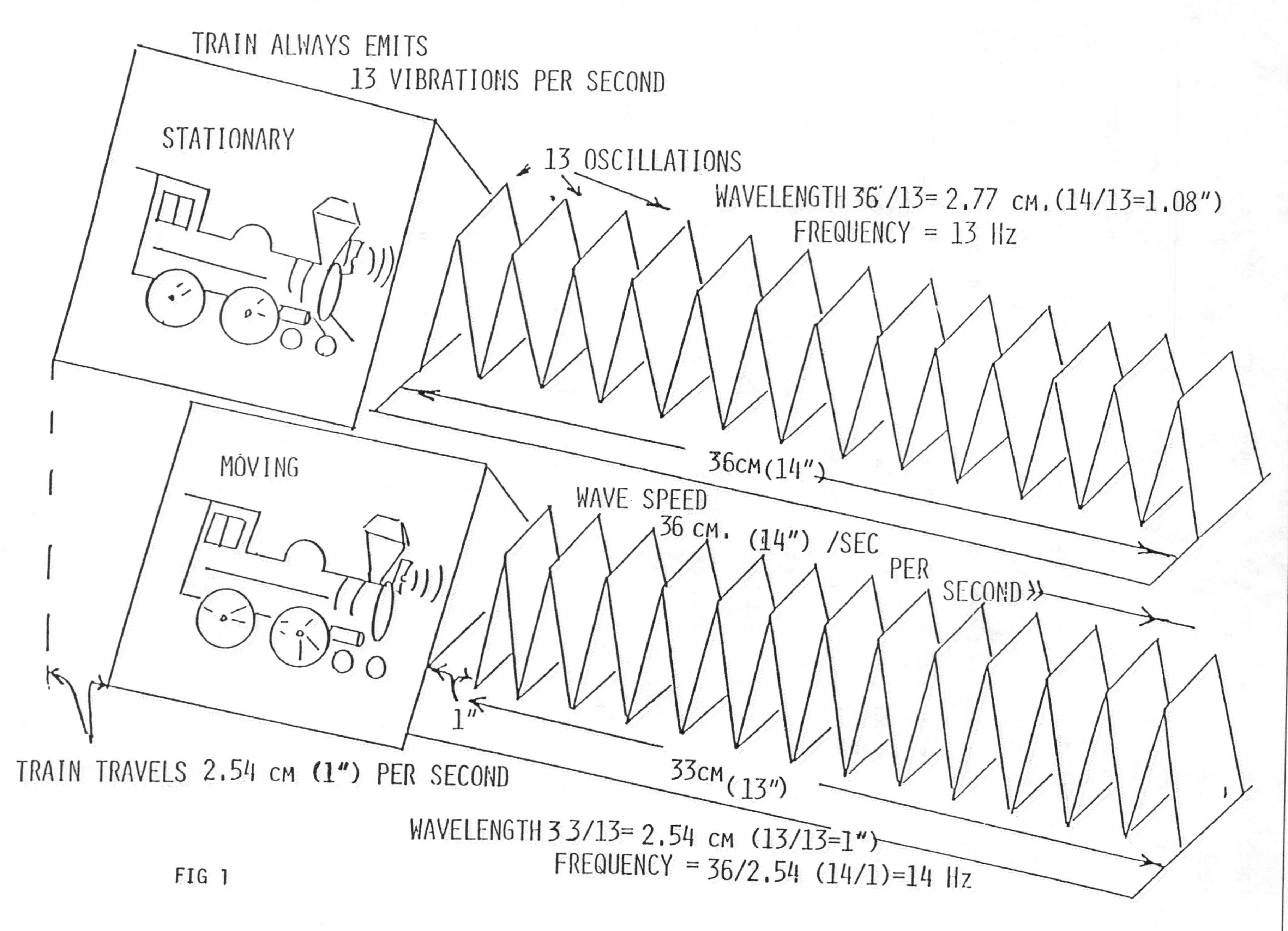
TRAIN ALWAYS EMITS
13 VIBRATIONS PER SECOND
STATIONARY
13 OSCILLATIONS
WAVELENGTH 36/13= 2.77 CM. (14/13=1.08")
FREQUENCY = 13 Hz
36CM(14")
MOVING
WAVE SPEED
36 CM. (14") /SEC
PER
SECOND
1"
TRAIN TRAVELS 2.54 CM (1") PER SECOND
33CM(13")
WAVELENGTH 33/13= 2.54 CM (13/13=1")
FREQUENCY = 36/2.54 (14/1)=14 Hz

FIG 1

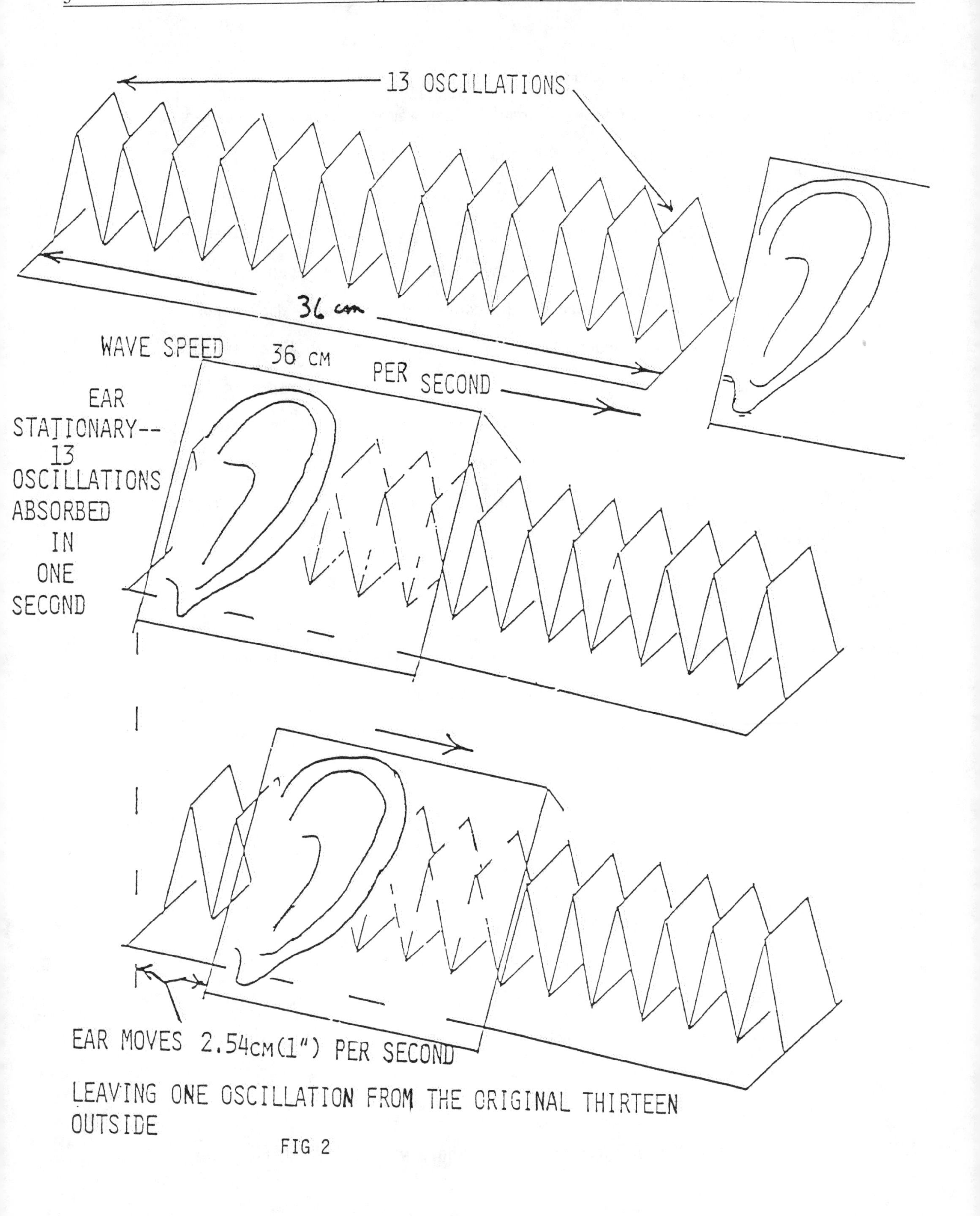
13 OSCILLATIONS
36 cm
WAVE SPEED 36 CM PER SECOND
EAR STATIONARY-- 13 OSCILLATIONS ABSORBED IN ONE SECOND
EAR MOVES 2.54CM(1") PER SECOND
LEAVING ONE OSCILLATION FROM THE ORIGINAL THIRTEEN OUTSIDE

FIG 2

Experiment 8.01 Molecular size

Equipment: Soap, chalk dust, wash bowl.

Procedure: Take the smallest portion of soap you can remove, with a finger nail, or one drop if it is a liquid, and dissolve it in a cup of water--about 500 cc. The soap will be approximately 1/10 gm. Now run a bowlful of tap water, let it settle until it is quite still, then sprinkle chalk dust on the surface. Take one drop of soap solution, and place it on the surface in the center of the bowl. What happens? The soap molecules reduce the surface tension, and the water surface, stretched like a rubber sheet with a hole suddenly punctured in it, pulls the chalk apart where the hole is. The limits of the hole are set by the soap molecules which form a layer one molecule thick. If you measure the diameter of the hole, the area which the soap molecules cover is given by $1/4\,\pi d^2$.

Qualitative: Why do you think the area covered by soap could not spread out forever?

Quantitative: Let the original mass of soap be .1 gm. Its volume will be about .1cc, diluted by 500 cc, of which we take one drop, about 1/2 cc. So we have $\frac{1/2}{500}$ x .1cc soap. The area it spreads is $\frac{\pi d^2}{4}$, so the thickness of the film is

$$\frac{\text{Volume}}{\text{area}} = \frac{\frac{1/2 \times .1}{500}}{\frac{\pi d^2}{4}} = \frac{4}{10{,}000\pi d^2}\ \text{cm}$$

approximately $\frac{1}{10{,}000d^2}$ cm

The maximum size of the film area is provided by its minimum thickness. This occurs when the film is one molecule thick. In fact, soap molecules are long and thin, and like to have one end attached to the water surface, so they are all parallel to one another, with one end in the water. For example, if the molecule were 100 Å = 10^{-6}cm, d=10 cm. If the open patch covers the whole wash bowl, it shows the molecular size is less than about 10^{-7} cm.

EXPERIMENT 8.02 Nuclear Cross-Section and the size of a Penny

Apparatus: Sheet of paper, pennies, pencil.

Procedure: The object of this experiment is to measure the size of pennies,(or marbles) without actually using a ruler, or any similar measuring device, other than statistics. Take as many pennies as are available (about 40) and lay them out at random, but more or less uniformly (i.e. not all in one corner) on a sheet of writing paper. Now, drop a pencil, point first, from a height of, say, four or five feet, onto the sheet of paper. Neglect all shots that miss the paper, and count separately those shots that hit pennies. (A blunt pencil works best!) Thirty or forty times are necessary.

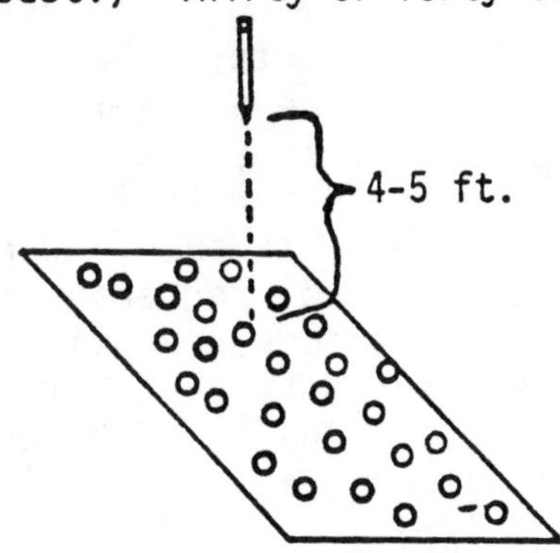

From this you can find the area of a penny.

Quantitative:

A sheet of writing paper is 8½ x 11 inches, an area of 603.22 square cm (93.5 sq. inches).

The pencil was dropped in a random fashion, and not aimed specifically to hit the pennies. Hence, the chance, or probability it hits a penny is proportional to the ratio of the area of the pennies to the paper. Let the area of penny be A. If there are m pennies, their area is mA. Let the total number of shots be N, and the number hitting pennies be n. Then

$$\frac{n}{N} = \frac{mA}{603.22}$$

and the area of one penny is

$$A = \frac{603.22n}{m\ N} \text{ square cm.}$$

The area of a penny is 2.835 cm^2 (.44 square inches).

How accurate is your answer?

The importance of this experiment is in explaining how the size of the nucleus is measured when it is inaccessible to a ruler. We fire nuclear particles at a target, in the same way we dropped the pencil on the paper, and see how many times they hit a nucleus, and how many times they miss. Knowing the number of nuclei per square centimeter of the target, we can find the area of the nucleus, or its cross section, in exactly the same way we found the area of the penny.

Experiment 8.03 Radioactive Decay--An Analogy

This simple experiment is an effective way to introduce students to the abstract concept of radioactive decay. It was suggested by Shirley Stekel of the Physics Department, University of Wisconsin-Whitewater. It requires beans or marbles of two different colors, although even pieces of paper of the same size but two different colors could be used.

Count exactly 100 red beans and place them in a paper cup or other small container. Have a handful of white beans available in another container. Each red bean will represent a radioactive atom and each white bean a stable daughter atom. The start of the decay process is simulated by taking ten red beans from the cup and replacing them by ten white beans to represent the decay of ten radioactive atoms to their stable products. Then, in a series of trials the contents of the cup is stirred each time and a randomly selected sample of ten beans removed. Each red bean removed is replaced by a white bean and the white beans that have been selected are returned to the cup so that there are 100 beans in the cup before each sample is taken. The number of red beans removed each time is recorded, and the number of red beans remaining in the cup is then calculated. Continue the sampling procedure until only about 20 red beans remain in the cup. Now draw a graph of the number of red beans remaining in the cup versus the sample number. A "half-life" can be determined if it is assumed that the samples were drawn at equal time intervals.

The theoretical curve for this process is an ordinary decay curve and the students' results are usually quite close to the theoretical values at first but tend to deviate as the number of red beans in the cup decreases. The half-life of the beans depend on the ratio of the sample size, S, to the original number of red beans in the cup, n_0.

The sampling process may be simply explained on the basis that 10% of the red beans decay,i.e.are removed from the cup and replaced by white beans, each time. In the case of radioactive atoms, we would say that 10% of the radioactive atoms (red beans) decay to stable atoms (white beans) in the time taken for each sampling. Thus, 10 red beans are removed the first time, leaving 90 red beans, and 10 white beans. From now on, we cannot predict the exact number of red or white beans we shall pick - we could pick 10 red or 10 white beans. However, since there is one white bean for every nine red beans in the cup, this is the proportion, and in fact the number most likely to be selected statistically. We will now have 81 red beans, and 19 white beans. The next throw removes 10x(81/100) or 8.1 red beans, leaving 72.9 red beans, so 7.29 will be taken next time, leaving 65.6, and so on.This calculation, and the average of five experiments, is shown in the figure. The increased accuracy of averaging five runs or sampling 50 out of 500 is worth it, if time is available

To relate this to the conventional decay formula, the number of red beans drawn in any given sample n is proportional to thenumber left, n.

$$S \quad n = n \qquad (1)$$

This leads to the relationship

$$n = n_0 \exp-(t/S)$$

where t is the number of the sample. This assumes the sample n is so small that we can integrate (1) . Since this is not in fact the case, the formula is only approximate. The half life (the time it takes for half the radioactivity to decay, or in our case, half the red beans to be converted to white) should be given by $T_{\frac{1}{2}} = S \ln 2$ which, for 100 beans sampled 10 at a time, would be seven samplings. If we went by time (as we do with radioactivity) and sampled once a minute, it would be seven minutes.

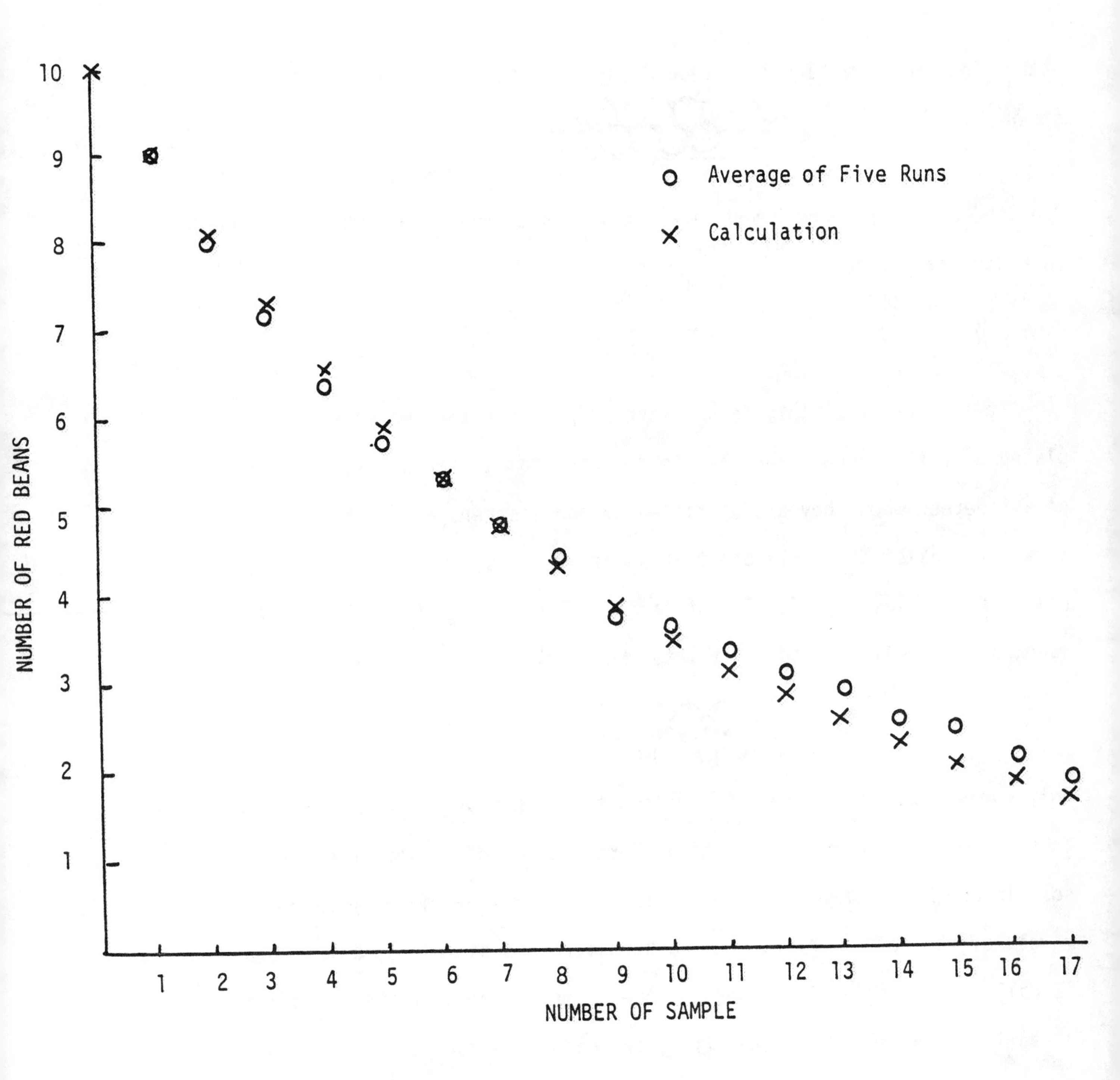
10
9
8
7
6
5
4
3
2
1
NUMBER OF RED BEANS
1 2 3 4 5 6 7 8 9 10 11 12 13 14 15 16 17
NUMBER OF SAMPLE
Average of Five Runs
Calculation

Experiment 9.01 The Packing Together of Atoms in Metals

Materials: marbles, paper, sticky tape

Procedure: Take a piece of tape, and attach five marbles, as shown, as close together as possible.

Attach four more marbles to a second strip, adjacent to the first, as shown

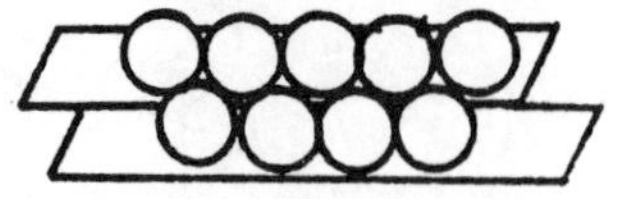

Continue doing this until you have a triangle of marbles similar to that on a pool table, so

It is clear that this is as close as possible that marbles can be placed on a flat sheet, and this is the way atoms, such as metal atoms pack together when they are attracted to one another. However, a metal does not consist of merely one flat sheet of atoms, but is solid. How does the next layer go down? Put marbles down on top of the layer you made, and you will see they fit into the cracks as shown. So there is

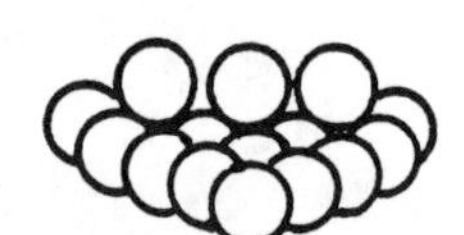

only one way to put two such layers together. However, now try putting on a third layer, and you will find there are two different ways you can do so. In one, each marble of the third layer will be directly above a marble of the bottom layer. Such a structure has sixfold symmetry, and is called hexagonal. In the other, the atoms of the third layer do not lie directly above atoms of either first or second layers. Such a system has cubic symmetry. To examine this, continue piling marbles on until you build a little pyramid as shown.

Such a pyramid, which is the way in which cannon balls are piled outside a courthouse, form a regular tetrahedron, the sides of which are all equilateral triangles, the corners all being 60^{0}, two thirds of a right angle, as you can see by holding the corner of a sheet of paper on one side.

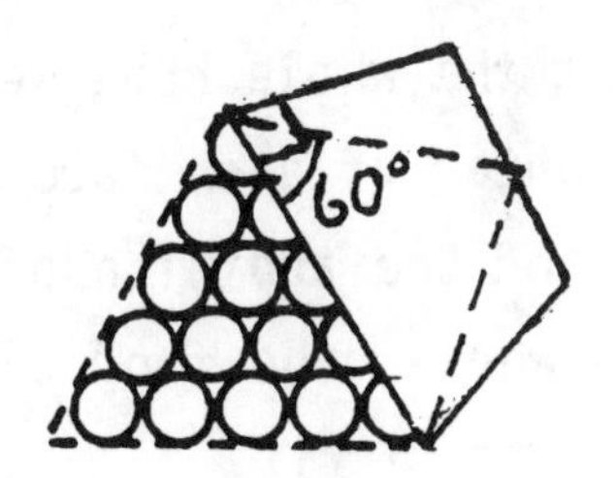

Three sides meet at the top, showing threefold symmetry about the corner.

Remove the marbles down to the bottom layer, and put down a second layer, composed of six marbles, as shown.

Then place one marble on top of the pile in the middle. If you measure the angle of the corner with a sheet of paper again, you will find it is 90^{0}, as are the corners of a cube. This shows that, the cubic symmetry we obtain by piling marbles is actually arranged so that one is building the corner of the cube, which has an axis of three fold symmetry passing through it, since again three sides meet at a corner.

Notice, if we label the layers from the top down which we used to build the tetrahedron, 1, 2, 3, 1, 2 where all the 1's and 2's lie vertically above one another, the corner cube would be 1, 3, 2 - both have the three layer scheme, but arranged in a different order.

The cubic packing we have examined is called "face centered cubic". We can see why by looking at the diagram below. "A" forms the corner of the cube we have built, and looking at the whole cube we have drawn, we see there are atoms at each corner, and at the center of each face of the cube. ABCD is the tetrahedral structure we built, and the addition of GFE turns it into the corner cube. See if you can pick out the whole cube from the second structure of cubic symmetry you built.

Qualitative Question: How do you think you could tell a cubic metal crystal from a hexagonal crystal? Do you think any other structures are possible besides cubic and hexagonal? Try with the little pile of marbles.

Magnesium, scandium, titanium and cobalt all have hexagonal close packed structures. Aluminum, nickel, copper, silver and gold all form close packed cubic structures.

Quantitative: Do you think the crystal size is the same if the same number of atoms are packed in a hexagonal or cubic array?

You can try this out, experimentally, by measuring the outside dimensions of the pyramid or the equivalent hexagonal structure.

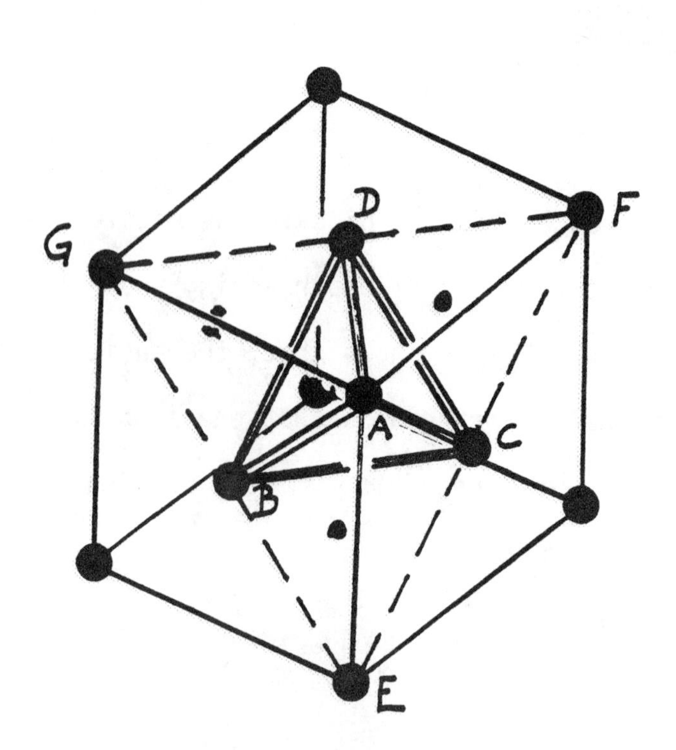

INTRODUCTION TO SECTION 10

Electrostatics

One of the problems in devising experiments employing cheap and easily available materials has been to find equipment for electrical experiments - particularly electrostatic experiments. The most satisfactory solution appears to be the (empty) aluminum soft drink (or beer!) can (12 oz) which can be used for inumerable purposes. Unlike the old steel cans, the aluminum may be cut easily with a pair of scissors, and without ruining the scissors. The edges of the cut metal are not as sharp as the steel cans, and the sheet can easily be bent with the fingers - yet it is robust enough, unlike aluminum foil, to retain the shape which it is given, even under stress. This item of garbage must definitely be added to our list of resource materials.

10.01 THE ELECTROSCOPE

To make an electroscope from an aluminum can, first, cut off the top and bottom, and slit the side, so that it can be opened out flat to give a sheet approximately eight inches by four inches (10 x 20 cm).

The electroscope is a device which measures the charge on itself, (and hence the voltage of whatever supplied that charge). The type described employs the mutual repulsion of a vane and its support, having charges of like sign.

Digression

At this point let us digress to examine one of the interesting features of the can itself. Aluminum drink cans, in general, are deep drawn - that is to say, the metal is squeezed in forming the can so that it runs from the bottom producing the sides of the can. X ray diffraction shows that aluminum crystallites align in a direction along the length of the can, rather than perpendicular to it. This affects the physical properties of the metal. To see this, cut two identical strips of aluminum from your sheet, say ¼ inch by 4 inches, with one cut along the "grain" (i.e. the length of the can) and the other perpendicular. Flatten the two strips carefully, then hang a quarter using sticky tape, from the end of each, and let them hang the same distance over the the table, as shown in figure 1. You will find the strip cut along the length deflects much less than that cut perpendicular to the axis of the can. This shows the coefficient of elasticity, (stress/strain), is larger for the strip cut along the length, in much the same way that wood has a higher coefficient of elasticity (Young's modulus) along than across the grain.

Returning from our digression, cut the shapes shown in figure 2 from your aluminum sheet. Now, bend the larger piece as shown in Fig. 3, and place the smaller piece, the vane, to balance on the pivot region A. The vane should be top heavy at first, and overbalance. Cut small pieces from the top of it until it just does not overbalance.

The electroscope must be insulated, so tape it to the bottom of an upturned styrofoam cup as shown. Humidity is the greatest enemy of electrostatic experiments. If your experiments do not work, it is quite likely because humidity has condensed on the insulators, allowing the charge to leak away. Hence, make sure the air is dry - in summer air conditioning will help, and in winter a hot dry room is best. Do not bring cold insulators into a warm room - moisture will inevitably condense on them to produce a conducting layer.

Since the sensitivity of the instrument depends on the position of the center of mass of the vane, each electrometer is different and requires a special calibration. However, the foil in which chewing gum comes wrapped also makes a suitable vane and it is very light, and of uniform thickness. Cut a piece of the wrapper, as shown in figure 4, making it as flat as possible, then fold over the top, and hang it (metal side inward) as shown on a support cut from the aluminum sheet with the dimensions given. We used the foil from Wrigley's double mint gum. The diagram shows the angle at which the foil will sit as a function of the voltage applied to the electrometer. It is not a linear scale, because the torque depends on the square of the charge in addition to its distribution as a function of the angle.

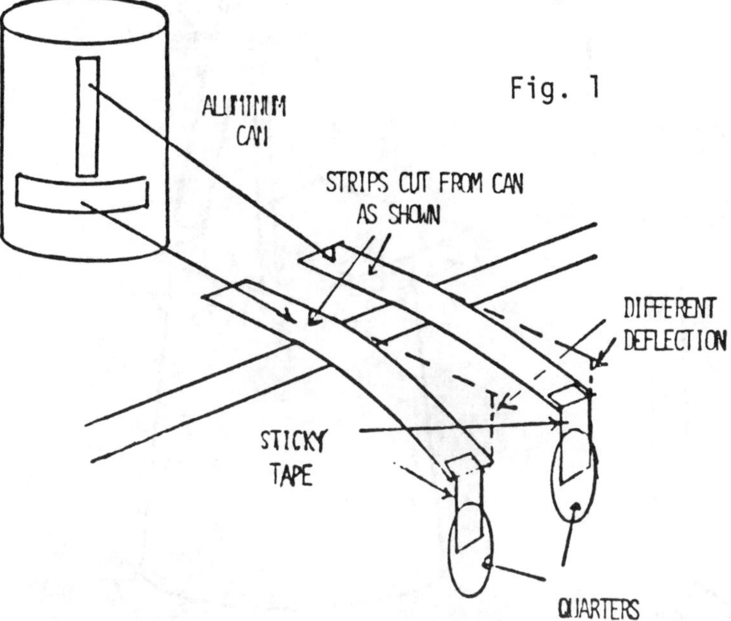

Fig. 1

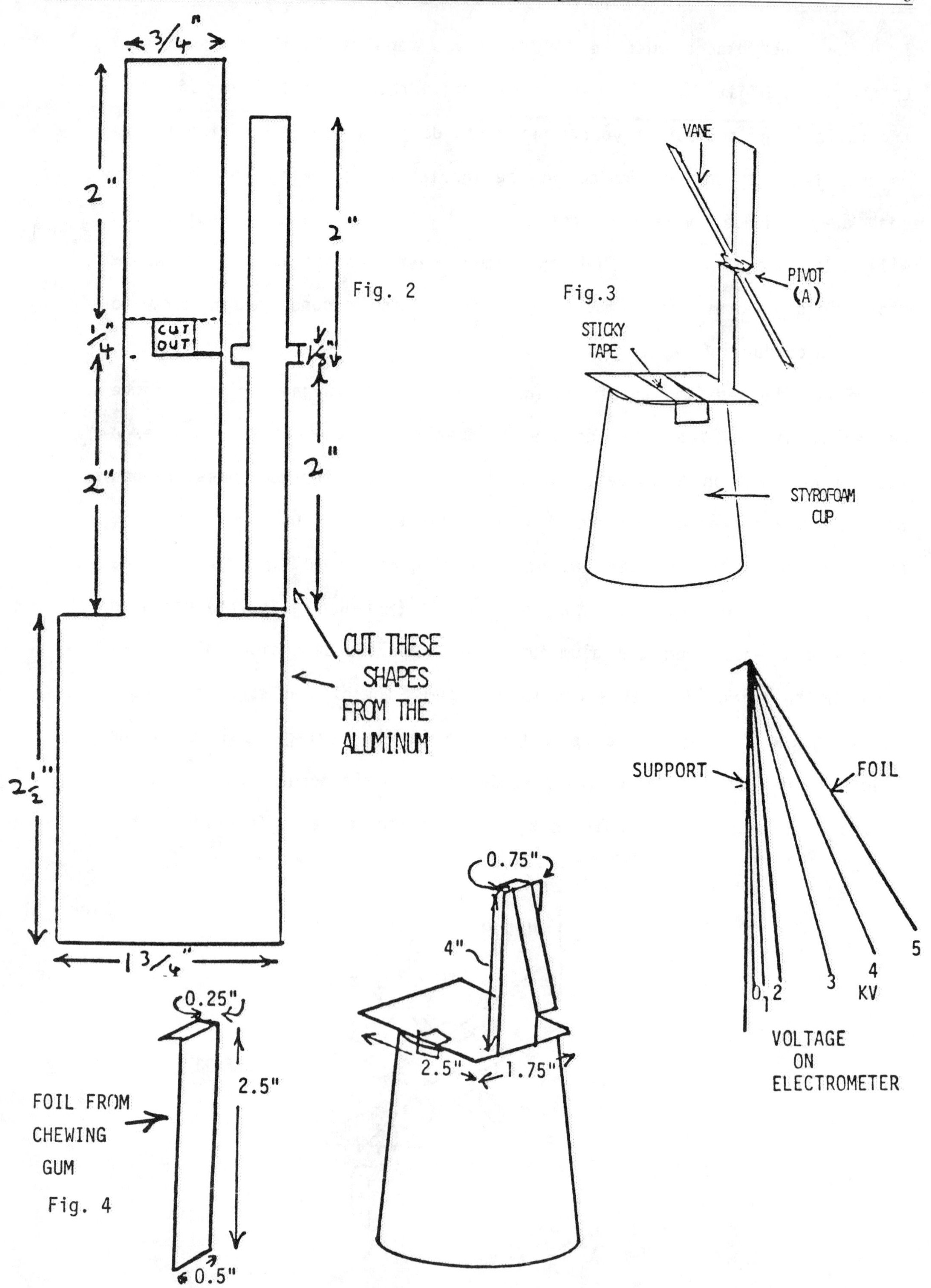
3/4"
2"
1/4"
CUT OUT
2"
2"
1/8"
2"
2 1/2"
1 3/4"
CUT THESE SHAPES FROM THE ALUMINUM
Fig. 2
Fig.3
VANE
PIVOT (A)
STICKY TAPE
STYROFOAM CUP
SUPPORT
FOIL
0
1
2
3
4
5
KV
VOLTAGE ON ELECTROMETER
0.25"
2.5"
0.5"
FOIL FROM CHEWING GUM
Fig. 4
0.75"
4"
2.5"
1.75"

10.02 The "Versorium" - a simple charge detector

It has been known since the time of the Greeks, that when amber, a natural yellowish fossil resin coming from trees, is rubbed, it will attract small bits of matter. From the Greek name for amber comes our work electricity, and the frictional process by which it is electrified is known as triboelectrification.

William Gilbert, a sixteenth century English scientist (and incidentaly Queen Elizabeth's physician) divided materials into "electrics," which he could electrify, and "non electrics" which he was unable to electrify, and which today we call insulators and conductors. It was left to Benjamin Franklin to suggest that there is really only one type of electricity, present to differing degrees in uncharged or neutral bodies. An excess of this in a body he called positive, and a lack of it negative.

Hang a drinking straw by a fine thread, or support it on a pin. Charge another straw by rubbing with a cloth, and bring it near the first. It will move. Such a device was invented by Gilbert who called it a "Versorium." Now make sure the first straw is charged with the same sign electricity as the second by rubbing with the same material. You will find the two repel one another, a simple way of demonstrating like charges repel. The versorium will tend to stick to the cloth used to rub it, showing unlike charges attract. (Since the charge on the straw is rubbed off the cloth, they must be charged of opposite sign.) If rubbing together A and B makes A+ and B-, and rubbing materials B anc C makes B+ and C-, then if materials A and C are rubbed together, A will become + and C - . Hence substances can be organised in a sequence called a triboelectric series, where a substance will be given a positive charge if rubbed by a second substance below it in the series, and correspondingly a negative charge if above it. The series (Table 1) drawn from the Smithsonian tables, is at best approximate - glass rubbed by rabbits' fur may not always be negative - it depends very much on the rabbit.

The electroscope you have built can be used to check this series, but to do so, and also for many other experiments, it is useful to have an "electrophorus," a device for putting a fixed charge on the electroscope, or other conductors.

Table 1

asbestos

fur (rabbit)

glass

mica

wool

quartz

cat's fur

lead

silk

human skin, aluminum

cotton

wood

amber

resins

Brass, Cu, Ni, Co, Ag etc.

rubber

sulphur

metals (Pt Au)

celluloid

India rubber

10.03 The electrophorus

The electrophorus demonstrates charging by induction, and the principle of electrostatic machines, such as the Wimshurst machine.

Cut the top off an aluminum can, and stick the botton end of it in a styrofoam cup, to act as an insulating support, as shown in figure (a). Now, charge a second cup by rubbing it on your jacket, your hair, or some suitable fabric. The cup will likely have a negative charge. Place the bottom end of the charged cup in the can, as shown in figure (b). The negative charge on the cup induces a positive charge on the inside of the can. Holding the can in one hand by the styrofoam cup in which it sits, touch the outside of the can to draw off the negative charge present there through your finger (Fig. (c)). Lastly, remove the charged cup. (Fig. (d)) The can will now be positively charged, and on touching the electroscope, charge will be transferred, causing the vane to rotate away from the support, because of the repulsion of like charges. Note, no charge was removed from the styrofoam cup - hence after discharging the can we can repeat the whole process, and transfer as much charge as we want to another conductor via the can, until the conductor is at the same potential as the charged can. This is the principle of electrostatic machines - only, instead of transferring the charge manually, it is done mechanically by the machine.

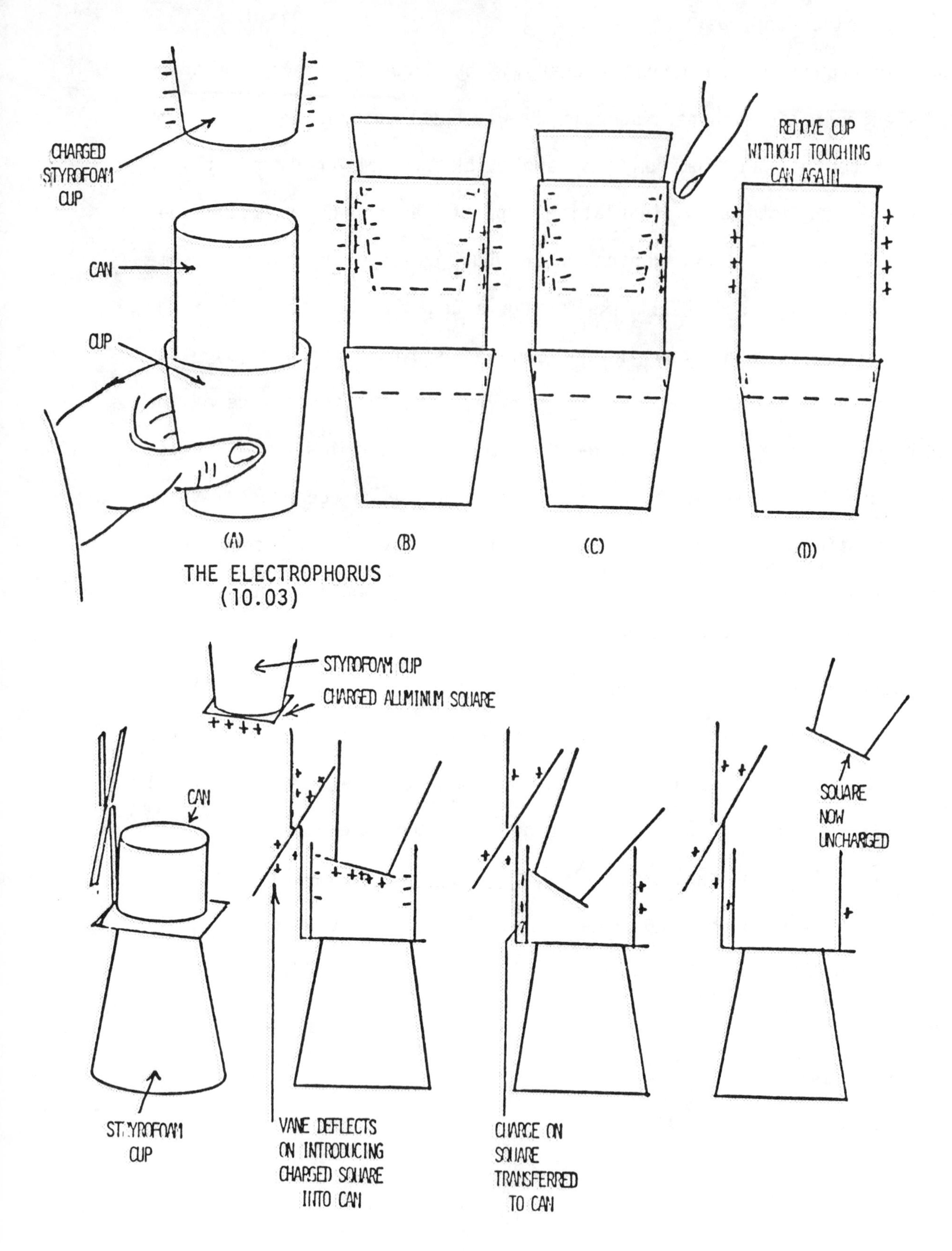

THE ELECTROPHORUS
(10.03)

FARADAY ICE PAIL EXPERIMENT
(10.04)

10.04 The Faraday Ice Pail Experiment

In 1843 Michael Faraday used an ordinary pewter ice pail for a number of electrostatic experiments, some of which showed there was no electrostatic field inside the pail, and therefore all the charge resided on the outside. We can repeat this experiment by using an aluminum can with the top cut off in place of the pail. Put this on the plate of the electroscope. Cut a piece of aluminum about two inches square, and stick on to the bottom of a styrofoam cup as shown in figure 6. Charge up a styrofoam cup, place the metal square on it and ground it, as you did before, with the electrophorus. The charge on the plate will be less than that on the can used previously because of its size. Now place the square inside the can on the electroscope, and touch the inside of the can with the metal square. If all the charge resides on the outside of a conductor, the square will be discharged and the charge run to the outside of the can. Remove the can from the electroscope and discharge it. You can now check that there is no charge on the square. We can now recharge the square as before, and repeat the process of transferring charge to the can on the electroscope as often as we want, raising the potential of the can much higher than that of the square before it is inserted in the can, until ultimately the can will spark over to ground, if the insulators are sufficiently dry. This is the principle of the Van de Graaf machine, but in this machine the charge is carried to the inside of the conductor via a belt.

10.05 Capacitance

Electrostatic charge is stored in a capacitor or condenser. The earliest form of this was the Leyden jar, a glass vessel with a metallic coating on the inside, and a second insulated from the first on the outside. What conditions are required to store the most charge?

To see this, take the charged metal square you have, and gradually bring it close to the plate of the discharged electroscope. Note that the vane diverges more and more the closer the charged square is to the plate of the electroscope - that is because the negative induced charge is drawn to the plate, leaving the vane and its support positive. Hence, to store charge the plates of the storage vessel should be as close together as possible. Now, overlap the charged square with the plate of the electroscope, keeping them the same distance apart. Again, when the overlap is small, the vane moves little, but with a large overlap it moves much, showing much charge has been drawn to the plate. So, the best capacitors have a large area, and a small separation between the plates.

Charge the electroscope with the charged plate until the vane is at about 45°. Now, try bringing up various objects which have been rubbed by other materials, and check out the triboelectric series. Put a can on the plate of the electroscope and discharge it. Rub a small straw with a small piece of paper or cloth (a sheet of toilet paper works well.) Put the paper in the cup - the vane moves out. Add the straw, and it moves back, showing the gain in charge by the paper was equal to that lost by the straw - but - make sure both are discharged to start with. Straws charge very easily.

10.05

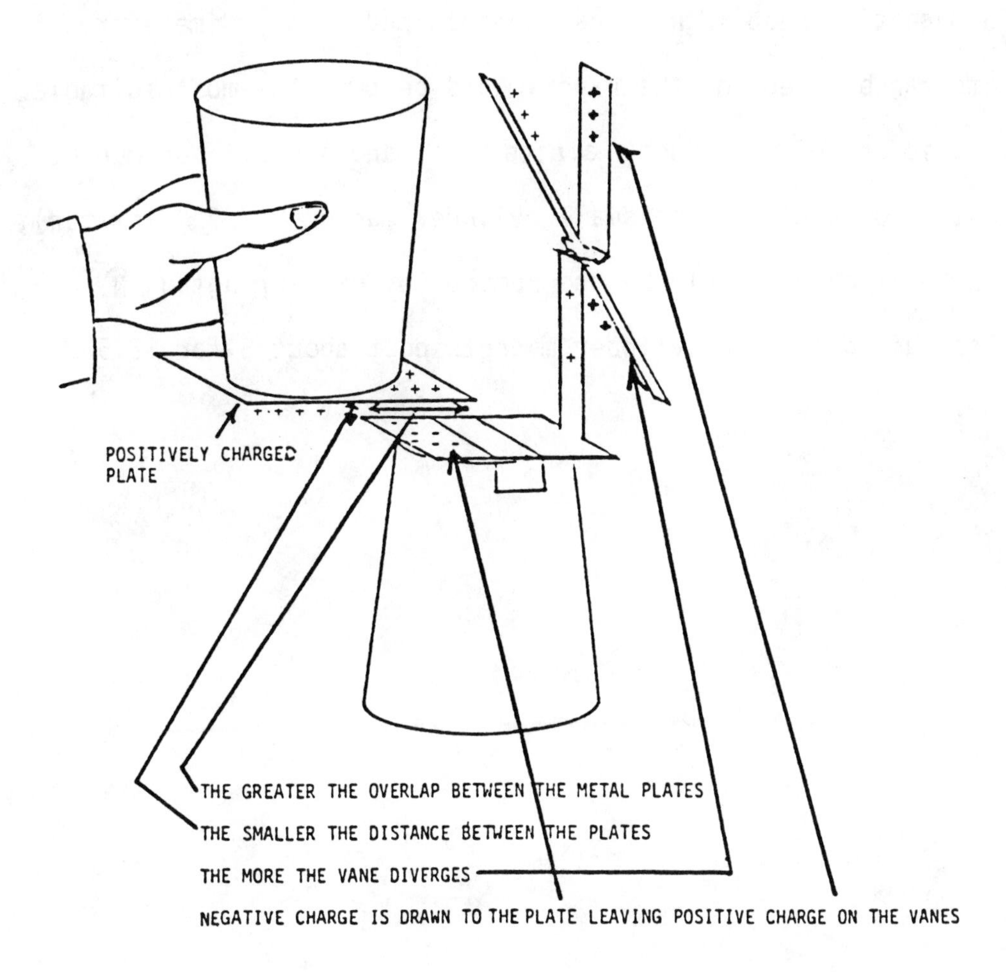
POSITIVELY CHARGED PLATE
THE GREATER THE OVERLAP BETWEEN THE METAL PLATES
THE SMALLER THE DISTANCE BETWEEN THE PLATES
THE MORE THE VANE DIVERGES
NEGATIVE CHARGE IS DRAWN TO THE PLATE LEAVING POSITIVE CHARGE ON THE VANES

INTRODUCTION TO SECTION 11

EXPERIMENTS IN MAGNETISM

Additional equipment is needed to perform experiments in magnetism, specifically a magnet! Cheap magnets can be obtained at any dime store, and such magnets can be used for the experiments below. The most suitable magnet, however, is one of the chubby alnico type, and a small bar magnet is probably best. For a class, the small cylinder magnets (2/3" diameter), or even the break-off magnets sold by the Edmund Co. of Barrington, N. J. 08007, are quite suitable. The cylinder magnets cost about 8 for $2.50.

Experiment 11.1 - Why is the North Pole South?

Equipment: Magnet, cotton

Procedure: Take about three feet of cotton, or unwind a thread from a piece of string the same length. Attach it to the center of the magnet. Hold the thread at the top end, and notice the magnet always points in the same direction. That end of the magnet pointing north is called the north-seeking pole, or more often, the north pole. However, as we shall see from experiment 11.3, the north pole of one magnet is attracted to the south pole of another. So, if we regard the earth as a big magnet, its south magnetic pole must be in the north, as shown. In fact, it is thought that the earth's field is provided largely by the circulation of conduction fluids inside the earth. Mark the North pole of the magnet in pencil with an N.

The way in which the force provided by a magnet can act through space mystified past scientists when there is obviously nothing joining the magnet and the object it attracts. They were looking for something like a connecting piece of string, or a spring. Our present day knowledge of atoms and electrostatic fields now makes us realize that even a piece of string is held together by electrostatic forces - concentrating our problem to what is such a field? Today, though we know much more about fields of force, the basic problem is still unsolved.

It used to be thought that a magnet breathed on by someone who had eaten garlic would remove its virtue. Luckily, this has proved untrue.

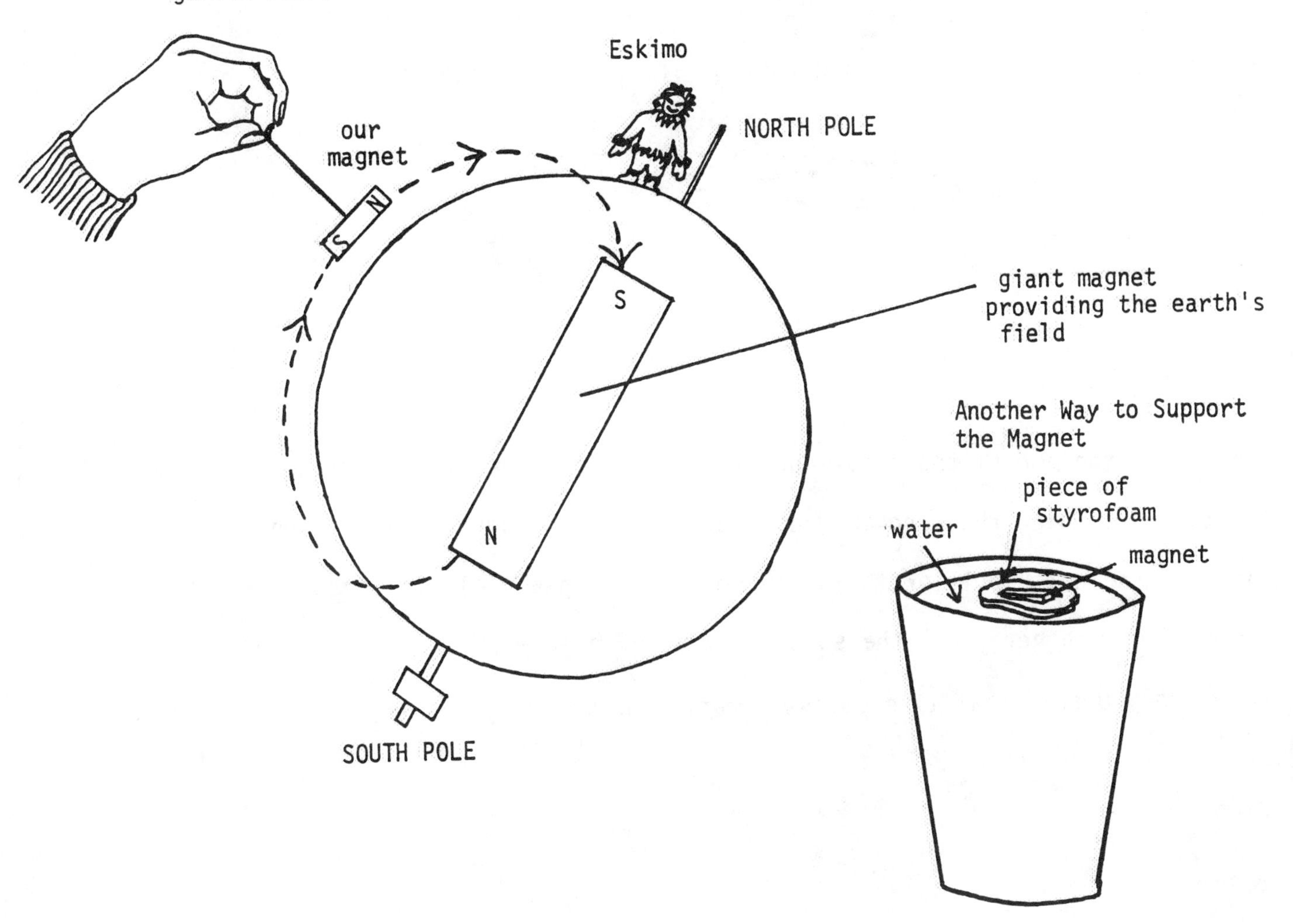

Experiment 11.2 Making a magnet, and using it as a compass.

Materials: cotton thread, or a fine thread untwined from the string, paper clips, magnet.

Procedure: Hold the paper clip on the table top, and stroke it from end to end, with one pole of the magnet, a large number of times, always in the same direction. Hang the clip from a thread, as shown, so that it balances

horizontally and see in which direction it points. Which is it?

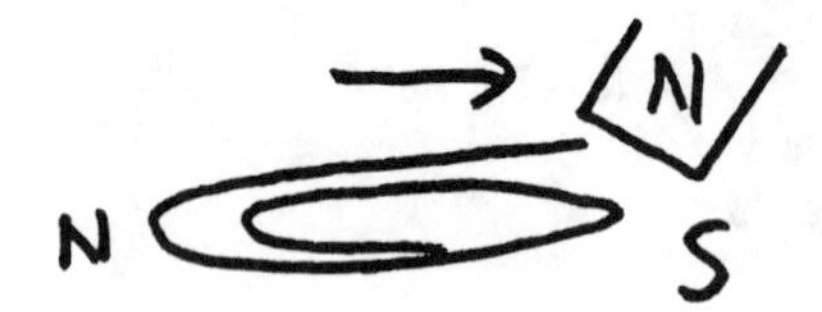

Note that the end of the clip where the stroke finishes has opposite polarity to the pole doing the stroking, as shown. Does the south pole of the magnet attract the north pole or the South pole of the paper clip? Does the south pole of the magnet repel the south or the north pole of the paper clip? Convince yourself that like poles attract, and unlike poles repel.

Experiment 11.3 - Lines of force around a magnet.

Equipment: Magnet, paper clip, cotton thread, pencil.

Procedure: The lines of force should follow curves whose formula is $\frac{\sin^2\theta}{r}$ = constant.

where θ is the angle the radius vector r makes with the axis of the magnet.

This experiment is to test this theory.

The curves are plotted on the next page. Place the magnet so the poles lie symmetrically along the line shown and tape it in place. It is best to orient the magnet North South, so that the earth's field affects the direction of the line of force least.

Support the magnetized paper clip using a piece of cotton thread, and hold it as far along the cotton as possible, as shown. A small piece of tape on the clip air-damps vibrations. Support the clip somewhere adjacent to the magnet, and gently lower it to the paper. Then draw a line along the edge of the clip just touching the paper. This is a field line, and should be parallel to the field loops drawn, as shown. Do this all around the magnet, holding the cotton closer to the clip as you approach the magnet, to prevent it being drawn in.

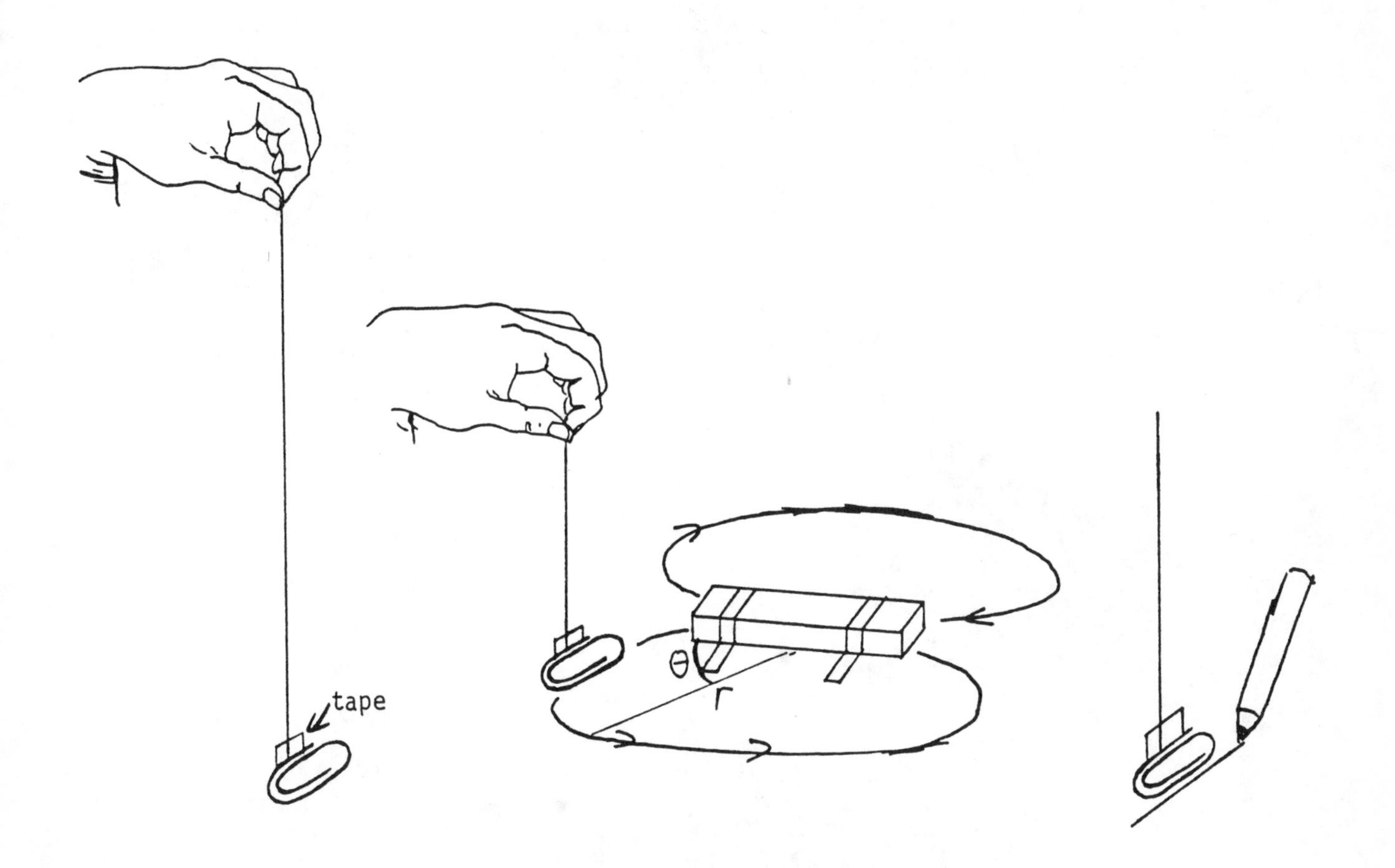

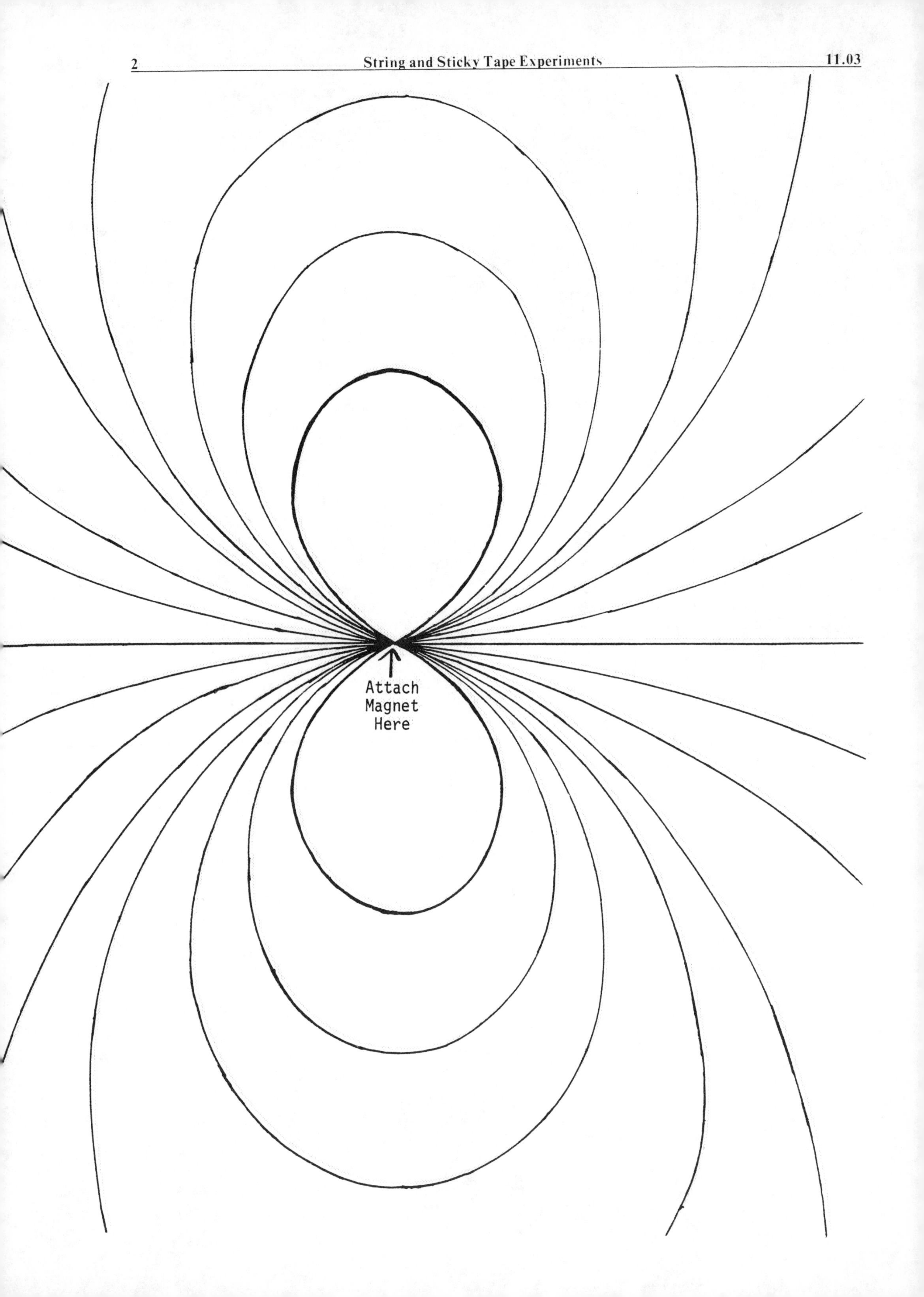
Attach
Magnet
Here

Experiment 11.4. Tangent Magnetometer, and Magnetic Strength of a Bar Magnet.

Materials: magnet, paper clip, cotton thread, paper protractor.

Procedure: Take the magnetized paper clip and hang it by a long thread from a table or chair to sit just above the floor.

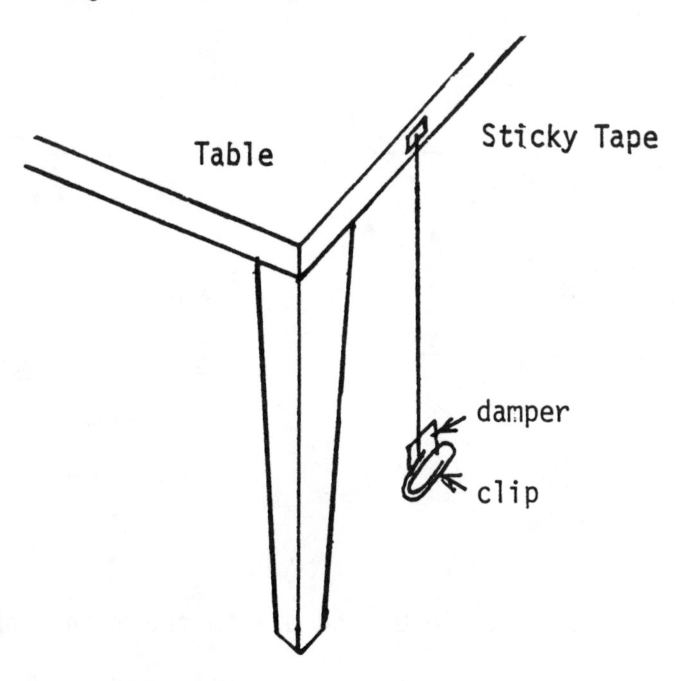

Make sure the thread is unwound. A piece of sticky tape attached to the clip will help damp out oscillations. Place the protractor beneath the clip, so that it is aligned at 0°.

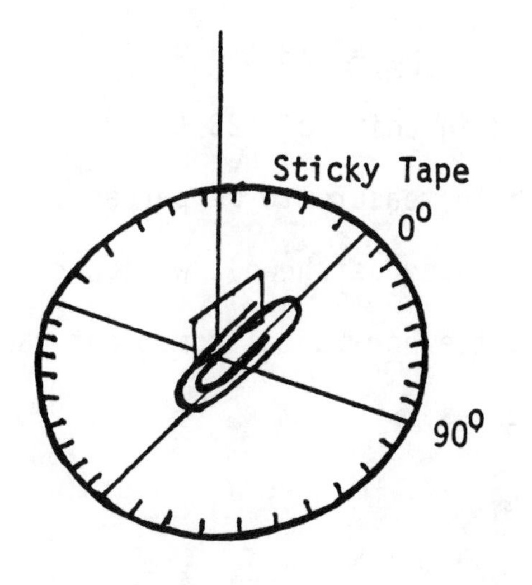

Bring up the magnet along the 90° line.

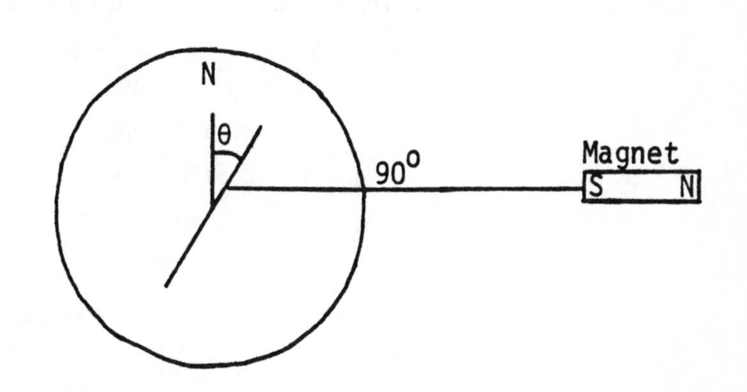

There are two forces acting on the clip. That produced by the earth's field H_e and the force due to the magnet, H_M. When the paper clip is sitting at an angle θ, as shown, the forces produce two couples, trying to twist the clip in opposite directions.

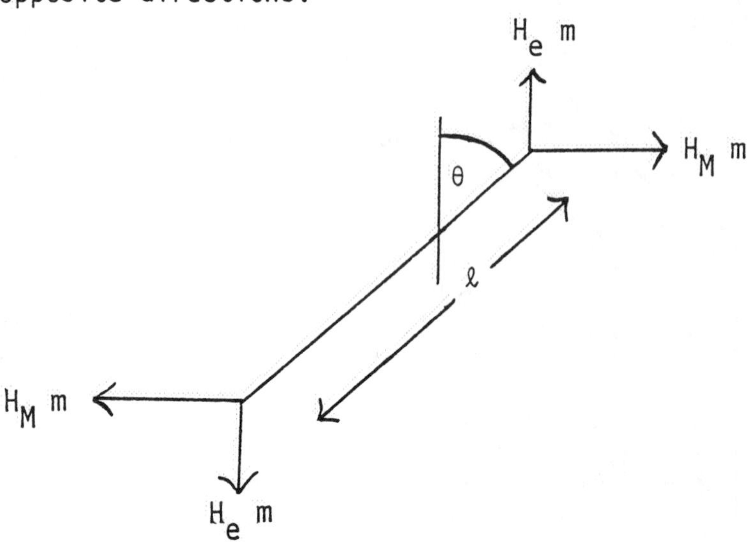

The couple due to the earth is $mH_e\ell \sin\theta$, and due to the magnet $mH_M\ell \cos\theta$.

Then

$$m H_e \ell \sin\theta = m H_M \ell \cos\theta$$

and

$$\tan\theta = \frac{H_M}{H_e}$$

Now, H_e is .23 gausses in South Carolina, and hence $H_M = .23 \tan\theta$. The outside of the protractor has been graduated in units of .23 tan θ, so, if we arrange the magnetic field we wish to measure to be perpendicular to the earth's field (as it is with the magnet placed as shown), we can read its value directly off the circle. In Florida, the horizontal component can be up to .29, and in Maine it is .14 to .16 gausses.

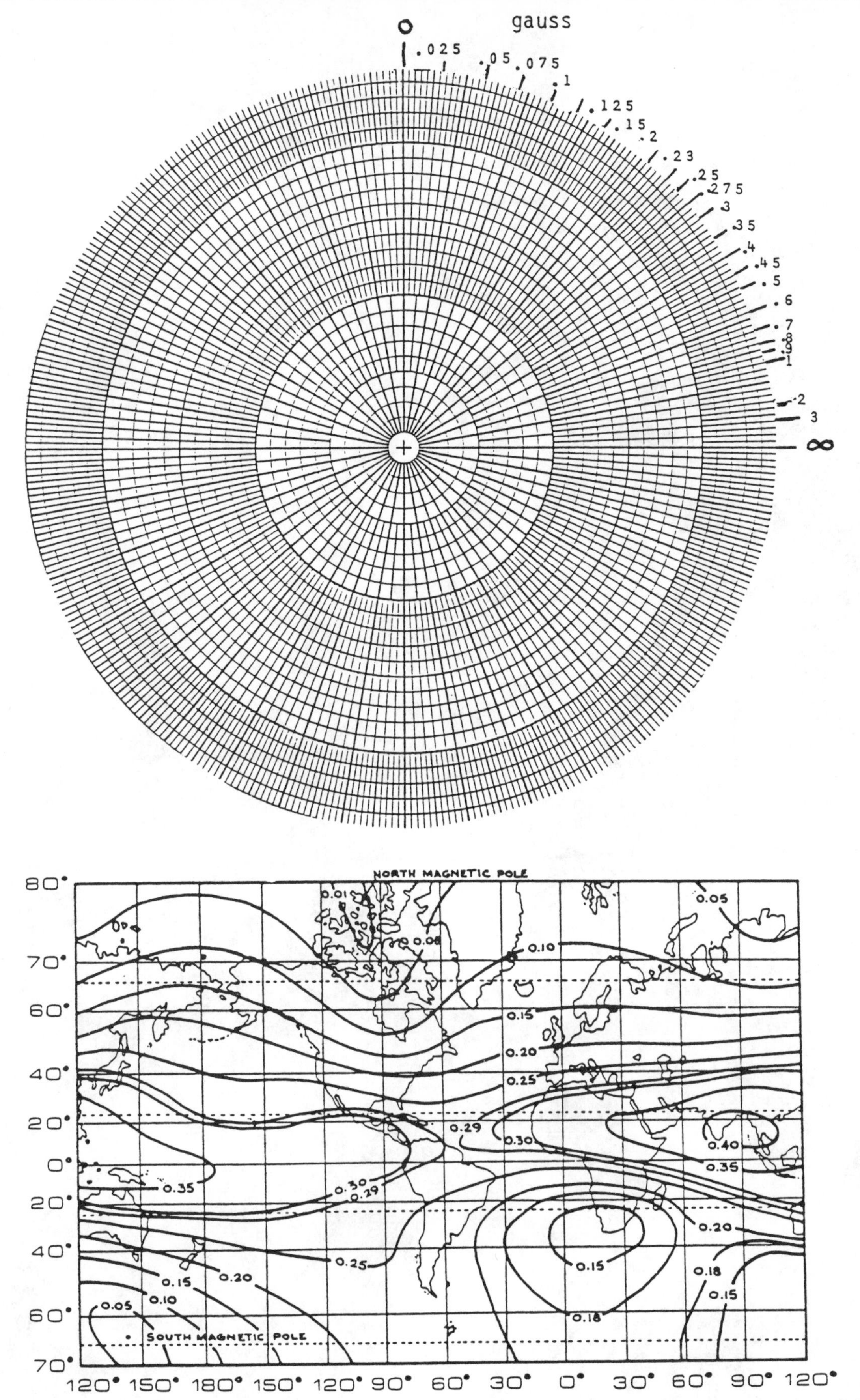

Map showing lines of equal geomagnetic horizontal intensity H (in 10^{-10} tesla or 10^{-6} gauss) for 1975 (U.S. Naval Oceanographic Office)

EXPERIMENT 11.5. The force law near a bar magnet.

Materials: Pencil, Magnet, paper clip and cotton thread.

Procedure: Place a centimeter ruler on the floor starting from the paper.

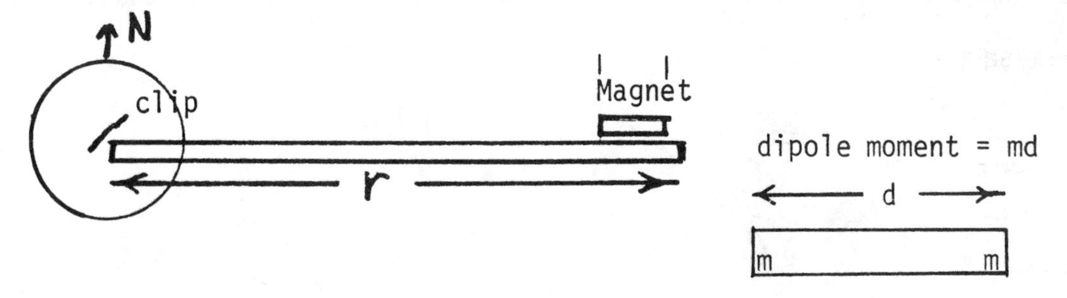

Now, measure the field strength as a function of distance from the magnet.

Distance from the magnet r(cm)	Field strength (gausses)	r^3	Magnetic dipole moment = Field Strength $x r^3$ (gauss cm^3)	$\frac{1}{r^3}$
5		125		.008
6		216		.00463
7		343		.0029
8		512		.00195
9		729		.00137
10		1000		.001
15		3375		.00029
20		8000		.000125
25		15625		.000064
30		27000		.0000370

Multiply the field strength by r^3, and record the result. Is it constant? If so, it shows the field strength is inversely proportional to the cube of the distance from the magnet. Alternately, you could plot the field strength against $1/r^3$ which would be a straight line through the origin, if the field strength $\propto 1/r^3$.

The field drops off from a bar magnet more quickly than from a point pole or charge ($1/r^2$).

The dipole moment of the magnet is the field strength xr^3 in units of Gauss cm^3.

(1 Weber/m^2 = 10^4 Gauss)

EXPERIMENT 11.6: Strength of bar magnets.

Materials: Magnet, paper clips.

Procedure: The pole tip strength can be estimated by hanging paper clips from the magnet as shown.

Hang the paper clips from the first one, then the opposite pole.

Can you hang the same number of clips from either pole? If so, this shows that the pole tips of a magnet must have equal, but opposite strength.

This is not an accurate experiment, because the paper clips have a certain amount of inherent magnetisation, but it does give a rough answer.

EXPERIMENT 11.7. The concentration of field in a magnet.

Materials: Magnet, paper clips.

Procedure: Try to hang paper clips at different points along the magnet.

whether it is a horse shoe or a bar. Where do the clips hang tightest? Where do they fall off? The field is concentrated at the ends of the magnet.

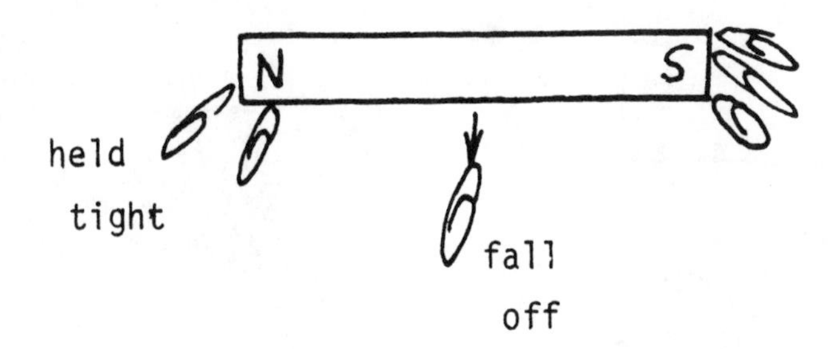

If you could break the magnet, you would find the field still concentrated at the two ends.

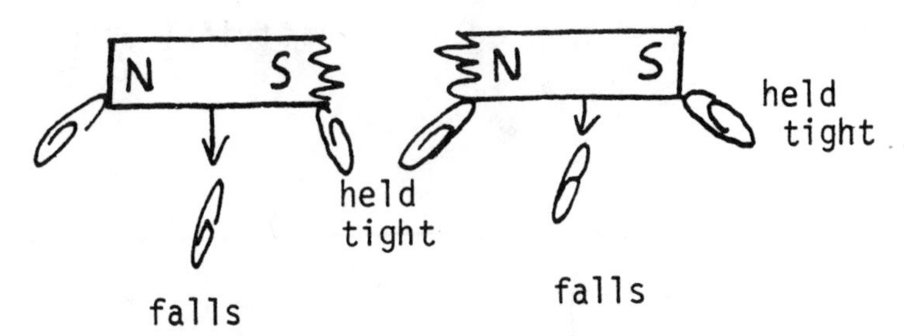

You may check this roughly by magnetizing two paper clips by stroking, and connecting them end to end, then try hanging a third, unmagnetized paper clip from them. It will only hang at the ends, but when the two are separated, it will hang from all four ends.

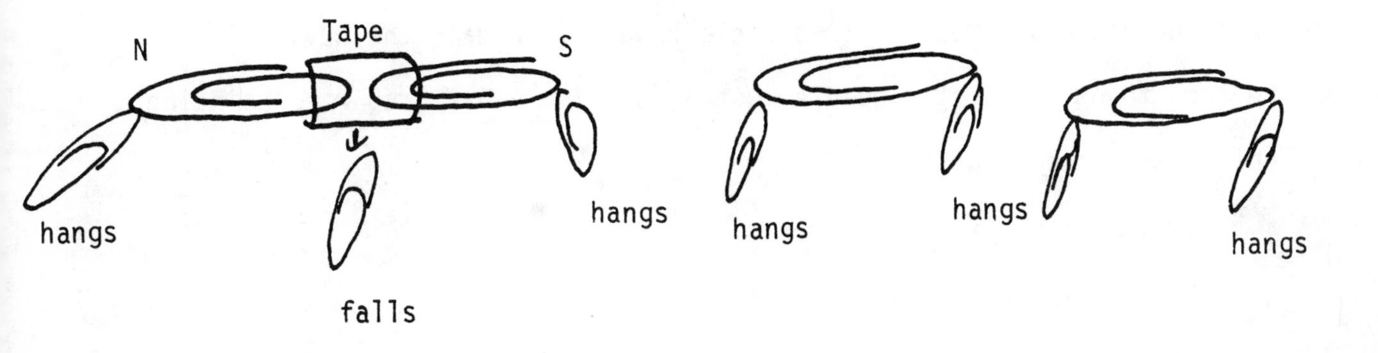

EXPERIMENT 11.8. The dip circle.

Materials: Paper clip, straw, piece of card, tape.

Procedure: Cut and fold the card (thick paper will do) as shown:

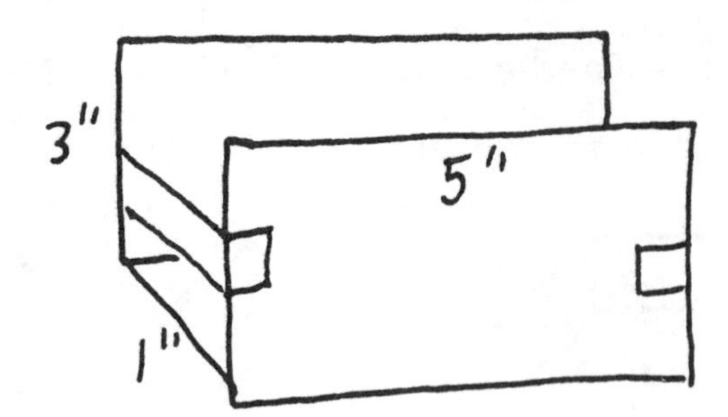

Unfold a paper clip, and magnetize by stroking.

Push the clip through a piece of soda straw two inches long.

Adjust the clip so that it appears as balanced and symmetrical as possible.

Place it on the stand.

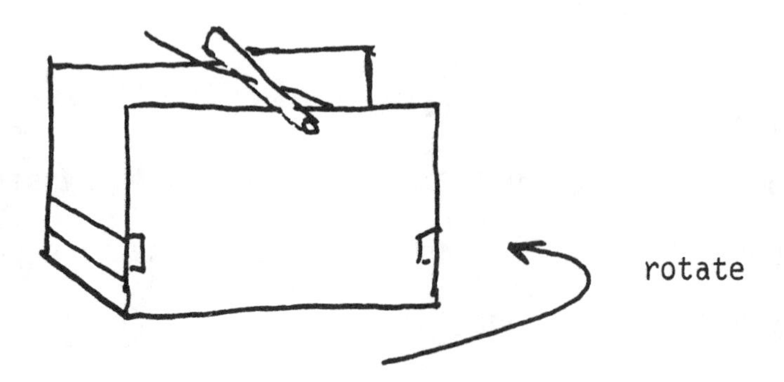

Rotate the stand on a table. The maximum angle of dip of the arrangement occurs when the wire points N.S. Measure this angle roughly. Rotate the device 180° and do the same thing. The clip needle must be carefully balanced when pointing E.W. to obtain accurate results.

How does the angle of dip correspond to the latitude?

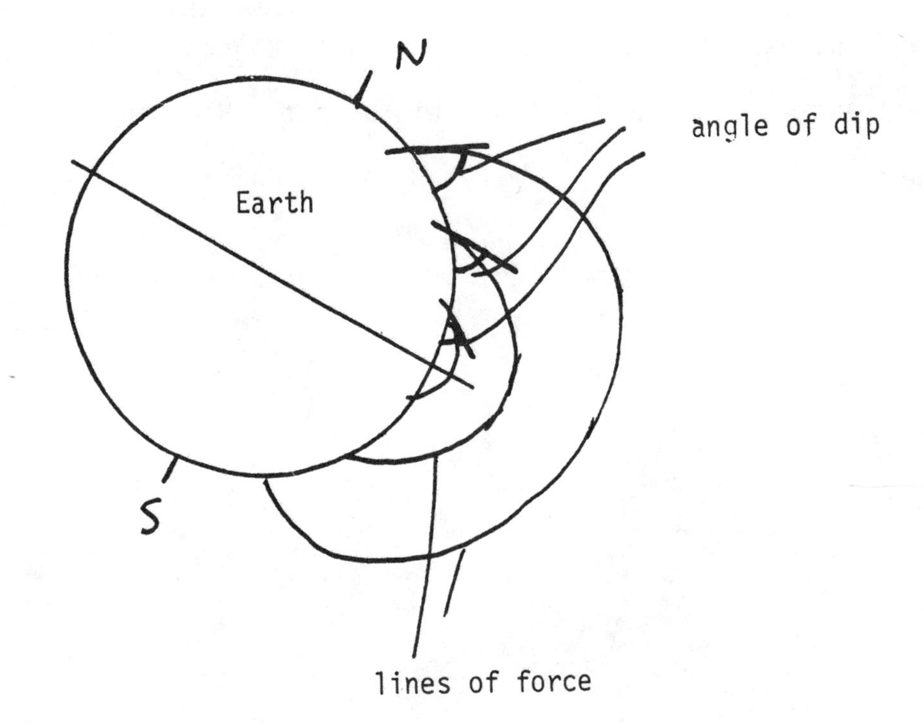

EXPERIMENT 11.9. What is magnetic?

Materials: magnets

Procedure: Find out what materials are magnetic by experimentally seeing which are attracted by the magnet. Try various coins, metal objects such as desks and shelves, knives, wooden and plastic objects. Make a list of the materials and whether they are magnetic or not. How many magnetic materials did you find?

Experiment 11.10. Penetration of magnetism

Materials: magnet, paper clip, various materials.

Procedure: Place sheets of different materials (paper, wood, iron or steel sheet from a desk drawer on stool) between the magnet and the paper clip. Which materials reduce the magnet's power, and therefore the ability of the magnetic field to penetrate? Make allowance for the thickness of the sheet of material.

You will find magnetism penetrates most materials except for those capable of magnetization themselves, such as steel sheeting. This is because they become magnetised in such a direction as to oppose the field.

Experiment 11.11 - The Mysterious Magnet

Take a small magnet, such as are used to hold papers to metal walls or the refrigerator, (1" x $\frac{3}{4}$" x $\frac{3}{16}$" is suitable) many of these have a $\frac{3}{16}$" hole, through the center of the flat face, which is good. Now, place a small ball bearing (about $\frac{1}{4}$" diameter) at the center (resting in the hole, if there is one). (If no ball is available, simply fold a paper clip as shown in the figure). Bring up an unmagnetized paper clip, as shown, touch it to the top of the ball and raise it. Will the ball bearing stick to the magnet or the clip? Most people would think it would stick to the magnet, but in fact, the bearing rises nicely with the paper clip. How does this mysterious event occur? We tend to think that it is the power of the field of the magnet which will cause an object to be attracted - but, for an induced magnet as with the ball bearing, it is the gradient of the field. To see this, imagine the bearing placed in a uniform field, no matter how strong - the induced poles are equal and opposite at each end, so there is no net force. Back to our little ball. The magnets holding papers to walls are magnetized with the poles on the flat faces - so the lines of force are emitted more at less uniformly from this surface as shown. These converge on the ball bearing, and even more on the paper clip, as shown - so the induced poles on the top of the ball bearing and paper clip are very large - causing the ball bearing to rise. If, now, you put the ball or folded paper clip against the edge of the magnet, as shown in the last figure, it will be impossible to remove with a paper clip, as previously, because the lines of force now go from one pole to the other through the ball or folded paper clip, without going through the other clip at all.

Arthur Schmidt , Physics Teacher, 23, 375 (1985).

John Ecklin, Physics Teacher, 25, 73 (1987).

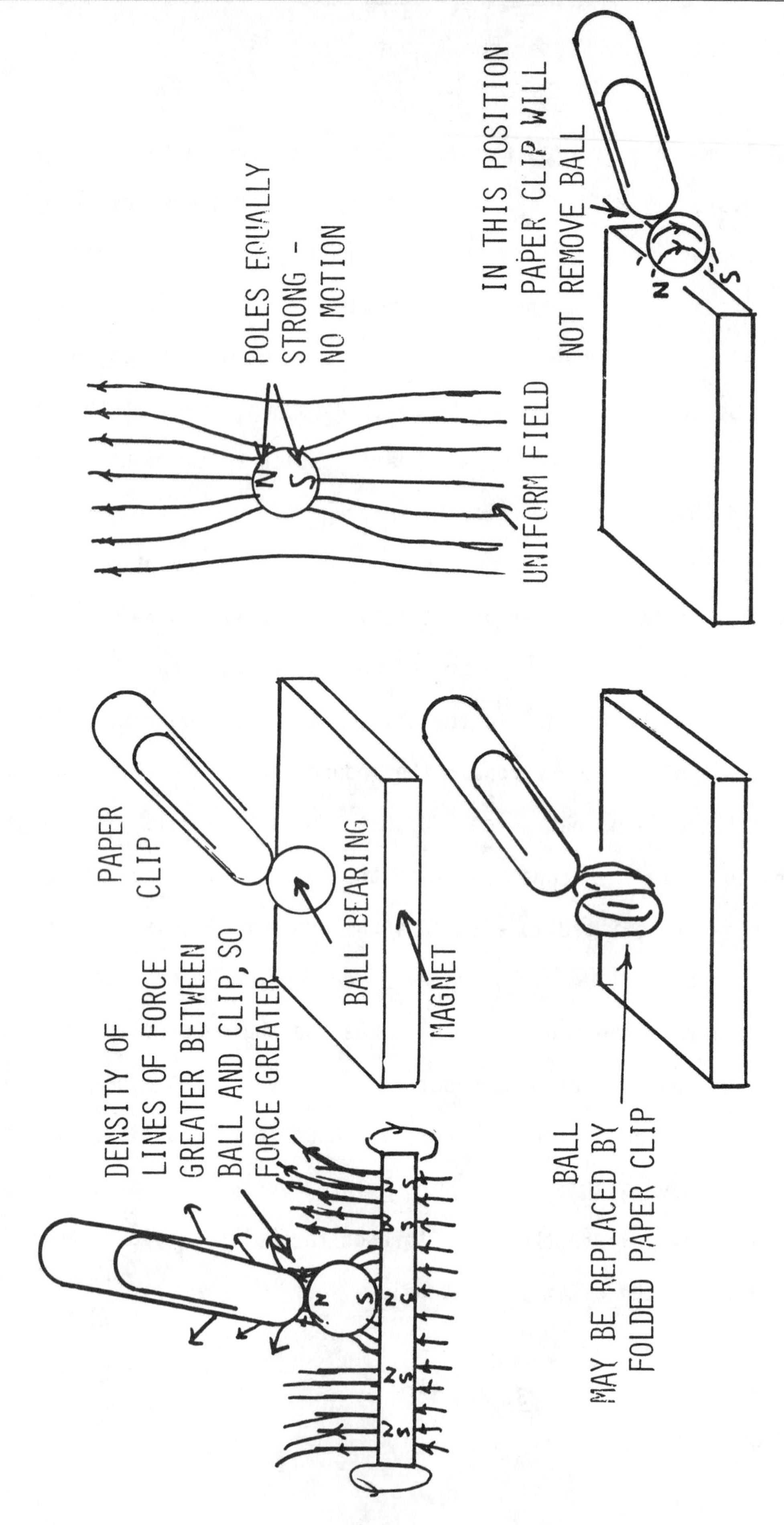
DENSITY OF LINES OF FORCE GREATER BETWEEN BALL AND CLIP, SO FORCE GREATER
PAPER CLIP
BALL BEARING
MAGNET
BALL MAY BE REPLACED BY FOLDED PAPER CLIP
POLES EQUALLY STRONG - NO MOTION
UNIFORM FIELD
IN THIS POSITION PAPER CLIP WILL NOT REMOVE BALL

INTRODUCTION TO SECTION 12

Experiments in Electricity

It is impossible to employ the same simple equipment for experiments in electricity as it is for mechanics, heat and light, for example.

The primary requirement is a source of D.C. Batteries are most commonly used, but are a poor idea, since, from one semester to the next, new batteries must be found, the old ones having died on the shelf. It follows, a power supply running off the 120V A.C. is much preferable.

The most suitable source we have come across is a model electric train supply. The HO gauge power supply provides D.C., variable from 0-12V, costing about $10.00. It has an overload button, and is practically indestructable. It can be used indefinitely, and cannot be drained flat. (Anything made to be used by ten year old kids must be virtually indestructible)

The second requirement is a source of insulated copper wire. Again, this can be purchased, but the heavy-current windings of a transformer, the motor of an old toy, any of these can be employed. Note:- #26 insulated wire is good .

The fundamental D.C. experiments concern the relationship between current and magnetism, and between potential and current, resistance, and inductance.

Experiment 12-1. The Magnetic Field Around a Coil of Wire

Materials: Dixie cups, copper wire (#26 insulated works well), paper clips, magnet, a fine thread or a hair, HO model train power supply.

Procedure: Place two cups, one inside the other, and tape them together over the lip, as shown:

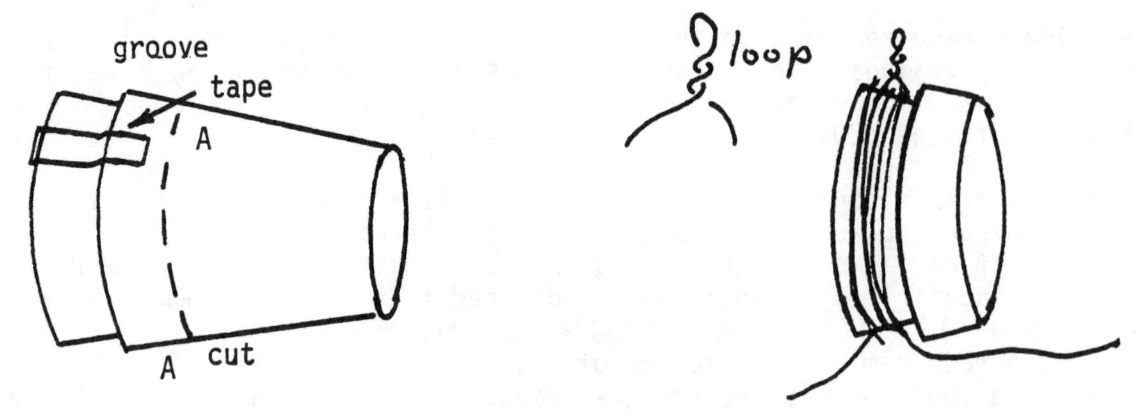

Now wind twenty turns of wire around the inner cup, in the groove, form a loop,as shown and continue to wind eighty more turns, fastening the coil in place with tape. Clean the ends of the wire and the loop by scraping with the scissors blade. Cut the cups along AA to form a coil of wire, and attach to the D.C. terminals of the power supply. Tape the coil, as shown, to a piece of card on the table. Place a sheet of paper on a book to come half

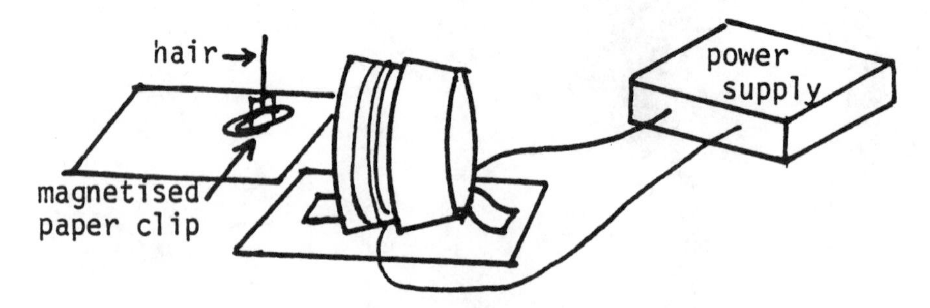

way up the coil. Now, map the field, as was done for the bar magnet (experiment 11.03) using the magnetized paper clip. The lines of force of the field are shown overpage. Do they agree with your results?

1. Increase and decrease the current. Are the lines of force dependent on the current? They should not be. The strength of the field depends on the current, but the direction of the lines of force,solely on the geometry.

2. Pass the current through the twenty turns, as well as through the 100 turns. Does the field change? Again, the number of turns should affect the magnitude, not the direction of the field.

3. Do the lines of force form complete loops? Trace one all the way round the coil to ensure that they do, unlike the bar magnet, where they run from the north pole to the south pole, outside the magnet.

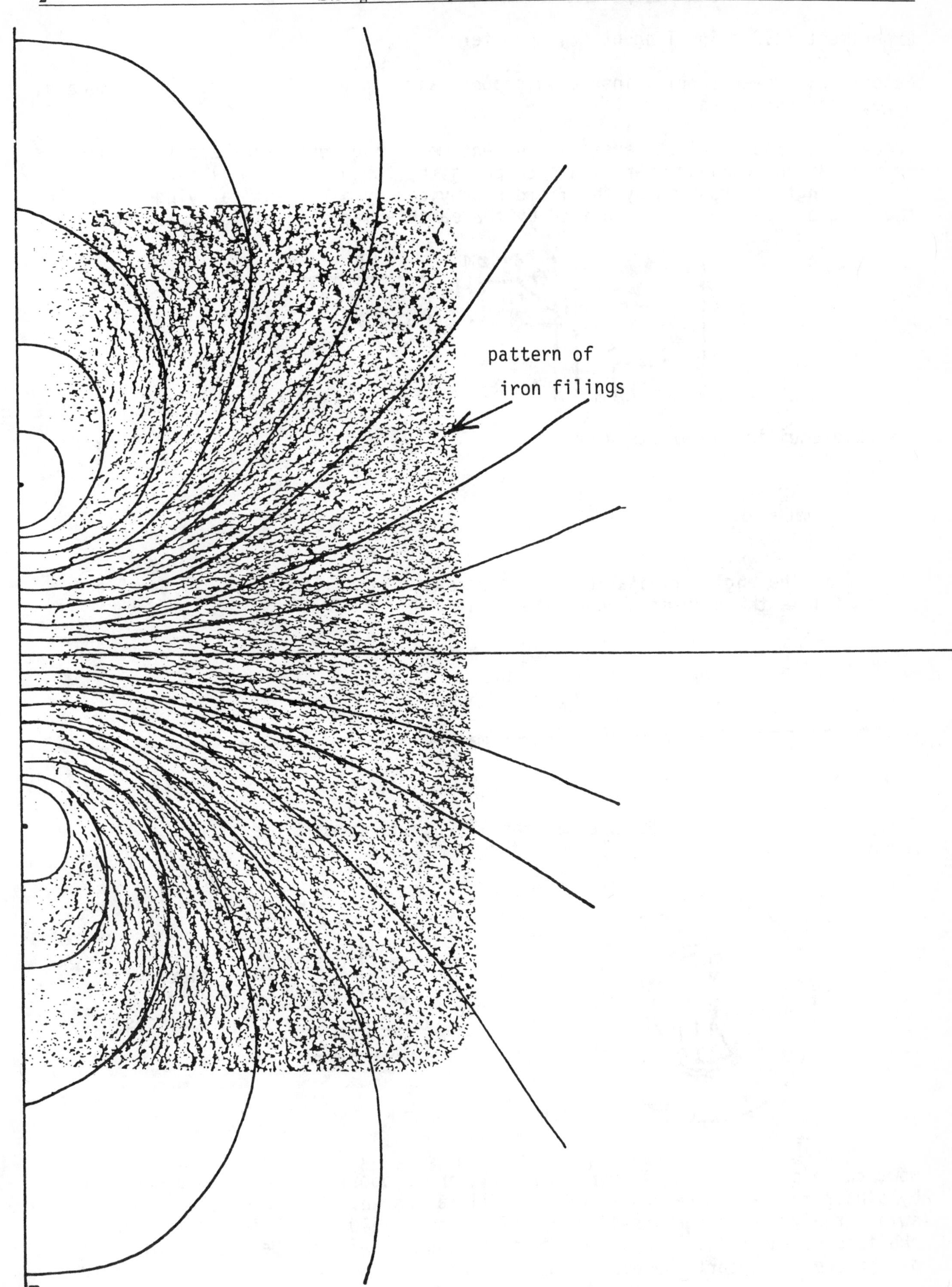
pattern of
iron filings

Experiment 12.2 The Tangent Galvanometer

Materials: Power supply, insulated copper, wire, paper clips, string, cardboard, tape, one long human hair.

Procedure: The simplest sensitive current measuring device one can construct is the tangent galvanometer. This compares the torque on a small magnet of pole strength m produced by the field H provided by a current carrying coil, with the torque provided by the earth's magnetic field H_e as shown

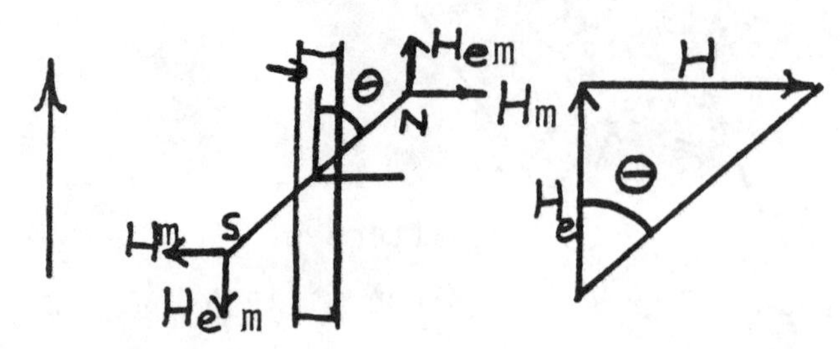

A stable equilibrium exists when

$$mH = mH_e \tan \theta$$

where θ is the angle of rotation of the magnet. It follows that $H \propto \tan \theta$ and since if i is the current through the coil

$$i \propto H \propto \tan \theta$$

For small angles (less than 10°) approximately

$$i \propto \theta$$

Take the coil of wire used in experiment 12-1, and tape it over the dial shown

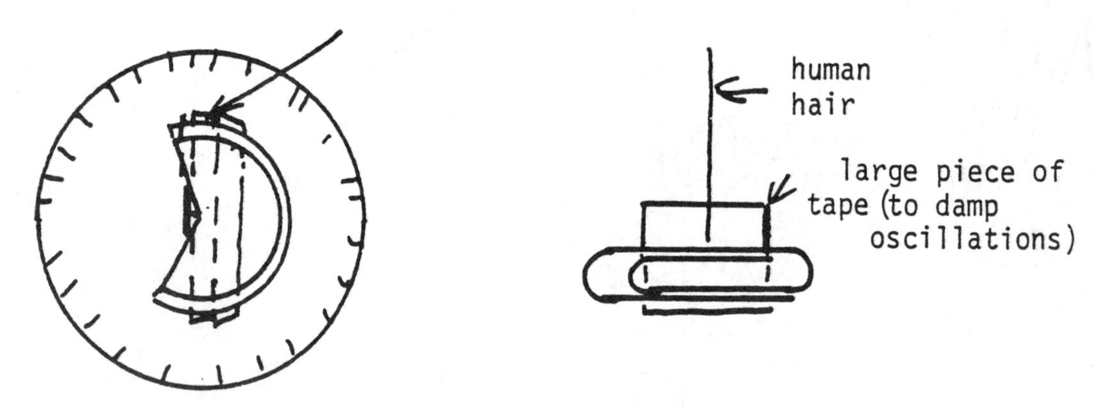

Now, cut a paper cup to fit over the coil, and support a magnetized paper clip by a hair to lie at the center of the coil, as shown. The paper clip should swing freely, and align itself with the earth's field, which is parallel to the face of the coil. The tripod from experiment 1-34 may be used in place of the cup to support the paper clip.

Pass a current through the coil from the power supply and notice the deflection of the paper clip. You can read the value of the current in this way.

You may vary the voltage supplied to the meter by using the control on the power supply. However, it may not be possible to reduce the current sufficiently to be read, even when only twenty turns of the coil are used. A resistance, as described in 12.3, must then be placed in the circuit.

Notice that, since we are comparing the current with the earth's field, which is sensibly constant, we build a calibrated meter by simply using a fixed geometry.

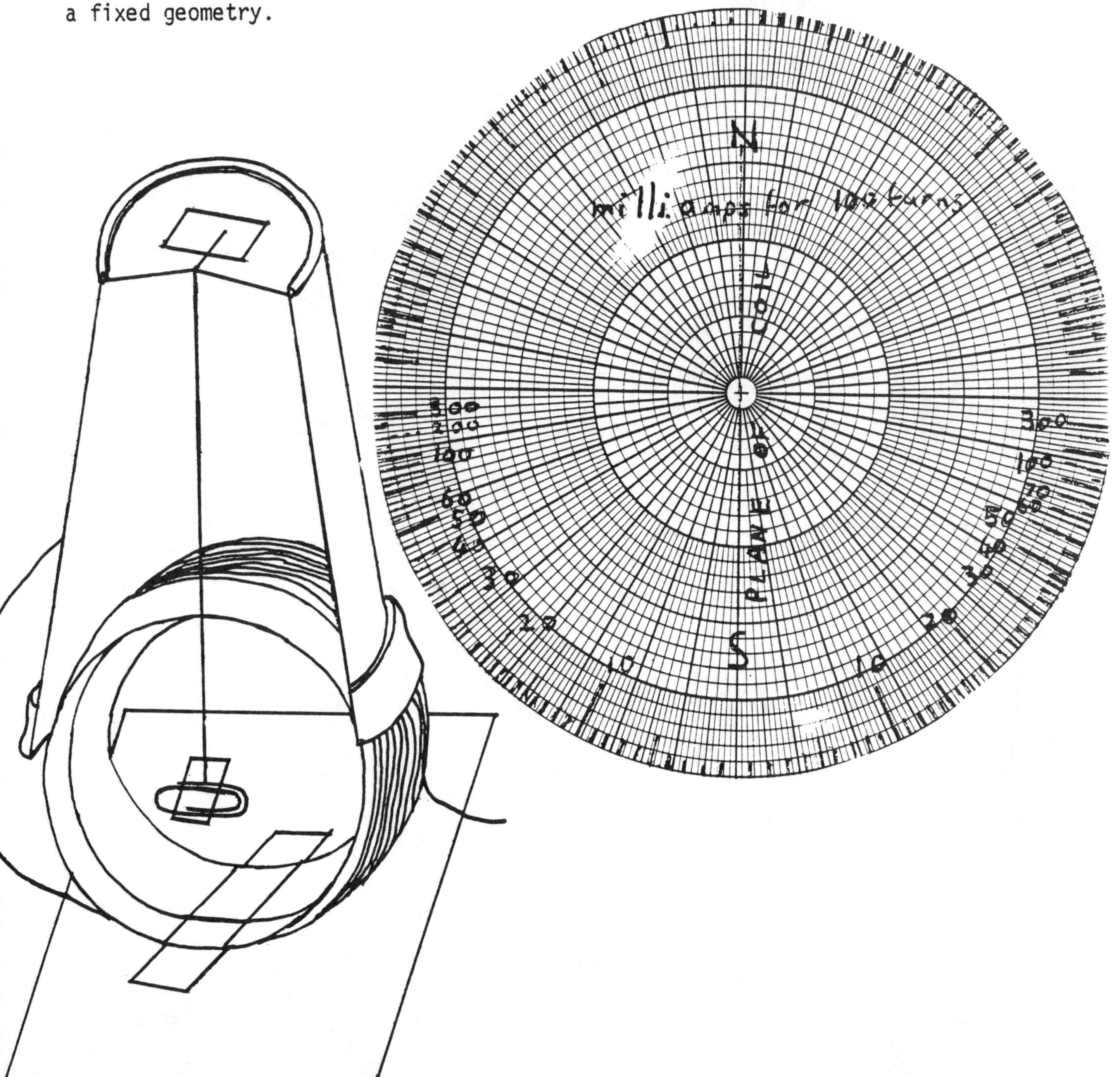

Experiment 12.3 Electrical Resistance, (series and parallel)

Materials: straws, paper clips, galvanometer, power supply, pencils.

Procedure: 1) Resistance of a conducting fluid: straighten a paper clip and push in to the center of a straw, using another clip.

Bend the straw, and fill with water.

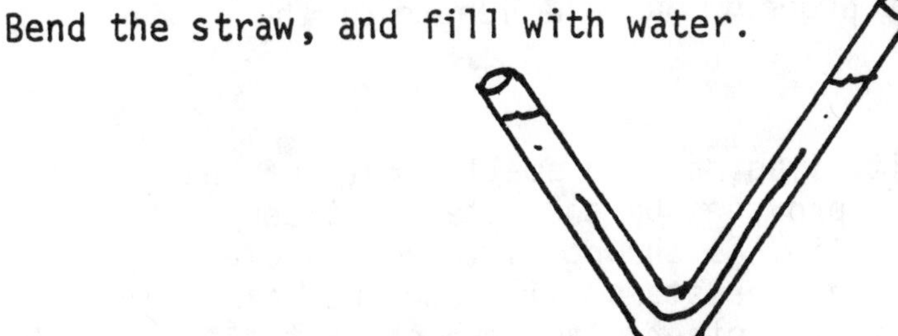

Straighten the large end of two paper clips: Put the two clips on the opposite sides of one end of the straw:

Connect the two clips between the power supply and the galvanometer.

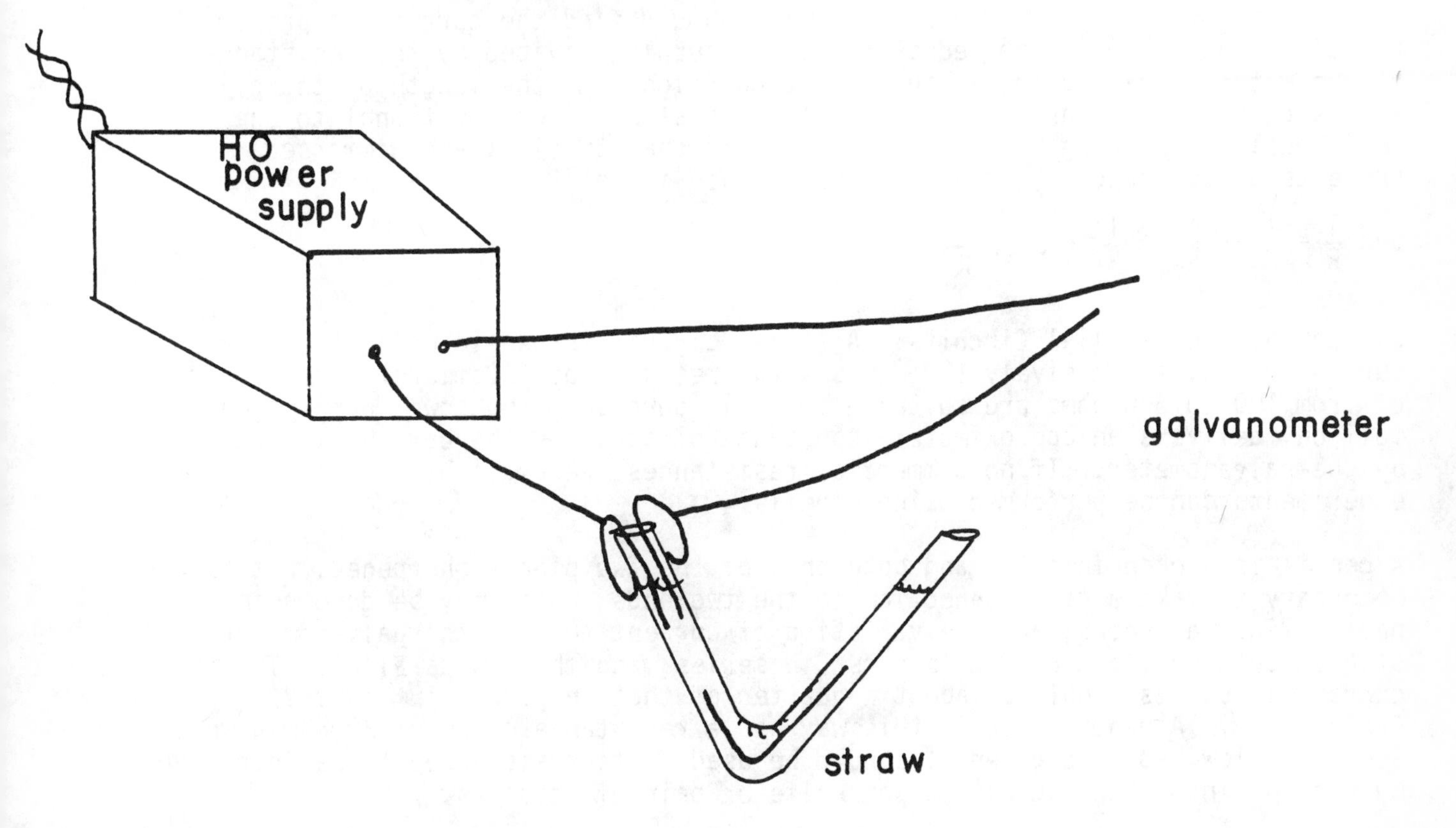

The resistance of the water now determines the current through the galvanometer. Measure the current, versus the length of the paper clip immersed in the water,

The resistance of the water is inversely proportional the length of the clip immersed--this is a parallel circuit.

To see why this is a parallel circuit, imagine each small length of one clip which is immersed has aresistance - provided by the water, between it and the corresponding part of the other clip, as shown. As the clips dip deeper into the water, more and more of the resistors will come into play. Since each runs directly from one clip to the other, they are connected in parallel.

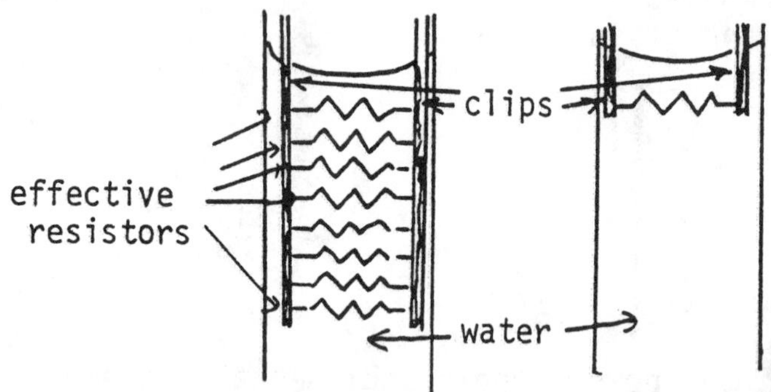

You should find the current is proportional to the voltage supplied by the power supply, divided by the resistance of the water. This ratio, in turn, is proportional to the length of the clip immersed, i.e. in a parallel circuit, the resistance is proportional to the reciprocal of the length immersed. Since, if the nth millimeter immersed presents a resistance R_n, the total resistance is not $R = R_1 + R_2 + R_3 + \ldots . R_n$

but $\frac{1}{R} = \frac{1}{R_1} + \frac{1}{R_2} + \frac{1}{R_3} + \ldots . \frac{1}{R_n}$

2) Series and Parallel Circuits: A series circuit is one in which the same current passes successively through several resistances. Commercial resistances of from 100 to 500 ohms are suitable for this purpose. The power supply, when full on, delivers an approximately constant voltage, and the current is measured by the galvanometer. If no commercial resistances are available, some interesting experiments can be performed using pencils.

A pencil is broken in half, and both ends of the two pieces sharpened. It is necessary to make a good connection to the two ends, which may be done using paper clips, as shown, and the variation of current through the galvanometer studied putting the two pencils first in series, and then in parallel. The current in series should be about a quarter of that in parallel. Is it? Explain why. A pencil used in this way has a resistance of about 20 ohms, which is rather low, so a long pencil should be used. Its resistance can be increased by cutting into the lead with a pen knife or pair of scissors.

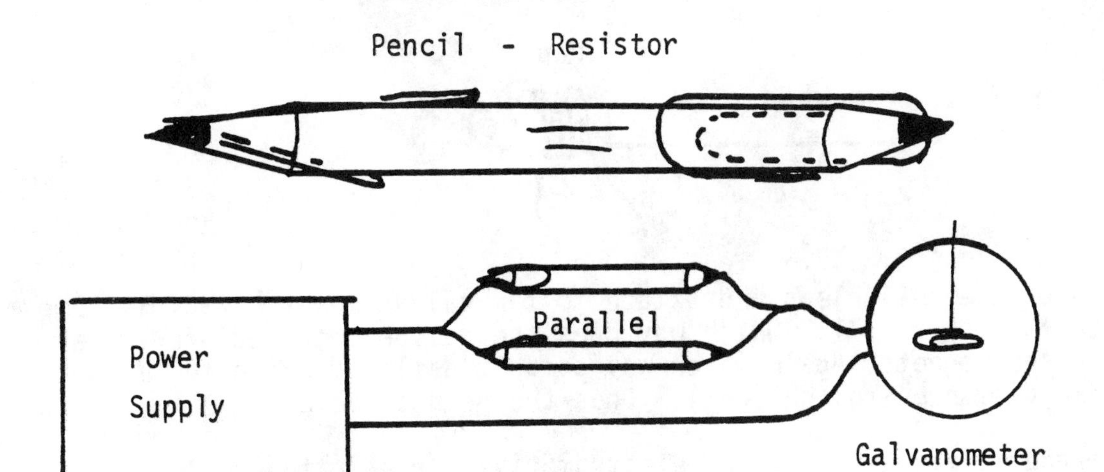
Pencil - Resistor
Parallel
Power
Supply
Galvanometer

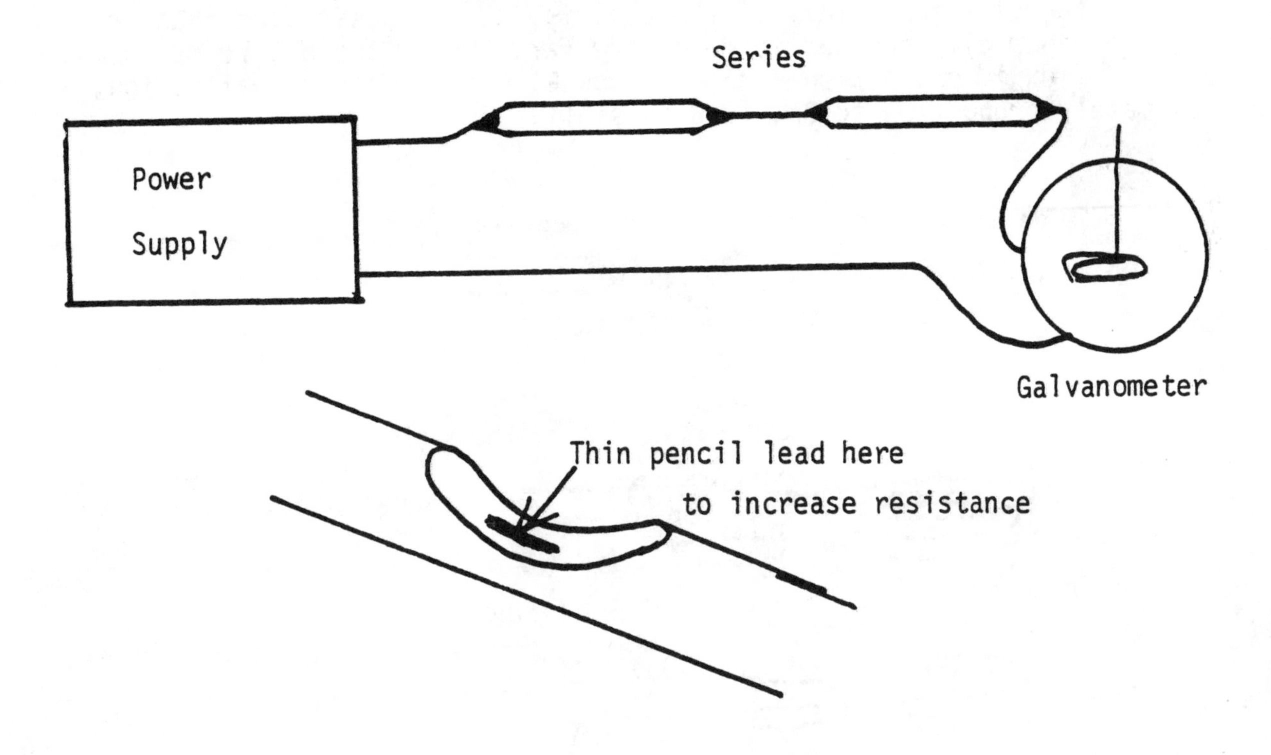
Series
Power
Supply
Galvanometer
Thin pencil lead here
to increase resistance

Experiment 12.4 Magnetic Induction

Materials: Copper wire, magnet, galvanometer

Procedure: Wind a coil of one hundred turns on a small paper tube, formed by rolling a sheet of paper around the magnet pole, and taping the ends.

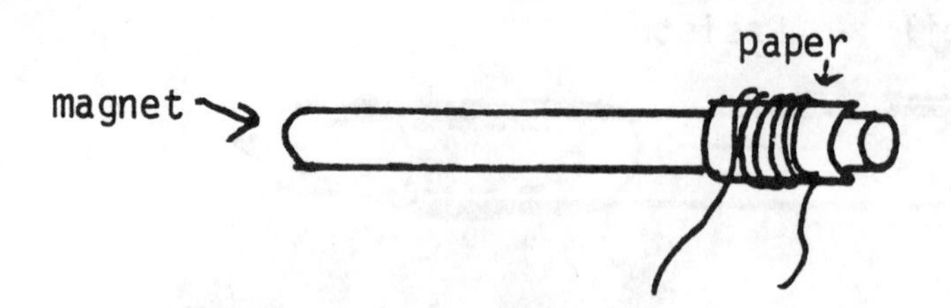

Scrape the ends of the coil clean and attach to the galvanometer. Rapidly remove the magnet from the paper roll. Which way does the galvanometer pointer move? Rapidly replace the magnet. Again which way does it deflect? Move the magnet to and fro, some distance from the coil. Does the pointer move?

The changing magnetic field induces an electromotive force in the coil which causes a current to flow through the galvanometer. Make sure the leads from the galvanometer are sufficiently long for the magnet not to affect it.

The deflection of the galvanometer is a measure of the change in magnetic flux passing through the little coil, and we can use this device as a flux meter. It only measures the flux accurately if the coil is moved in less time than it takes for the paper clip pointer to move very far. Under these circumstances, the change in flux is proportional to $(1 - \cos \theta)$ where θ is the deflection, or approximately proportion to θ^2, if θ is small.

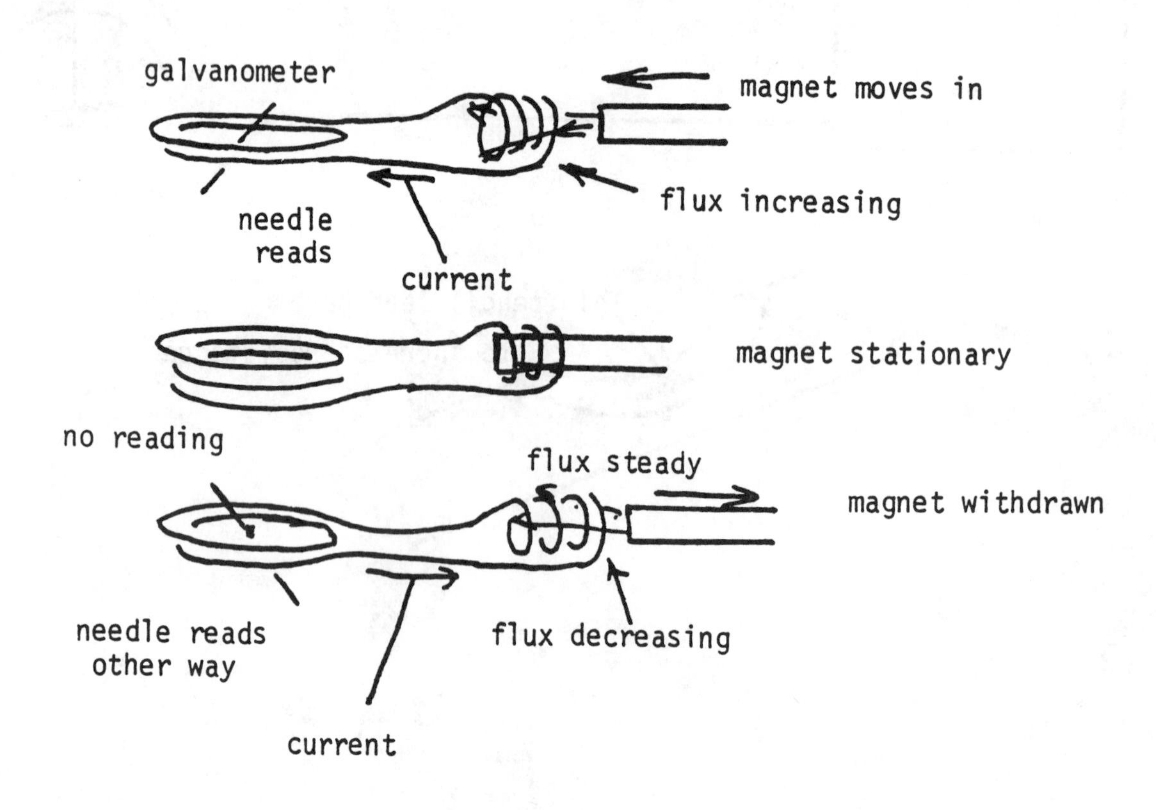

Experiment 12.5 Mutual Inductance

Materials: Insulated copper wire (#26 is good), cups, tape, galvanometer

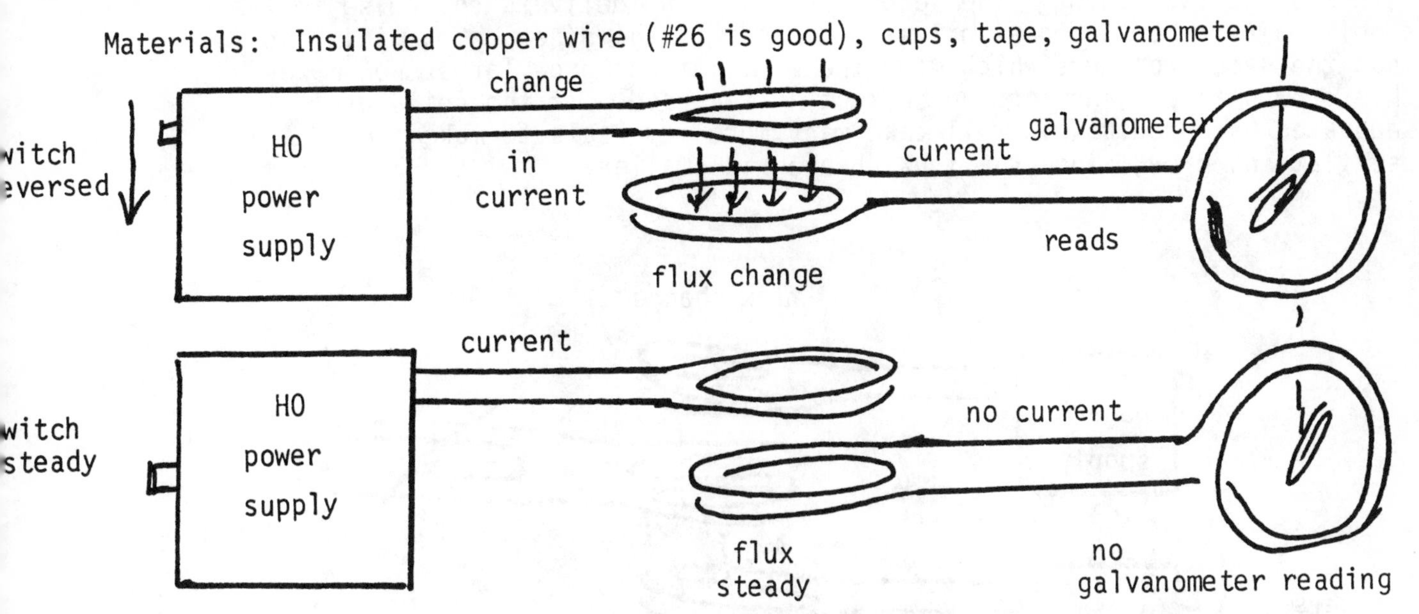

Procedure: Wind one hundred turns of wire on two cups, as described in experiment 12.1. Wind a second coil of twenty turns on a second pair of cups, make a loop, and wind eighty more turns. Scrape the ends of the two coils, and loop. Connect one to the DC on the power supply, and the second to the galvanometer. Place the two coils face to face, as close as possible. Now, turn the reversing switch on the power supply, and notice how the galvanometer responds, swinging to a certain angle, before returning to zero. The maximum deflection of the galvanometer corresponds to the pulse of charge flowing through it. The galvanometer responds to the charge rather than the current. Used in this way, it is called a ballistic galvanometer, and the deflection depends upon the charge.

On reversing the switch, the magnetic field established by the first coil is reversed, and since the same field passes through the second coil, the reversal of the field induces an E.M.F. in the second coil, which in turn induces a current in the galvanometer coil. Reverse the power supply switch the other way, and notice the deflection of the galvanometer is reversed. Now, separate the two coils, by about one radius and notice how the deflection is reduced on reversing the power supply, because the field looping the second coil is reduced.

Connect the power supply to the twenty turns at the loop instead of the hundred. On reversing the field, does the galvanometer deflection increase or decrease?

If the deflection of the galvanometer is too small to read with one reversal of the DC power supply, you can deliver a series of reversals to build up large oscillations of the galvanometer needle which can easily be seen. The principle is the same as delivering judicious pushes at the right time to a child's swing to build up large oscillations.

The needle should be set oscillating very slightly, then the switch reversed in the middle of each swing. The oscillations will either get larger, if the pulses are delivered in the right phase, or they will get smaller, if you are reversing one way when you should be reversing the other. Each reversal reverses the flow through the secondary coil, and delivers one pulse to the needle. Try reversing one coil with respect to the other. You will find that the same procedure which made the oscillations grow larger now makes them grow smaller, showing the current is reversed. If the coils are separated by a radius, it will take many more reversals to make the oscillations grow, since the flux change is smaller.

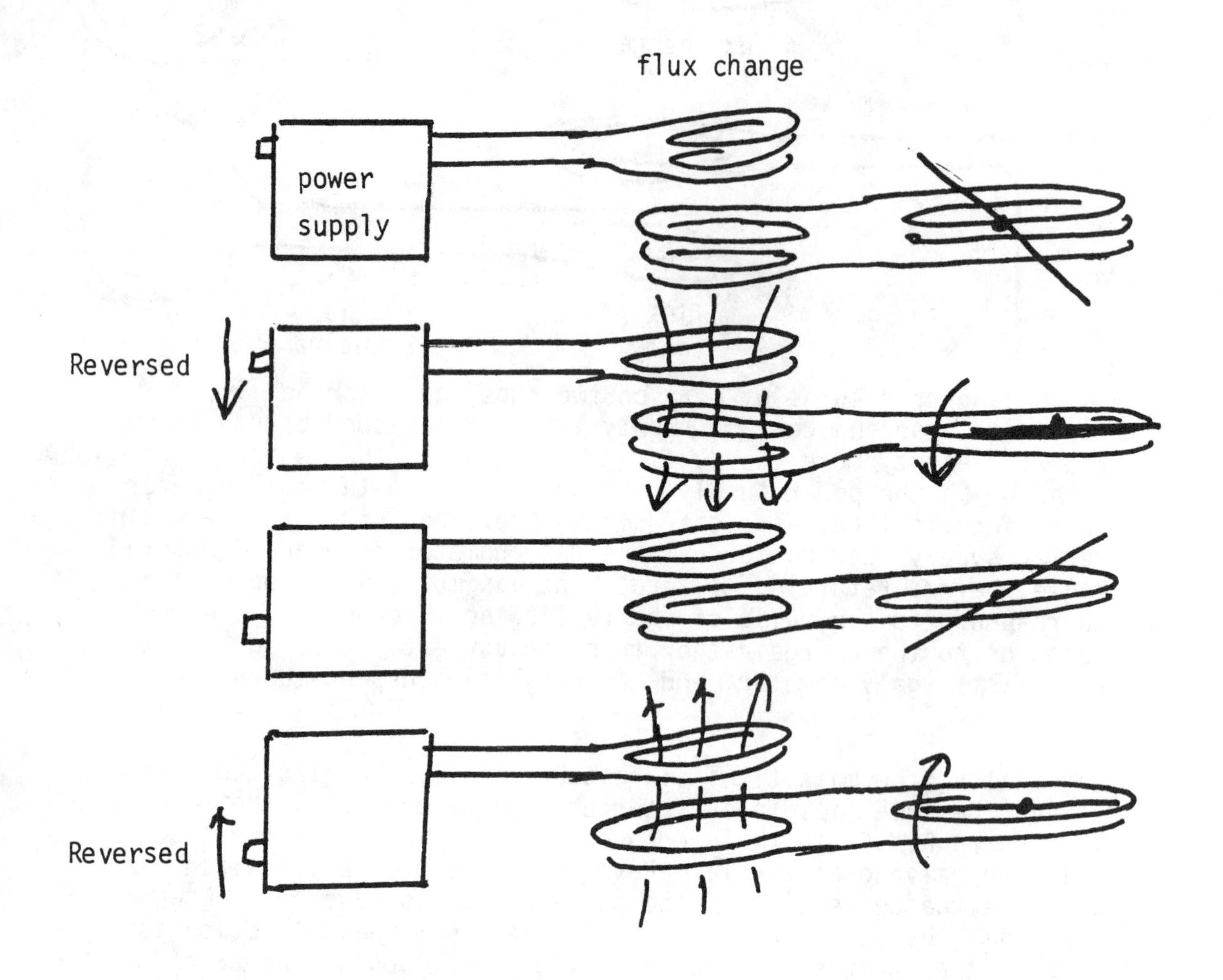

EXPERIMENT 12.6 Electrolysis

Materials: power supply, insulated copper wire, straws, paper clips, water, aluminum foil

Procedure: When an electric current is passed through water, hydrogen is liberated at the negative terminal and oxygen at the positive.

$$4H_2O \rightarrow 4(H^+ + OH^-) \rightarrow \begin{cases} 4H^+ + 4 \text{ electrons} \rightarrow 2H_2 \\ 4(OH)^- - 4 \text{ electrons} \rightarrow O_2 + 2H_2O \end{cases}$$

Hence there should be twice the volume of hydrogen liberated as oxygen. Most metal cathodes are oxidised, and hence no oxygen is released, only rare metals such as platin um not being attacked. Aluminum, however, forms a thin, insoluble oxide coating, and, once formed, it protects the metal beneath.

Cut two pieces of aluminum foil about one inch by three inches. Fasten one to the inside of a cup filled with water using a paper clip. Attach this to the power supply using copper wire with the ends scraped clean. Fasten the other foil, using a paper clip and wire, to the other pole of the power supply, and turn it on. Submerge the foil in the water, as close to the other foil as possible without touching. Note bubbles form rapidly on one foil, and more slowly on the other. They form rapidly on the cathode, where hydrogen is released, and the oxygen bubbles form more slowly on the anode, since only half as much oxygen as hydogen is liberated.

Better still, you can use quarters in place of the aluminum foil as electrodes since the silver is not attacked by the water.

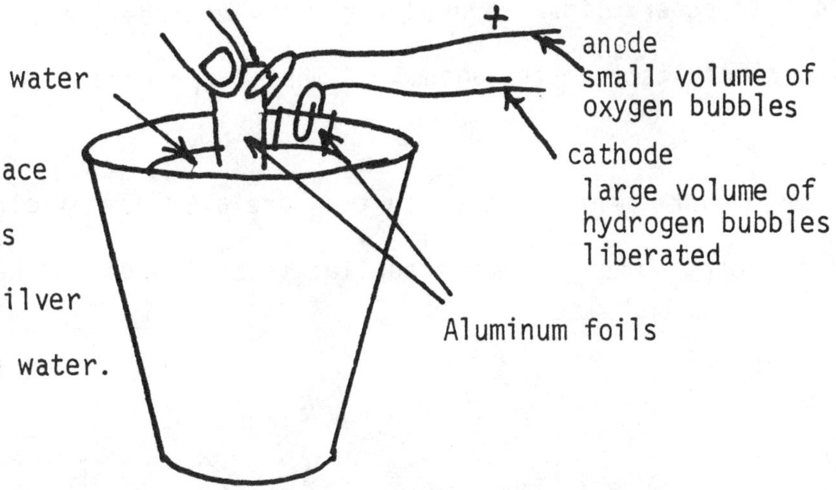

water
+
−
anode
small volume of oxygen bubbles
cathode
large volume of hydrogen bubbles liberated
Aluminum foils

Cut the bottom end of a straw, as shown

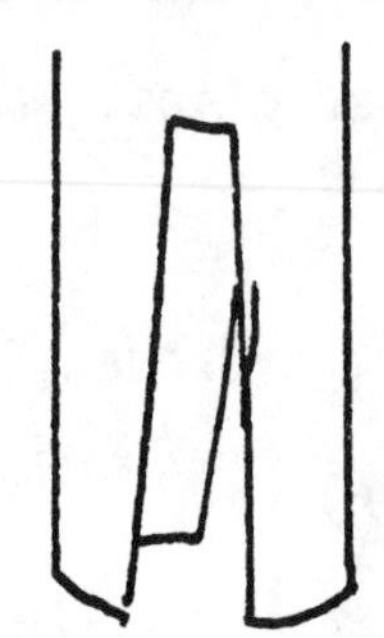

Fasten foils over the piece cut-out, using paper clips. Submerge in the cup of water, suck up water to fill the straw, fold over the end and tape shut, as shown. Connect to the power supply. The oxygen and hydrogen liberated can be collected together.

Tape

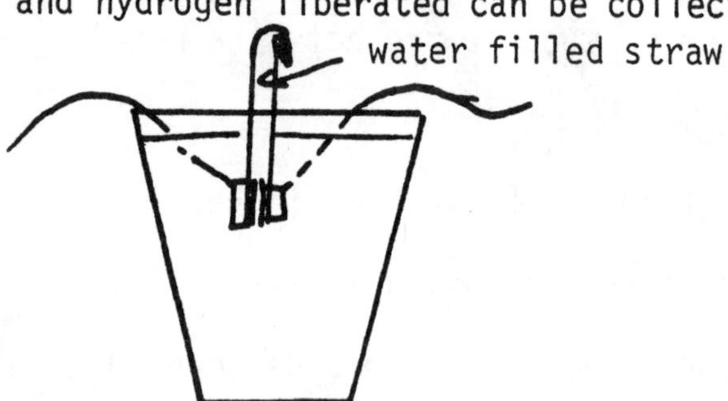

Light a match, and hold over the end of the straw as it is opened. A flame, or slight pop should result from the explosive mixture of oxygen and hydrogen.

Replace the aluminum electrodes by placing bare copper wire in the water. Note hydrogen is given off, but no oxygen-----instead copper oxide is formed. If paper clips are used for the electrodes, a brown discoloration of the water results from the anode, of what is probably iron oxide or hydroxide.

A simple experiment is to place two bare aluminum electrodes close together in soapy water. A match applied to the bubbles makes a nice pop.

EXPERIMENT 12.07. Forces between parallel conductors

Materials: This experiment needs no wire! --Only aluminum foil and a battery or other current source.

Method: Now for a simple experiment to show that currents in the same direction attract, and in the opposite direction, repel. Cut a strip about 1 cm wide and 70 cm long from a roll of thin household aluminum foil. Fold it in the middle, and tape the two ends to the edge of a table, as shown in Fig. 1, then run the fingers down the two adjacent parts of the strip to bring them as close together as possible without touching. Cut two more pieces of foil to connect the ends attached to the table to the opposite sides of a battery or power supply. As the connection is made, the two parts of the strip will repel and move apart slightly, since they carry current in opposite directions.

For a large audience, the strip can be folded and laid loosely across an overhead projector, taping the ends of the strip to the edge of the projection plate. The motion is not as large, because of friction with the transparent plate, but the magnification makes it easily visible. To show that currents in the same direction attract, the top ends of the folded strip are taped together and attached to one terminal of the supply, and the fold at the bottom end attached by a separate foil strip to the other terminal of the supply. This time, the two parts of the strip will move together as shown. It is possible to use either a d.c. or a.c. supply in this experiment, but connecting directly across the 110 V supply is definitely not recommended--you will either blow a fuse or melt the aluminum!

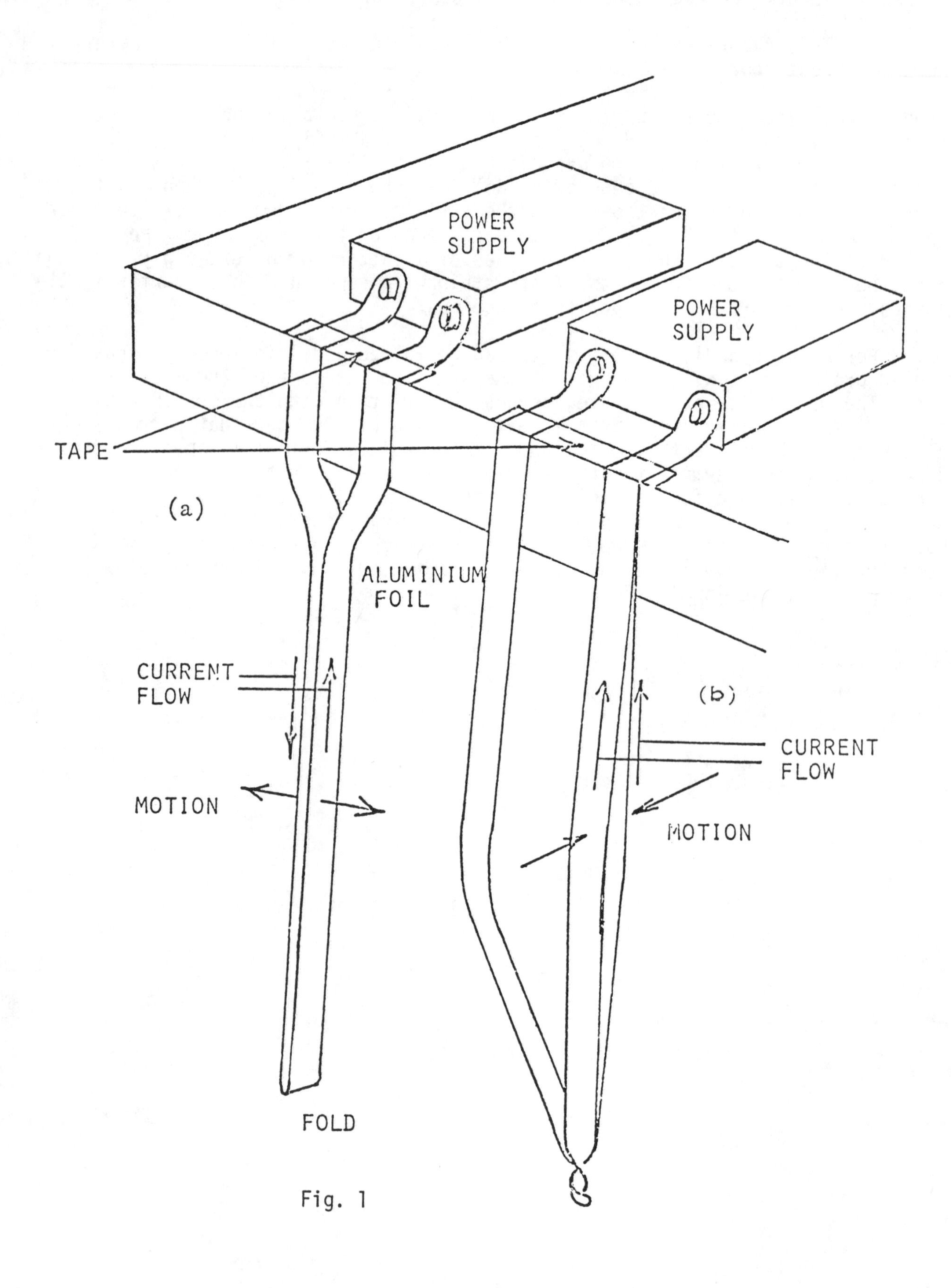
POWER
SUPPLY
POWER
SUPPLY
TAPE
(a)
ALUMINIUM
FOIL
CURRENT
FLOW
MOTION
FOLD
(b)
CURRENT
FLOW
MOTION

Fig. 1

Experiment 12.08 The Current Balance

Materials: #26 copper wire, aluminum foil, or two coins, HO power supply, paper clip.

Procedure: Bend a piece of copper wire 40 cm long to the shape shown. This is the moving portion of the balance. A paper clip acts as a counter weight.

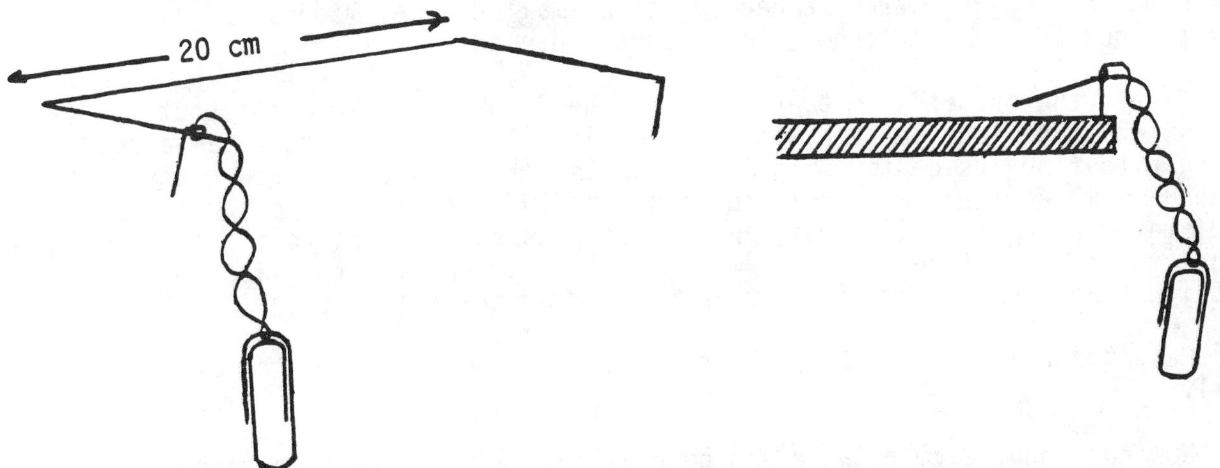

Clean the ends, and support the device on two coins or pieces of aluminum foil. A paper clip is used to attach one foil to a copper wire which runs under the conductor of the moving portion of the balance to the power supply. The other coin is attached to the other terminal of the supply.

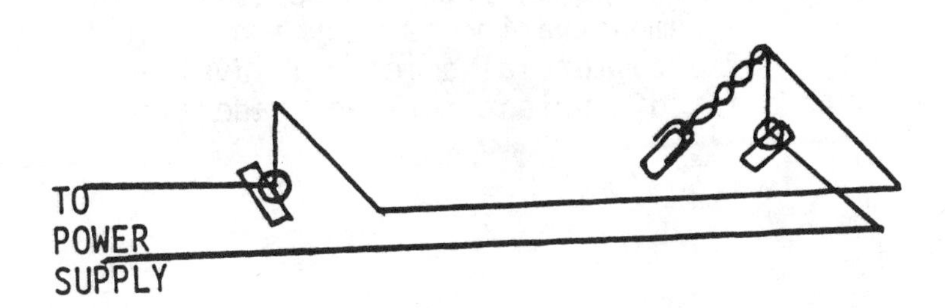

Turn on the DC current and note the deflection of the wire . Since the circuits are opposed, there is a repulsion. It is difficult to get an accurate measurement with this device. The current must be read by a meter, and the balance calibrated by a standard weight, which could be a postage stamp, for example, noticing the weight giving the same deflection as the current.

Experiment 12.9. A Current Balance using a magnet

Materials: magnet wire, alnico magnet, straws, tape, needle.

Method: Here is a simple experiment designed by E. J. Wenham for the British Nuffield Physics course (Nuffield Physics Teachers' Guide, Year 1 and 2, Longmans 1978). It requires a small alnico magnet (about 1.2 cm long and .4 cm square) a drinking straw, a needle and a small channel made of card, such as the outside of a matchbox, to support the needle.

Fasten the magnet near to the end of the drinking straw with about 4 cm of sticky tape (cut to half width). See Fig. 1. Balance the straw across your finger to find its center of gravity, and stick the needle through the straw about 1 mm farther away from the magnet. Wind a coil of wire about 25 mm diameter with about 20 turns, and fix it with sticky tape to the table top close to the end of the channel, which is made by cutting the matchbox top so that it stands about 5 mm higher than the top of the coil. (Fig. 2).

Rest the drinking straw on the channel with the magnet on the axis of the coil.

Now cut about 2 cm of wire and bend it into a u to act as a counter-balancing rider on the straw. Use sticky tape to hold a second straw vertical just by the end of the balanced straw--and mark the position of the balance on it. If correctly connected into a series circuit with say, two 1.5 V cells and two 1.5 V lamps, the magnet will be pulled into the coil. When balance has been restored by sliding the rider along the straw, the current balance can be used to check whether or not the current is the same at all points in the circuit by connecting it between the lamp, between the cells, and between the cells and the lamps. This demonstration is very important in understanding current electricity. If a commercial ammeter is available the balance can be calibrated with a set of current markings made along the straw.

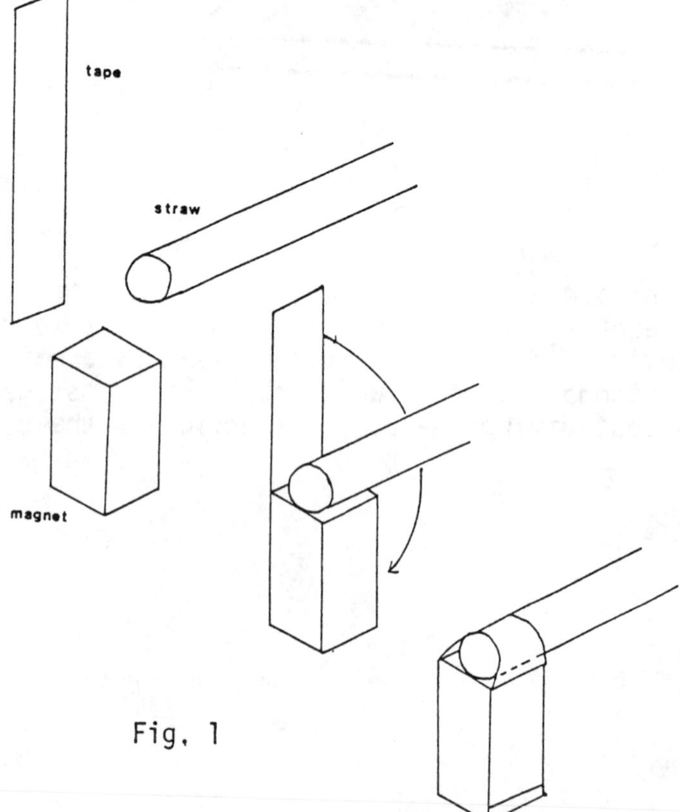

Fig. 1

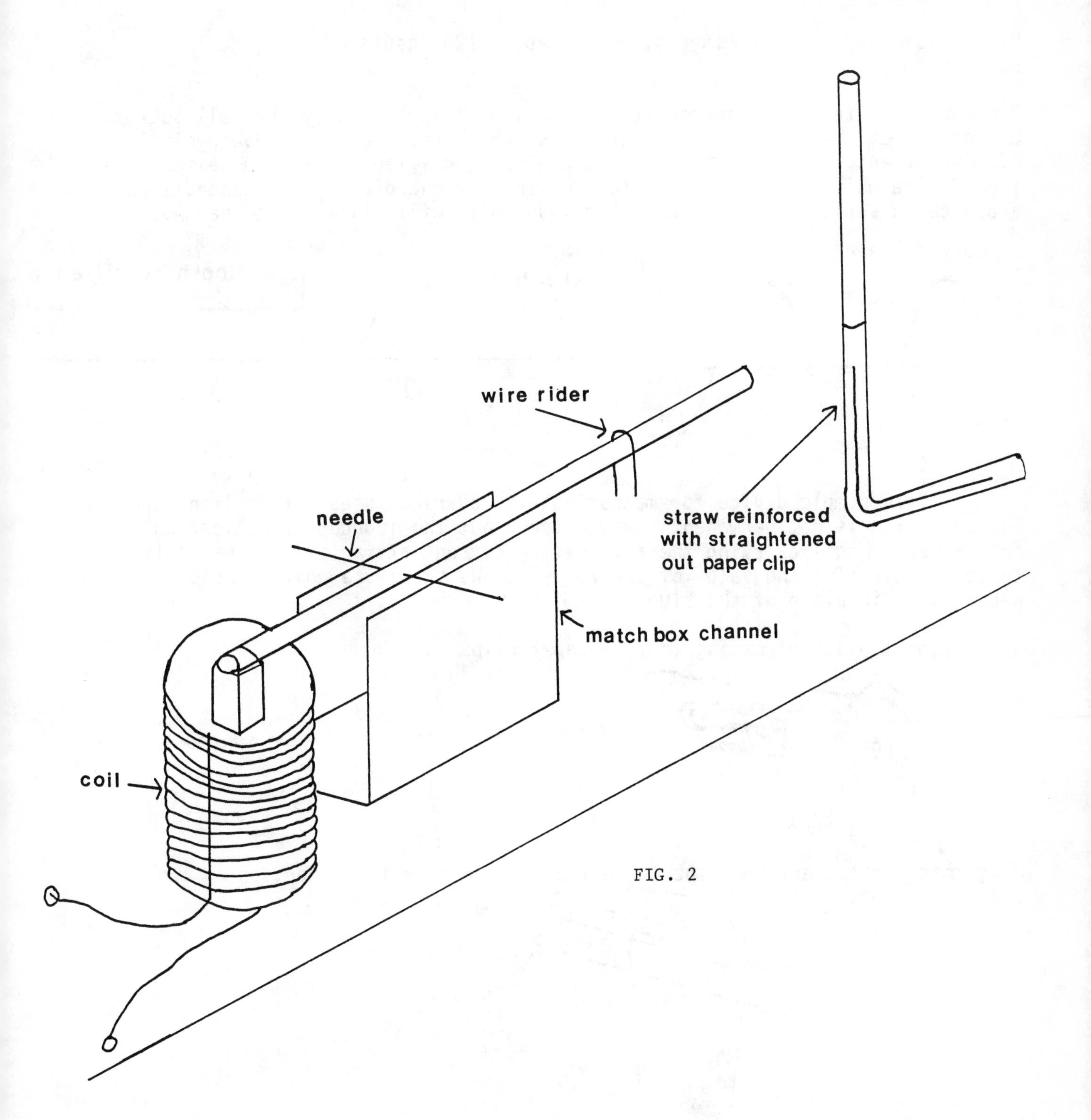
wire rider
needle
straw reinforced
with straightened
out paper clip
match box channel
coil

FIG. 2

Experiment 12.10 - A simple method of measuring alternating current, and the transformer.

Materials: paper clips, string, thick paper, #26 insulated wire, HO power supply

Procedure: Alternating current (A.C.) such as is provided by the wall socket, cannot be measured using D.C. instruments, which read zero, the average value of the current. However, the current can be "rectified" by using a semiconduction device which allows current to flow in only one direction. Hence, such a device in series with the tangent galvanometer will allow it to read A.C.

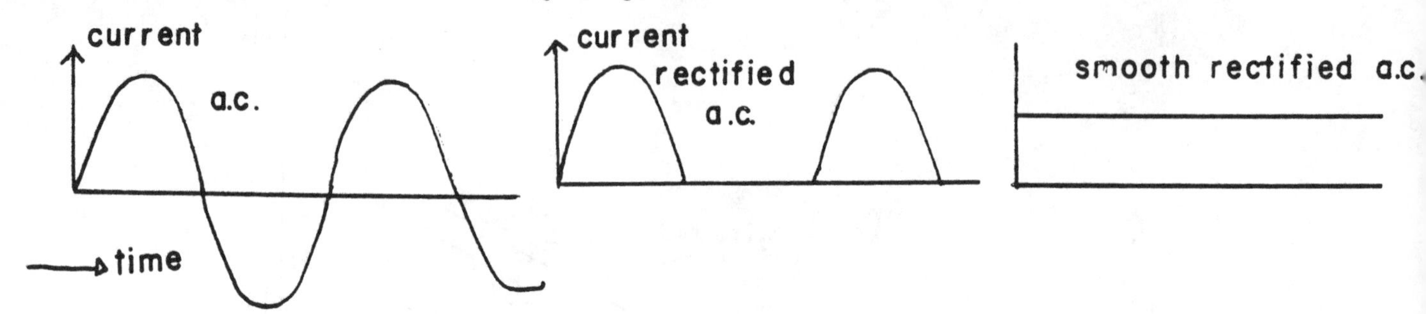

A very simple device for measuring A.C. currents uses a soft iron slug. The soft iron is not permanently magnetised, but the induced magnetisation from a nearby coil carrying the alternating current attracts it, causing it to move indicating the value of the current. We shall use unmagnetised paper clips in place of the slug.

Tape a piece of string to four paper clips, as shown.

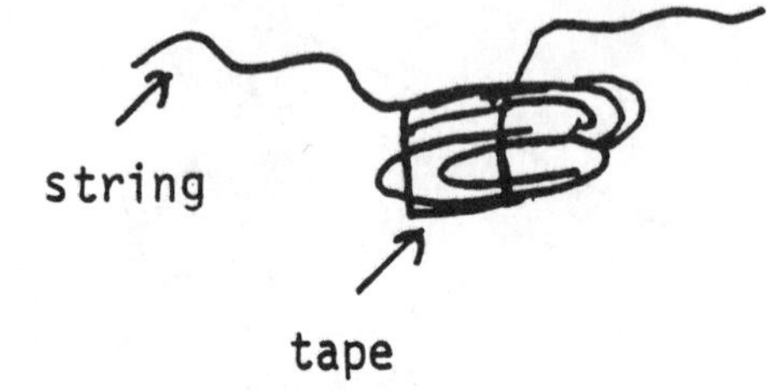

Hang this arrangement from the table top to come near the floor, using two pieces of tape.

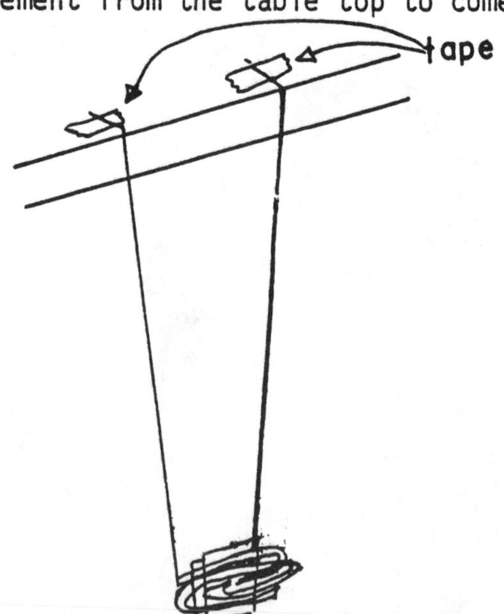

The height of the clip may be adjusted by pulling the string through the tape on the table top.

Now, roll a piece of card, or paper, into a cylinder large enough into which the four paper clips can slide easily, and wind a coil of 100 times on it. Place the cylinder on the floor, taping it so that the clip can slide in and out without twisting the sides. To check the device, connect the coil to the A.C. outlet of the power supply, and notice how the clips shoot into the coil.

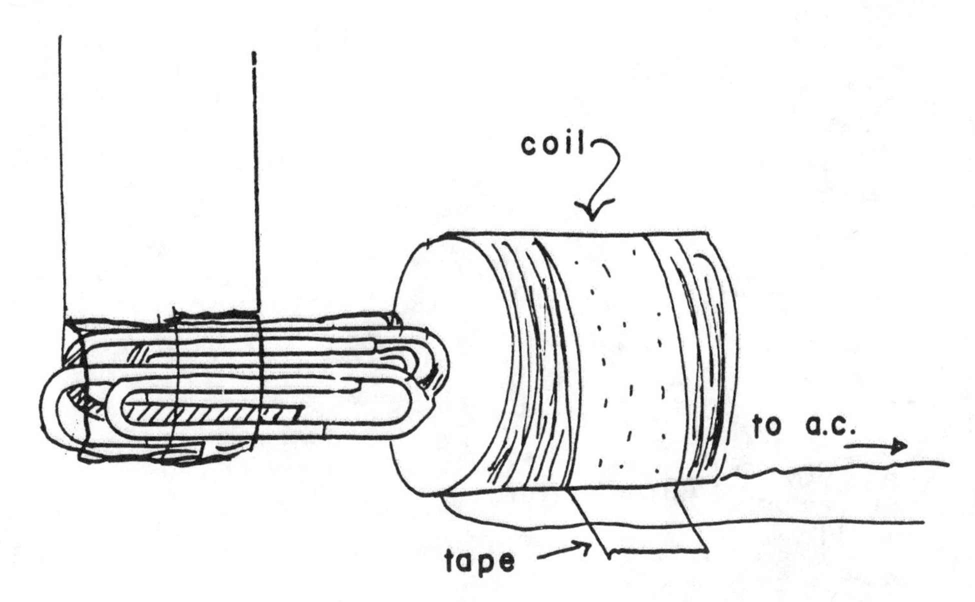

Now, to make a transformer, tape fifteen paper clips together. This will form the iron core of the transformer. Wind one hundred turns of wire on this. Tape up this coil. Now wind the secondary coil of one hundred turns on top of this, tape it down and wind a further one hundred turns on top. You can wrap a paper cylinder over the first hundred turns if you have difficulty in the turns sliding off the end of the coil.

Now, connect the one hundred turns of the secondary to the coil fastened to the floor, and fasten one end of the primary to one terminal of the 18 V A.C. outlet from the HO power supply. Observe the hung paper clip as you touch the other end to the other terminal. It will move, indicating that an A.C. current is induced in the secondary coil, passing through the detector coil also. Connect the second winding of the secondary in series with the first hundred turns. Note that, when connected so the turns oppose between the two secondary coils, i.e. are in opposite directions, no current is generated. With them in the same direction, this is not so--is the current larger, or smaller than with 100 turns? Why?

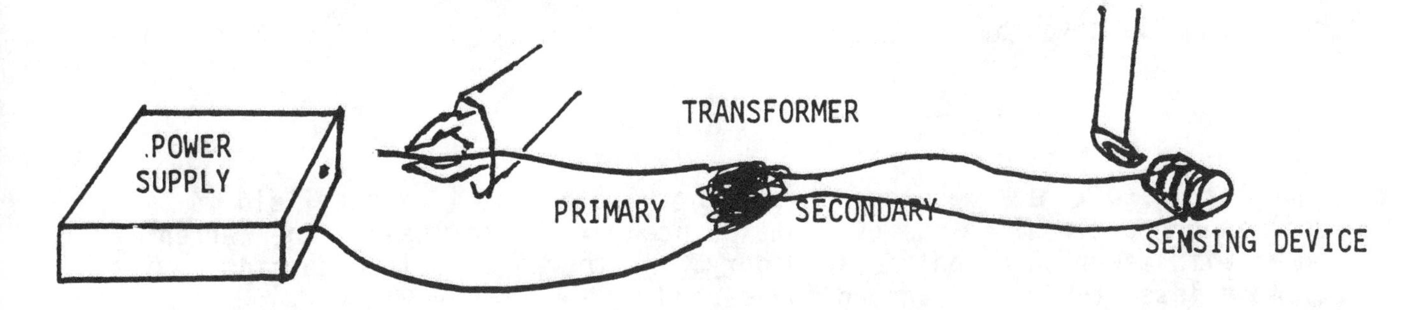

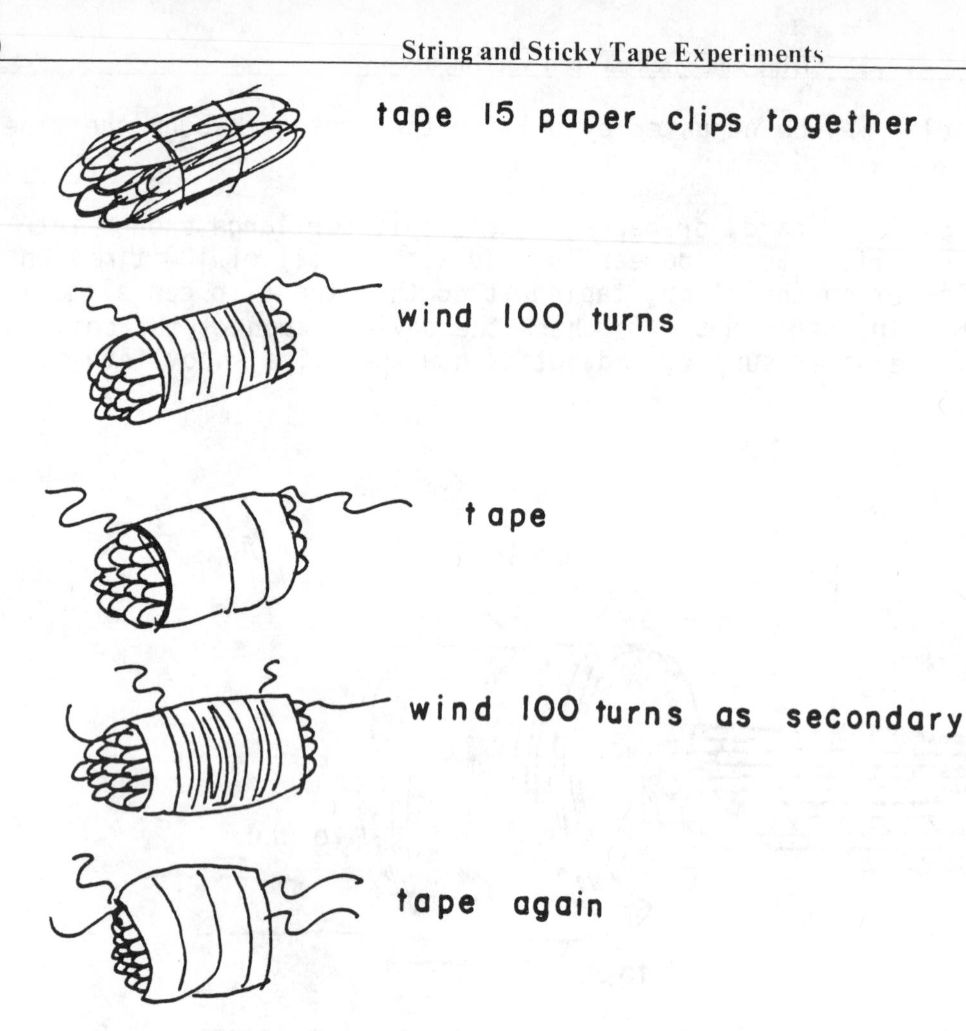

In a transformer, the ratio of the currents in the primary, i_p, and secondary, i_s, is given approximately by

$$\frac{i_s}{i_p} = -\frac{n_p}{n_s} \quad \text{for current circuits}$$

where n_p, n_s are the number of turns in the primary and secondary, respectively. The product of voltage and current is constant for both primary and secondary

$$V_p i_p = V_s i_s$$

so the voltage is given by

$$\frac{V_s}{V_p} = \frac{n_s}{n_p}$$

Now, our little coil measures the current in the circuit since the field in the coil is proportional to the current. Hence, excluding losses, the current is higher with fewer turns on the secondary, and the paper clip detecting device should move less, the more turns on the secondary.

Experiment 12.11

The simplest electric motor

Materials: two rubber bands (large), small ceramic magnet, two paper clips wire supports, #22 enamelled coper wire, sand paper (or knife), a D (or C) battery.

Instructions: The diagram is almost self-explanatory. A coil of wire of ten or twenty turns is wound on a former - the D cell itself will do for this. The two supports (straightened paper clips will do) have loops at the ends - they can be formed round a pencil - one or two rubber bands hold these wires and the magnet on to the battery as shown, the wires making contact with the terminals, the magnet being on top. If preferred, a "v" shaped support, as shown in the figure may replace the loop. The ends of the coil must now be scraped, or sandpapered on one side only, with the angle to the coil as shown (coil vertical) so the torque on the coil is in one direction only. The device is put together and the coil set spinning. A number of further suggestions (from Scott Welby's article) are:
Does the motor have a preferred direction of rotation? Why?
Does reversing the battery or magnet reverse the direction? Why? How does the shape of the coil (armature) affect its operation?
What is the optimum number of turns of wire for best performance?
Put an ammeter in series with the coil to measure the current. Estimate power output.

1. Physics Teacher 23, 172 (1985) (Scott Welby)
2. Physics Teacher 17, 308 (1985) (Rudy Keil)

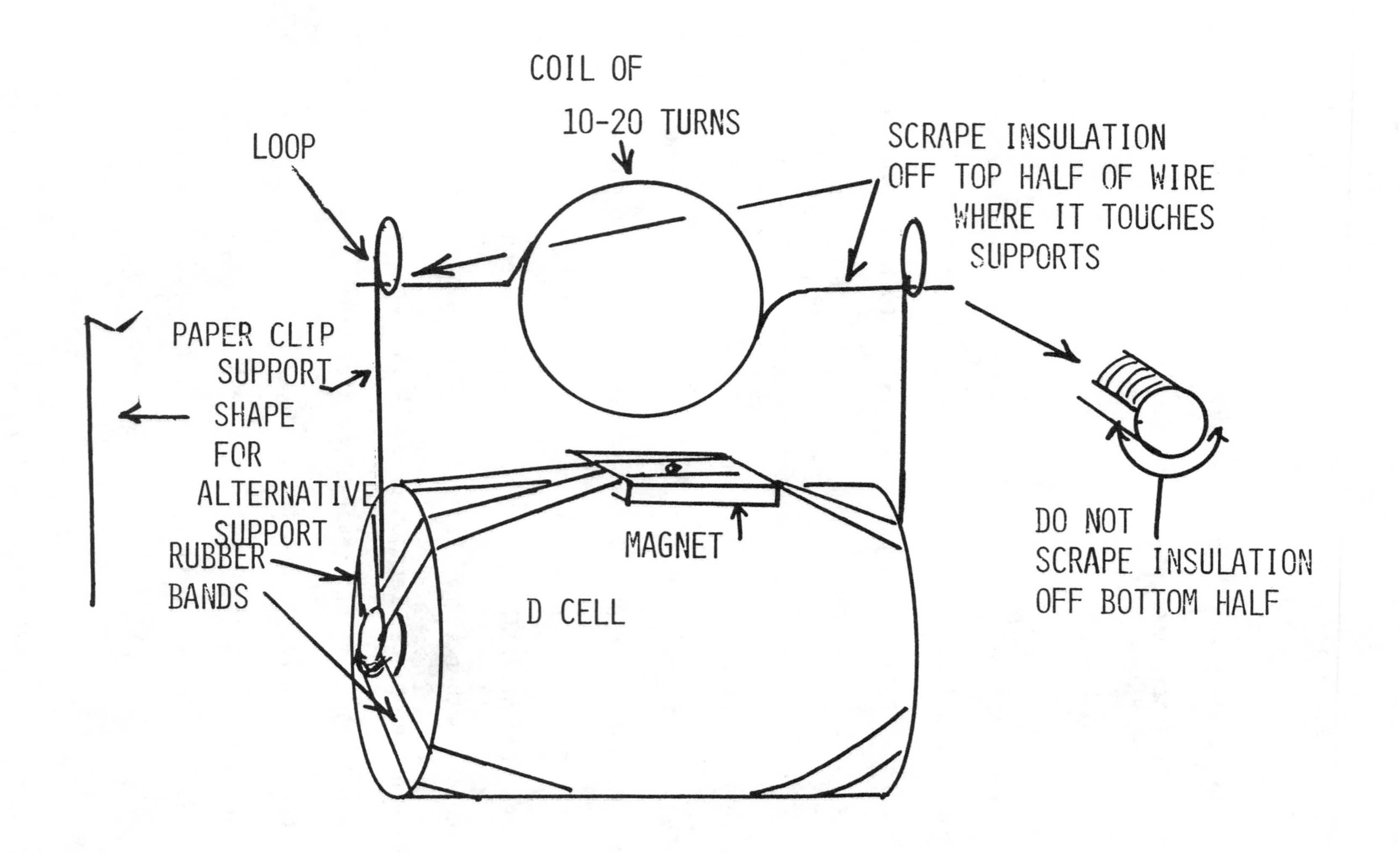
COIL OF
10-20 TURNS
LOOP
SCRAPE INSULATION
OFF TOP HALF OF WIRE
WHERE IT TOUCHES
SUPPORTS
PAPER CLIP
SUPPORT
SHAPE
FOR
ALTERNATIVE
SUPPORT
RUBBER
BANDS
MAGNET
D CELL
DO NOT
SCRAPE INSULATION
OFF BOTTOM HALF

Experiment 13.01 Benham's Top - an experiment in psychophysics.

Materials: straw

Procedure: Cut the disc below from the sheet, poke a small hole at the center using a pencil point, then push a piece of straw about two inches long through the hole. The disc can now be spun as a top.

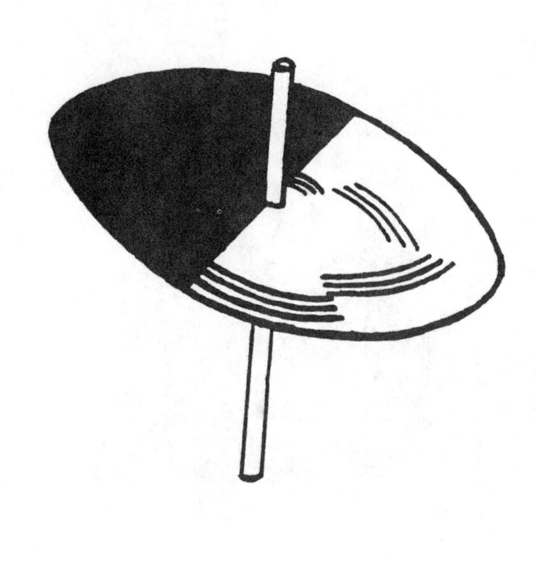

Spin the top clockwise, and note how the outer rings appear darkest, and the inner rings colored, even though they are black and white. Now, reverse the rotation of the top and note how this time it is the inner rings which appear dark, and the outer rings appear lighter and colored.

Qualitative: Why is this? Our eyes adapt to darkness in a variety of ways. After looking at a bright light for a short while, the nerves sending pulses to the brain become "tired" and cease to respond as well. On the other hand, the cells looking at the dark are "dark adapted" and give a much larger response to light. The cells which respond to different colors recover at different speeds, so a rapid change from black to white may allow one color sensor to give its full output, but not another. The effect is to tint the disk that color which gives its output most rapidly. Spinning the disk at different speeds can lead to different colors--very slow, and all the sensors give full value. Very fast, and only a random selection respond, leading to grey.

Although the general principle behind this experiment is understood, the details of the physiological process is not well known at all.

A similar illusion was invented by Bidwell (1897). When rotated clockwise at about six revolutions per second with the slot above a colored object, the object appears its complementary color.

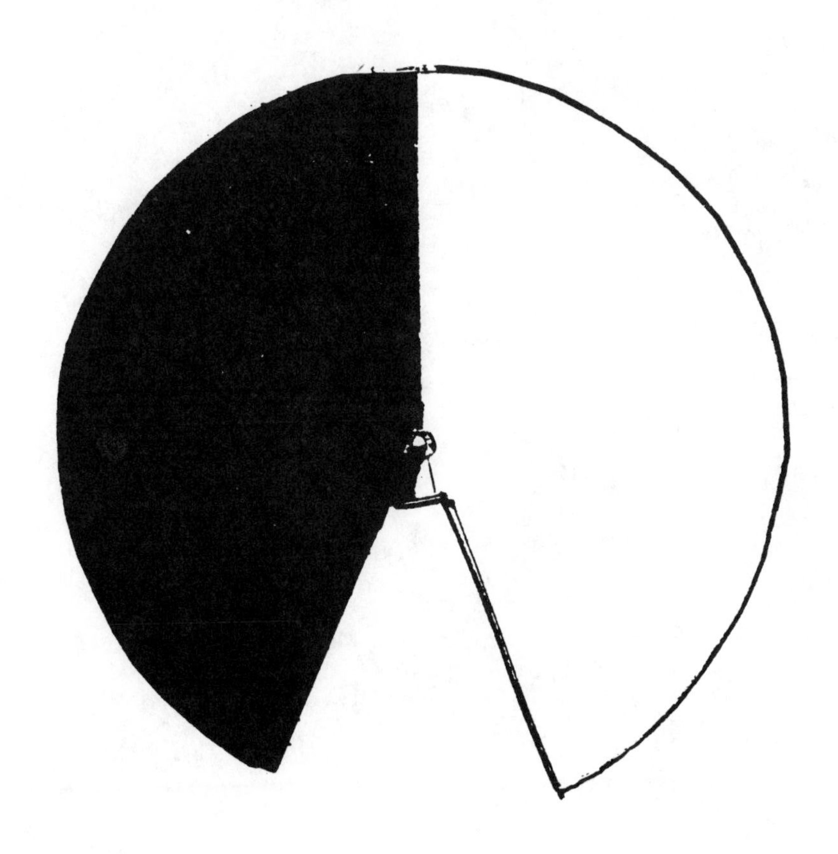

Experiment 13.02 Physical Perception - Optical Illusions

Materials: Scissors, hot and cold water, cups

Procedure: The importance of sensual illusions lies not only in the amusement they provide. They are also important in helping to understand the mechanism of perception, and in teaching us we can't always believe our senses.

1) Illusions and Distortions Due to Fatigue or "Adaptation"

Among the most common sensual illusions are those which arise when we suffer from prolonged stimulation. Here are some ways of adapting the senses to give distortions.

1. Weight

After a heavy book is carried for several minutes at arms length, the arm feels light, and may involuntarily rise up several inches.

2. Temperature

Put one hand in a cup of hot water, the other in a cup of cold water. After a few minutes or so, put both hands in a cup of tepid water. Although both hands are now in water of the same temperature, the one that was in the hot water feels it as cold, while the other feels it as hot.

3. Taste

Sweet drinks taste gradually less sweet. Try keeping a sweet soda drink, such as a cola, in the mouth for a few seconds and then taste fresh water. It will now taste distinctly salty.

4. Loudness

After hearing a rock concert, the outside sounds quiet.

5. Velocity

Distortion of apparent speed is common while drivinq. A car moving at 30 mph seems to be moving almost ridiculously slowly after an hour's continuous driving on a super highway. Traveling by train, where the forward motion is smoother than by car, a common sensation of drifting backwards occurs when the train stops at a station. Cut out the diagram on the following page; push a pencil through the center and rotate it by rolling the pencil between the palms of the hands. Better still, spin the figure on a record player. If, on rotating, it appears to expand, on stopping, the whole figure seems to shrink.

6. Brightness

If you stare at a lamp for a few seconds, then look at a white wall or sheet of paper, the effect of adaptation to white and black will be seen quite dramatically. Adaptation to the bright light will produce a corresponding dark area on the wall or sheet. The brightness and colors of such after-images are pretty well understood. Fixation on the lamp reduces the

sensitivity of the retinal light receptors on which this area falls. On looking away, we see the region corresponding to the part of the retina which has lost sensitivity by being exposed to bright light as darker, simply because it transmits less signal to the brain. The frequency of nerve impulses from this region is reduced, just as when in fact we look at a darker region of an object. This adaptation also applies to colors. After looking at a greeny-blue colored object, on looking at a white sheet we see the object, but in its complementary color red. The eye is adapted and no longer sensitive to green.

2) Optical Illusions

Optical illusions can be divided into two groups. One set is purely physiological. It depends upon the physical response of the eye to stimuli--a simple example of this is the vanishing of an object whose image lies on the blind spot. Look at the cross in the figure below with the left eye, and move the page back and forth. The dot will vanish at a point when its image falls on the blind spot of the retina. Correspondingly, the cross vanishes if you look at the dot with the right eye. Our brain fills in the absent visual response with its surroundings.

The second type of optical illusion is psychological, depending upon our association of ideas and visual patterns which we have developed over the years, and would, for example, be absent, in the case of a new born infant. Such illusions are those which arise from perspective -- two lines drawn to converge at a point, as shown in the figure below, look like railway tracks running away from us to infinity at the "vanishing point".

Often optical illusions are a combination of the psychological and physiological. For example, the television and motion picture industries are based on such illusions. Successive pictures in sequence of time are flashed on the screen at the rate of 16 to 25 per second, and because the physiological response of the eye is not sufficiently rapid to separate them, - the so-called "persistence of vision" - we see them as continuous motion and not as individual pictures. This effect is physiological - however the illusion of depth in the scene itself is psychological and based on our ideas of perspective mentioned above, which we have built up through looking at objects from different vantage points, and inferring their solid appearance from a knowledge of one view. Our mind puts in what the perspective does not provide. Interest in such perspective illusions has been recently revived by M. C. Escher, and several new perspective illusions have recently evolved - such as the perpetual staircase and the impossible prongs starting with two, and ending with three, and the triangle - each corner of which is a right angle, shown overpage. They depend upon the fact that the eye does not comprehend the whole of a picture at one time, but only sees and interprets a small portion.

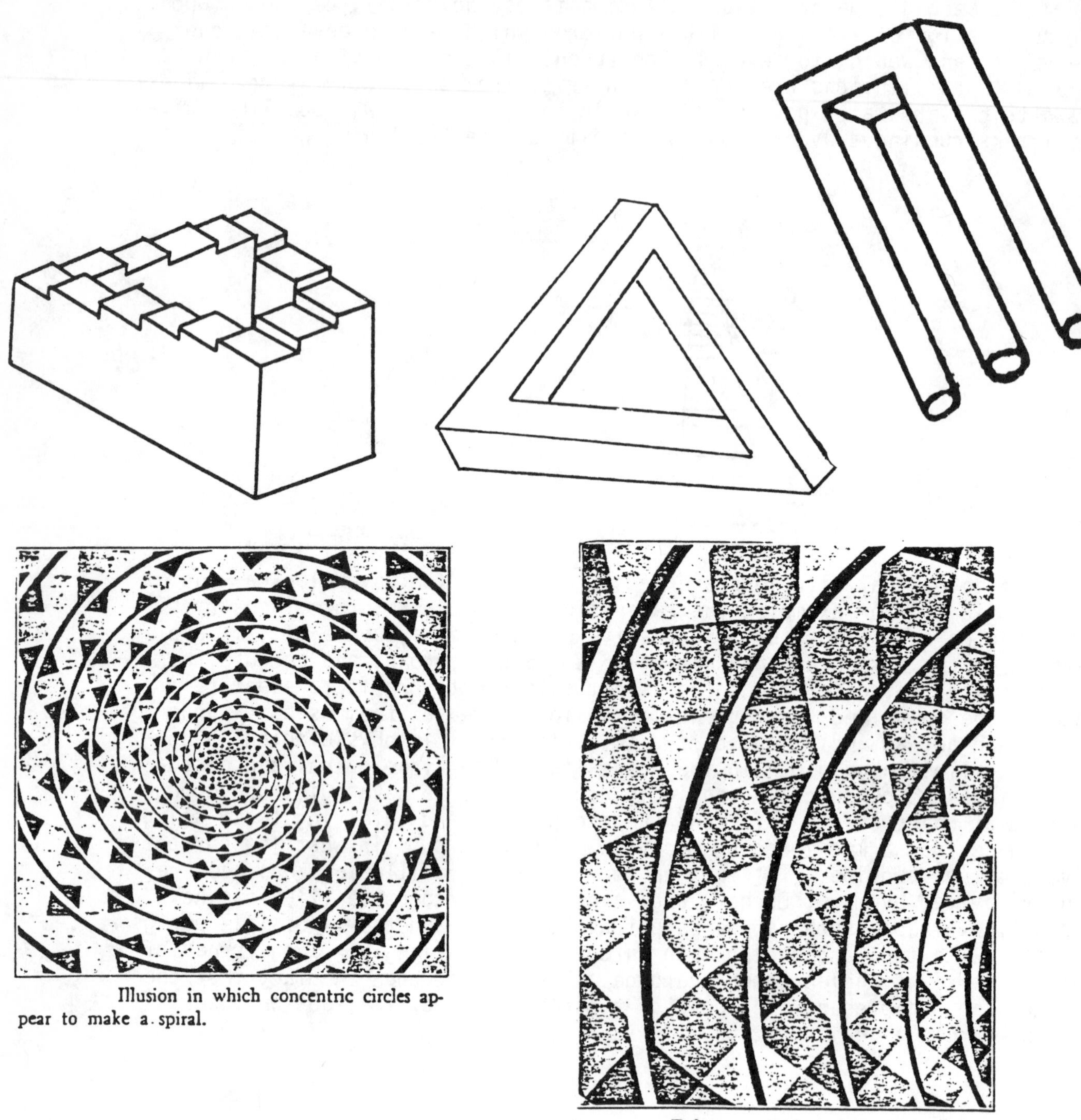

Illusion in which concentric circles appear to make a spiral.

Enlargement of preceding figure, showing true nature of concentric circles.

In the nineteenth century, eminent scientists such as Zollner, Hering, Jastrow, Poggendorf and even Helmholtz suddenly evinced interest in explaining illusions. Nearly always, they failed to realize illusions are more than mere laboratory curiosities, being the outward expression of human perception and judgment. Interest vanished for a while until in the 1920s the field of transactional psychology was born. and men like Adelbert Ames devoted their lifetime to visual perception.

Illusions are most difficult to categorize in any logical way. R. L. Gregory's book, "The Intelligent Eye", provides the best insight.

The earliest of the so-called "hatched line" illusions: is Zollner's paradox. (a) Although the long lines are parallel, the small cross hatching lines produce the very real illusion of making the vertical lines alternately diverge and converge - they do not appear parallel. Zollner discovered this on some dress material. Ward's illusion (b) is a variation of this, the parallel lines bowing in, and in Hering's illusion (c) the lines bow out.

In Poggendorf's illusion, (d) the visual question is whether the left hand line is a continuation of the upper or lower line on the right. All these illusions are based on the misleading effects of lines - the rest of the pattern other than the lines must be considered. The eye can only interpret relative effects - even a straight line is interpreted with respect to its surroundings. The illusions (e) have a similar basis - the eye being led inward by converging lines.

Oscillatory Figures

It is clear from the forgoing illusions that the brain interprets patterns of the eye in terms of external objects. A fundamental problem is how the eye splits the pattern into objects and surroundings. The figure-ground illusions of (f) Rubin's faces - vase is a case in point. Is this two faces (objects) looking at one another, or a vase? Boring's ambiguous "My wife - and my mother-in-law" is similar (g) The ear of the pretty girl becomes the eye of the mother-in-law.

Schröder's staircase illusion (h) is similar - are the stairs the right way up or upside down?

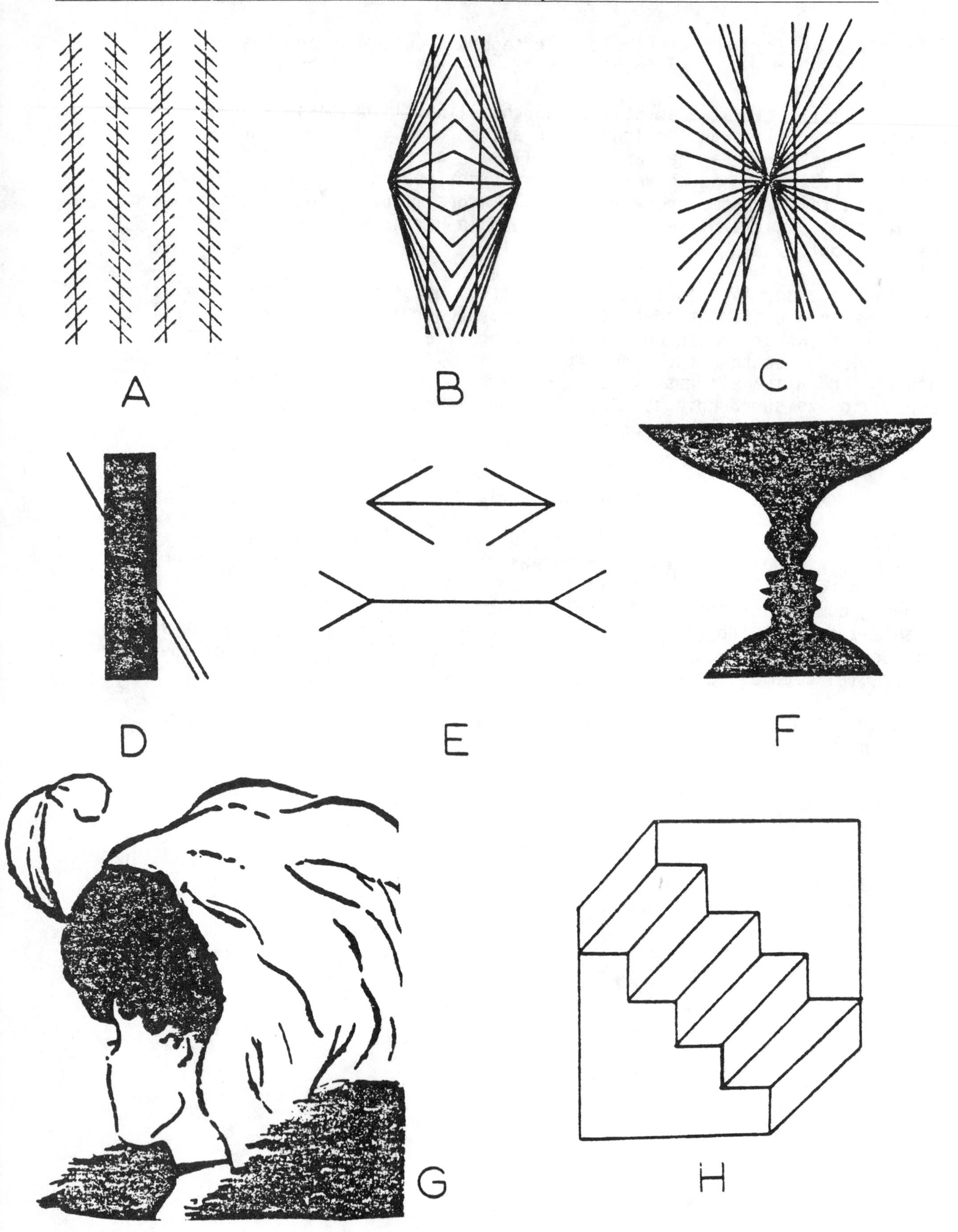
A
B
C
D
E
F
G
H

Experiment 13.03. Moiré Patterns

Materials: Scissors

Procedure: Perhaps you have noticed that when window insect screens overlap, patterns are produced which change wildly with a slight movement relative to the screens. These patterns are called moiré (pronounced mwareh) patterns. Two sets of railings on a bridge, or two combs with different teeth spacings also produce such patterns. The distance in which the pattern repeats is large, if the difference in spacing between the teeth is small.

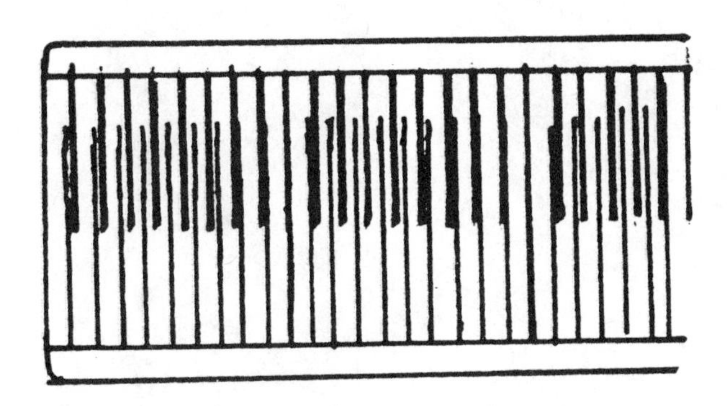

Some other patterns are shown below, and over page. Cut out two similar patterns, turn one over, and place it on top of the other pattern, so the two ink surfaces are in contact. Hold the two overlapping patterns to the light, so you can see through the two sheets of paper at the moiré pattern produced.

Move the patterns about on top of one another to see how they move. Describe the patterns produced. Examine the straight line patterns, and try to explain why the moiré lines are perpendicular to the pattern lines.

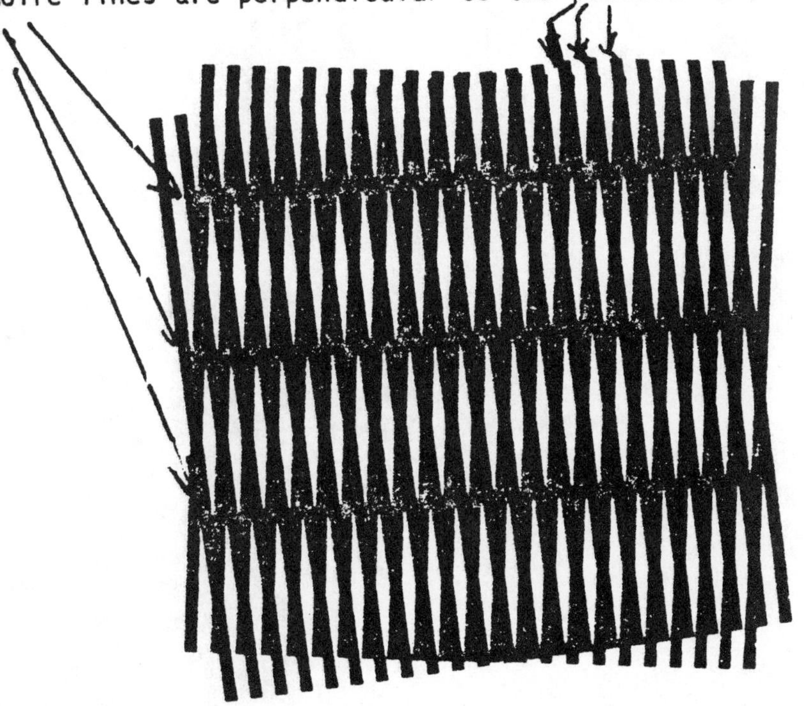

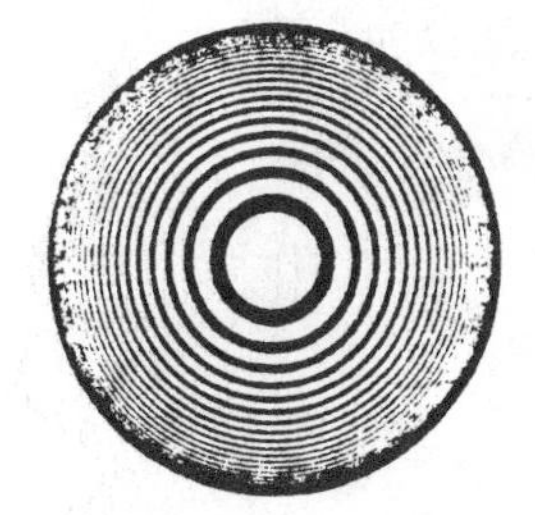

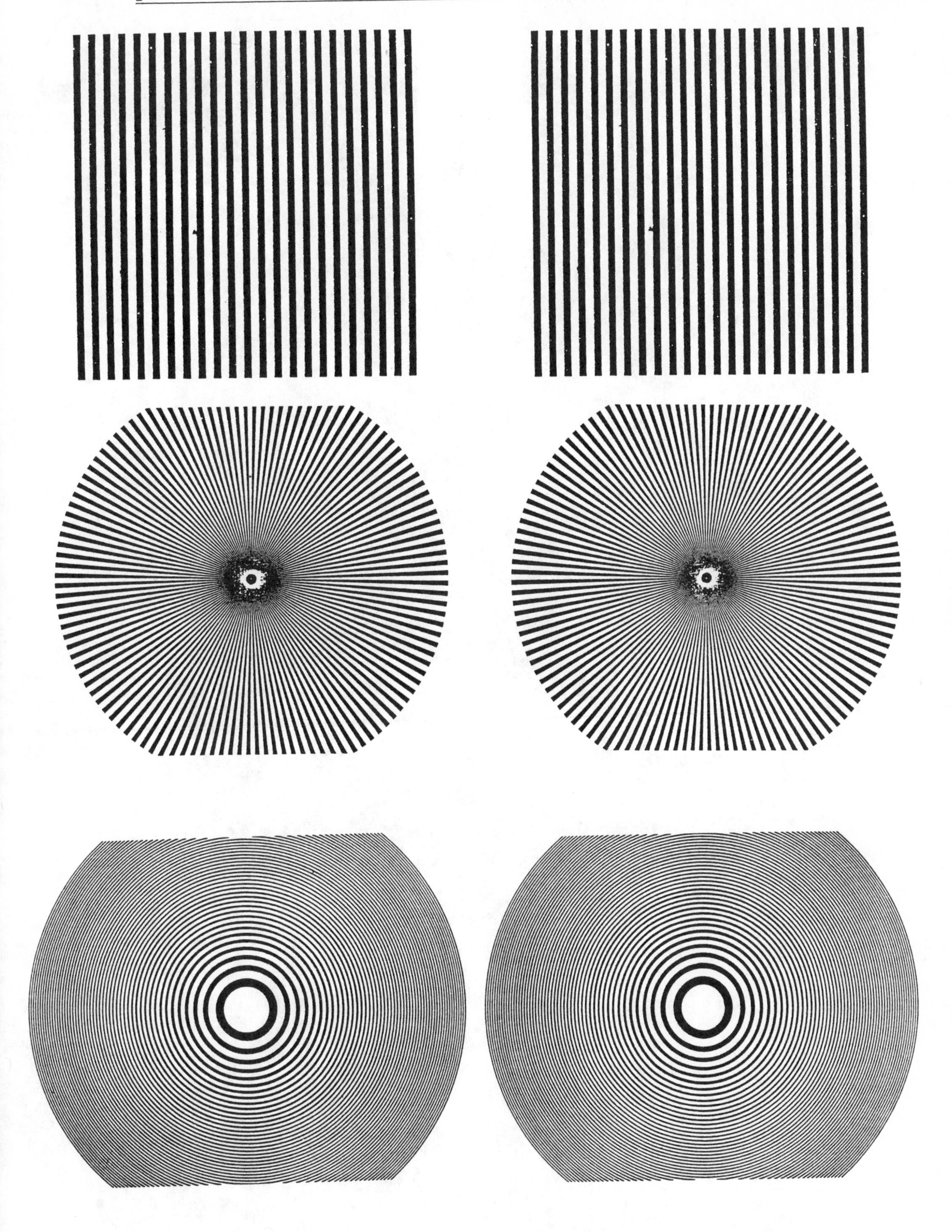

EXPERIMENT 13.04 Persistence of Vision

Materials: pencil, scissors

Procedure: Motion pictures and television depend on the fact that for a continuously flashing light, successive flashes become indistinguishable if the repetition rate is sufficiently rapid. So, the eye sees continuous action if a succession of pictures (each a little different from the last, because of the motion of the objects depicted) is placed rapidly before the eyes. The physical or physiological reason behind this is the time it takes the rod and cone sensors of the eye to recover. During this period the eye cannot perceive a complete new picture.

The earliest efforts at representing motion in this way occurred in the early and mid nineteenth century, and you will construct a typical device, called a "phenakistoscope". (All such devices had high-flown pseudo-Greek names, ending in "scope"). Cut out the black disk, and carefully cut the slots marked. Now, cut out the other disk. Take a new (so that it is long enough) sharpened pencil and push the point through the center of the black disk, continuing to push it until just before the other end. Now push the point through the disk with the girl on it. You are now ready to make the device work. Hold the pencil between the fingers and palms of your hands, as shown, and put them together, so the pencil rotates rapidly, peeking at the pictures through the slots simultaneously. You should see the girl skipping. After reading the instructions and questions, cut out the disk overprinted on this page. This shows an electron going round the nucleus in an atom. It gives out a photon of light, dropping to a lower orbit, and the photon travels to the next atom where it is absorbed.

Try to answer the following questions.

1) Estimate how many frames (pictures) a second are required to avoid jerky motion, such as you see in old-time silent movies, which have too few frames per second. As you rotate the disk, get a friend to count "one, two, three" for each rotation, from a mark on the edge of the disk. Now, a second friend can look at a clock, and see how long the "one, two, three" took. To give some idea, silent movies run at sixteen, sound movies at twenty four, and television thirty complete frames per second.

2) Reverse the slotted disc so that the white side faces outward. Does the device still work? Why ?

3) The slots in the disk are very narrow, so that you only see the picture about a tenth of the time--the rest is dark. In movies, the time the screen is illuminated is much longer than this--at least half the time. Enlarge the slots to find out why it is necessary that they be so narrow in the phenakistoscope. Why is this?

4) Why does the writing on the disk shown on this page not affect the picture you see?

PENCIL
POINT
HERE

Experiment 13.05 The Ames Window

Materials: Scissors, thread, straw, sticky tape.

Procedure: The object of this simple illusion is to show students how much they assume about an object from its superficial appearance. This is the reason, in physics, why we are so careful about the assumptions we make, and frequently go back and reexamine them. The experiment was devised by Adelbert Ames, who spent much of his life designing similar illusions.

Cut out the two printed window frames, and the sections within each frame where the glass would normally sit. Now, attach them back to back, using sticky tape around the edges, or glue, so that they exactly overlap. Attach a piece of thread to the top so that the window hangs level, as shown, and hang it from the ceiling, or other convenient support, so that it is about eye level. Wind up the thread so the window rotates slowly, and observe it with one eye from some distance. The window gives the appearance of oscillating too and fro, and not of going completely around, as in fact, it is. Our mind assumes the window has a rectangular shape, not the distorted perspective view it really has, so we feel we are seeing an oscillating rectangular frame, rather than a rotating frame of curious shape.

Now attach a straw through the window, as shown in the second figure. As we watch with one eye again, the window appears to oscillate, but the straw seems to have a curious motion, penetrating the window at times. It is very difficult for the eye to interpret this, but we still cannot overcome the feeling that the window oscillates, even with this additional evidence.

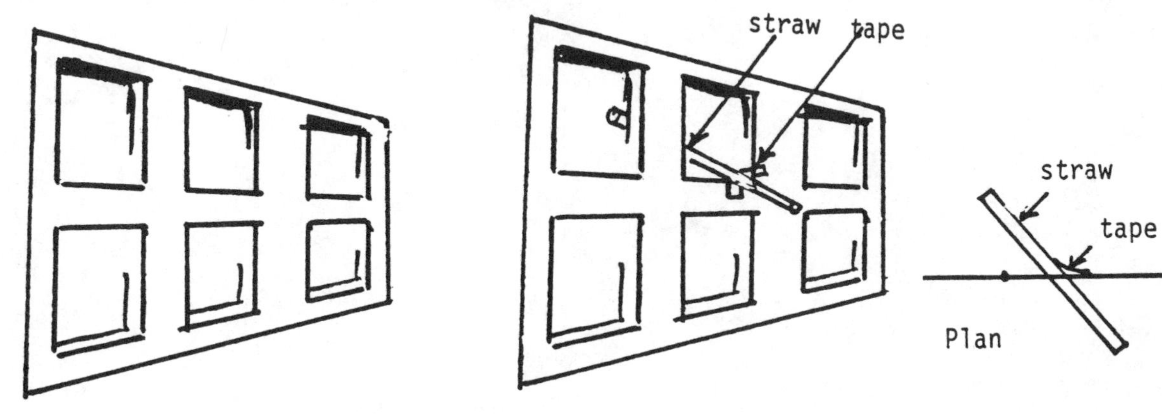

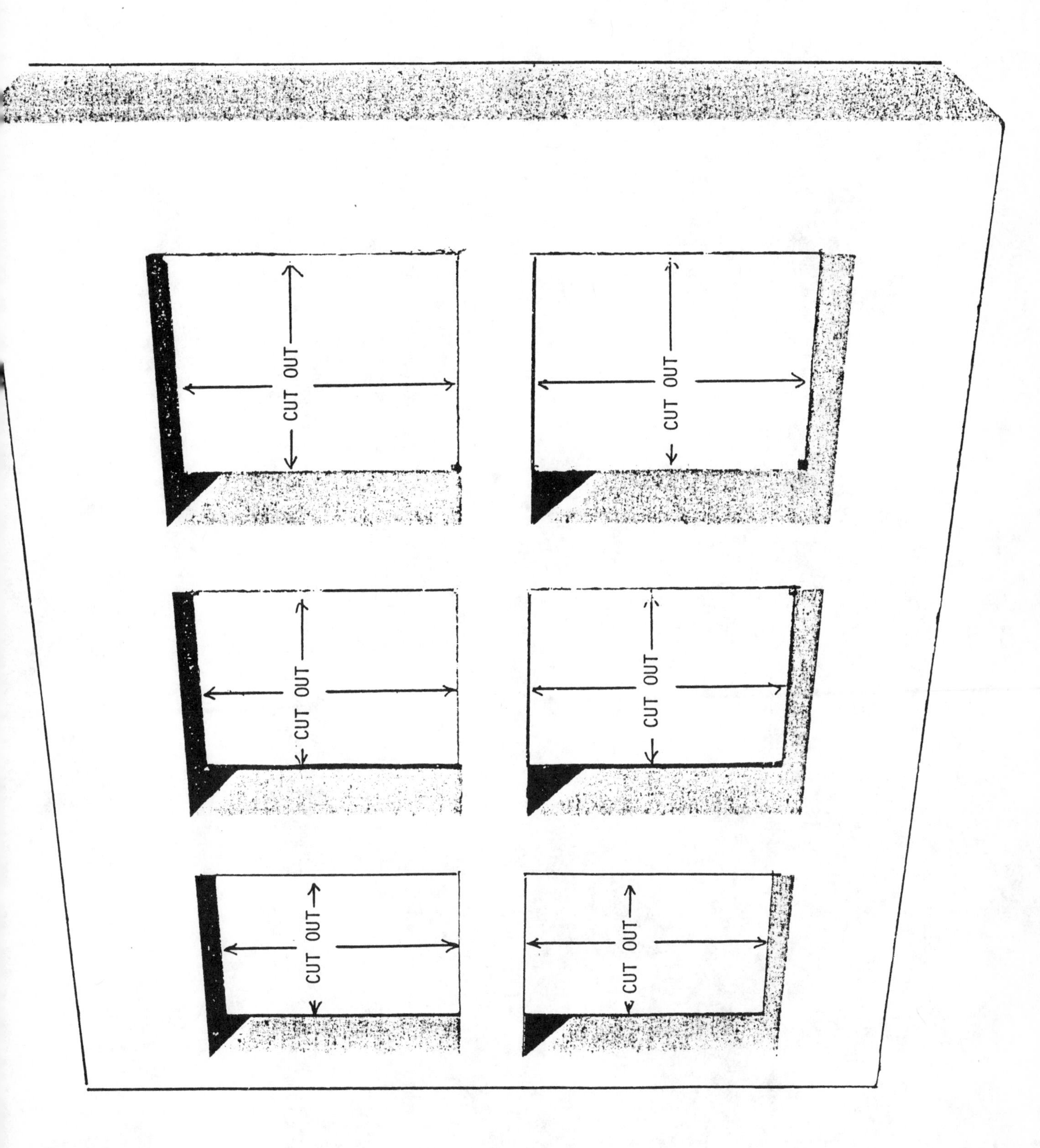
CUT OUT
CUT OUT
CUT OUT
CUT OUT
CUT OUT
CUT OUT

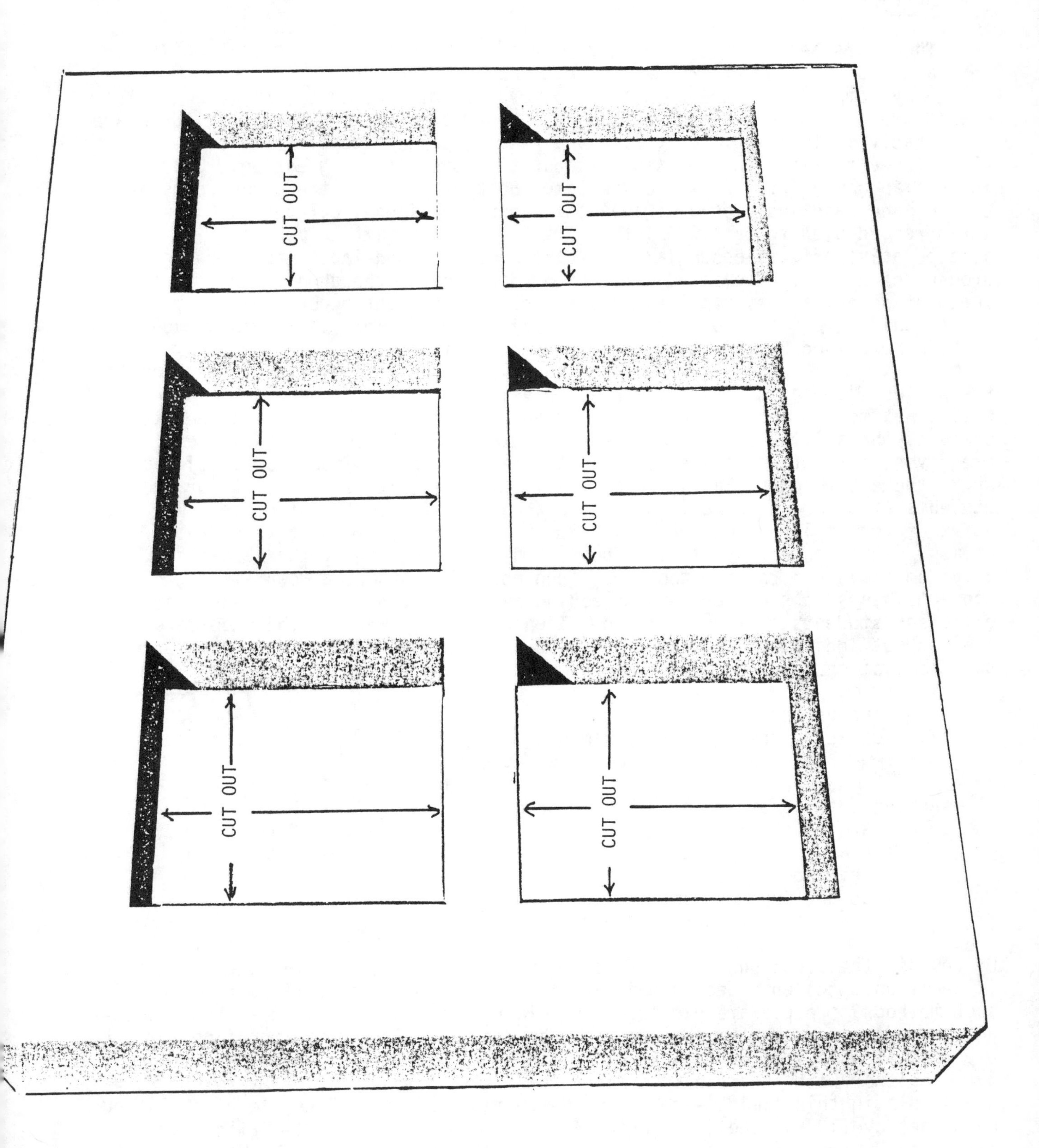
CUT OUT
CUT OUT
CUT OUT
CUT OUT
CUT OUT
CUT OUT

Section 14

Introduction

There has been a move afoot among child psychologists for quite awhile to change some of our attitudes toward games. In the past, games have been primarily competitive, the object being to beat the opponents. Any cooperative spirit arose through one team combining against the other. Not all children are so combatively inclined, however. As a result, games have been devised which allow for a spirit of cooperation without the concomitant agression. One such is the lap game, (p. 172 of the New Games Book). In this, the group forms a circle, and each individual sits on the lap of the individual behind. The world record used to be 1,500 lapsitters. Of course, if one person falls over, a cooperative effect--essentially a soliton wave of non-lap-sitters--moves around the circle. The speed with which this occurs, though not equal to the speed of sound, is nevertheless very rapid. The thought occurs that many such games could be devised to demonstrate physical principles. This was brought home by Ruth Howes and James Watson, of Ball State University, demonstrating physical principles using the students themselves. The point was made that, in a large class, interaction on a personal basis, not merely between the professor and the student, but also between the students themselves, is virtually non-existent. Furthermore, such large classes are given, more often than not, in vast auditoria not designed for physics use, where there are no apparatus storage facilities, and it is difficult for the students at the back to see. At Ball State, they invite the students down, and a few of them mill around, bumping into one another, pretending they are "gas atoms". This contact sport is evidently much appreciated by the students, provided it is not carried too far. Then more students come down, making closer contact, hands on shoulders, but breaking away to put hands on the shoulders of different students to form a liquid. Then, they put hands on the shoulders of the nearest individual and stick--a solid. Ruth Howes mentioned several other such simulations, which serve to

1. wake up the students
2. act as demonstration experiments
3. require no setting up time or equipment

Acting out really drives home the point.

I hear and I forget.
I see and I remember.
I do and I understand.
Chinese Proverb

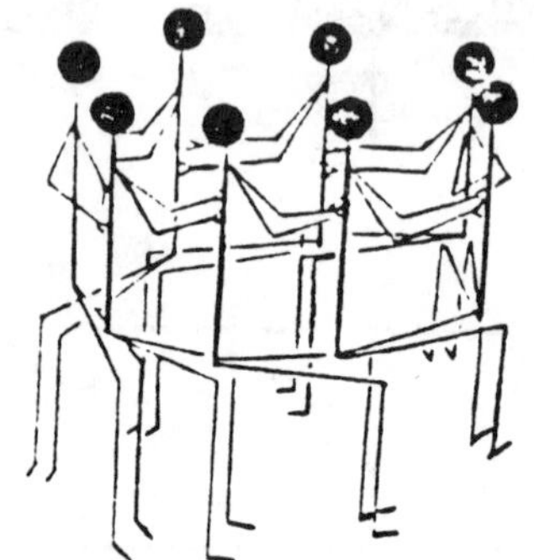

THE LAP GAME

Of course, the question arises as to how much is physics and how much game, but it seems an excellent idea in moderation. Which type of physics lends itself to such methods? People are particles, so clearly, gas kinetic theory is suitable--and in fact, many quantum phenomena can be described (short of tunneling through a potential wall.) I would like to give here some of the ideas we have tried out, together with the principle which it aims to demonstrate. Several books are a help in this field, such as The New Games Book, by A Fluegelman, published by Doubleday (1976); The Cooperative Sports and Games Book by T. Orlick, published by Pantheon (1978); and Learning through Movement, by P. H. Werner and E. C. Burton, published by Mosby (1979).

Experiment 14.01 Pirates Treasure Game (Vectors)

It is notorious that pirates always bury their treasure beneath some beach on a desert island, and then provide a nearly incomprehensible map to find it. The object of this game is to provide a suitable vector description for finding the treasure, using paces as the unit of length. It should be remembered that the pace was the unit employed by the Romans, whose professional pacers' sole job was walking between towns to measure the distance. So we can start by saying "take three paces north (or toward the blackboard, or whatever), and four paces west. Count the paces directly back to where you started, and continue beyond the same number,"--until you arrive at the desk with the treasure in it. The aim is to combine distances vectorially, to get where you are going to. One can extend this to three dimensions by going upstairs.

Captain Kidd
His
&
Treasure
Directions
to Calculate its
Exact position
From a latitude of
34° 52′ 10″ N, & a longitude 78° 32′ 17″ W
(or thereabouts) Starting from
the Schoolroom door, take six
paces directly forward, then turn
right, take eight paces. Count the
number of paces direct back to the
door, continue in the same sence as
many paces to locate the TREASure

Experiment 14.02 The Three-meter Dash (Kinematics)

To emphasize the difference between velocity and acceleration, we have two races--one a dash of three meters (ten feet), and the other a more normal length up to 50 meters, or whatever is available. The students who accelerate fast are not necessarily the ones who can do well over distance.

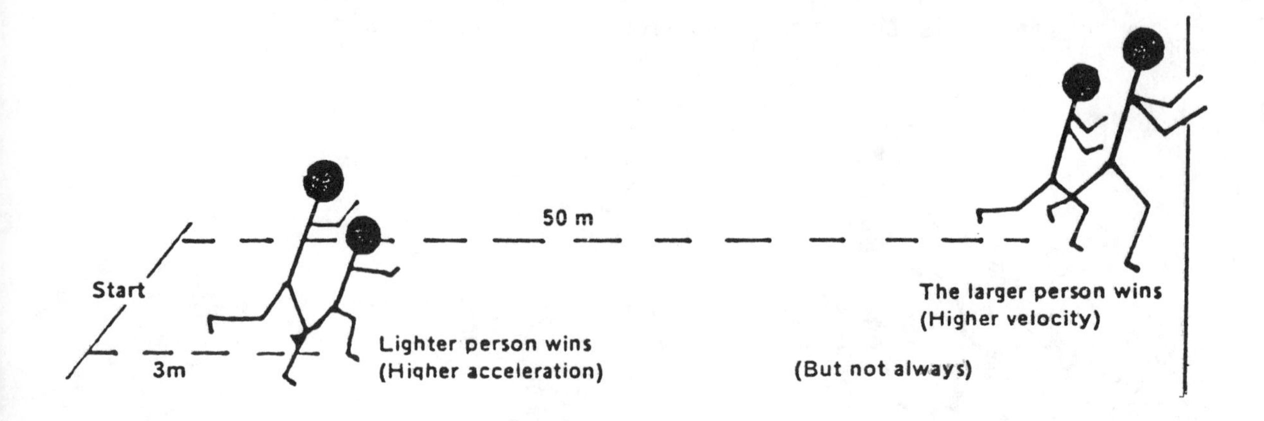

Experiment 14.03 The Knee-bend Game (Energy and Power)

This is due to Dr. J. Johnson. The participants do a knee bend, and the distance from some suitable part of their anatomy to the floor is measured. They then stand up, and the same measurement is made from this position. Most people know their weight, so the work done, mgh,mass times the acceleration due to gravity times vertical distance risen, is easily calculated. For example, if your mass is 50 kgm (110 lb) and the distance risen on standing is 60 cm, the work is 50 x 9.81 x .6 = 294 Joules. The work performed on rising is not regained on sitting--unlike a bicycle running downhill, we do not store the potential energy on doing a knee bend--it is lost as heat. Some student is bound to have a watch with a second hand, so the next portion of the game is to see how quickly you can do ten, twenty, fifty knee bends. The power is then the rate of doing work. If you do 40 per minute, in the above example, the power would be 294 x 40/60 = 196 Watts. (Joules per second). Generally, the rate of doing knee bends is about the same for men or women, but women weighing less, their power is also correspondingly less. One can also perform a similar game running up and down stairs and measuring the power required--however, one should avoid giving older students heart attacks.

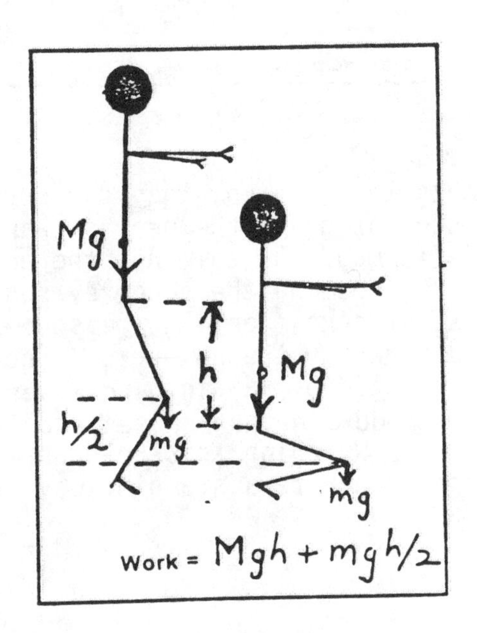

Experiment 14.04 The Wave Game

This one is great fun. Students stand in a line, fairly close to one another, and put their hands on the shoulders of the individual in front. The one in the front of the line rests his hands against a convenient wall. The last in the line gives a hearty push to the one in front, who (to avoid falling) pushes the one in front, and so on. When the front is reached, a push is given against the wall, and the compressive wave travels toward the back. About the only problem in this game is attenuation of the waves--a really good push is needed to avoid this. To simulate reflection at an open end, the last person in line pulls the shoulders of the individual in front, who pulls the shoulders of the next in front, and so on. The front of the line, on being pulled back, and having no one to tug on, falls back and "reflects" the rarefaction as a compression. This is not as self-generating as the first part. The wave will rapidly attenuate unless positive feedback is inserted--each student, on being pulled back, must make a conscious effort to pull back the student ahead. I have found, when the students see what is going on, that it makes understanding a difficult concept much easier--and enjoyable!

Transverse waves can be simulated by the last student pushing the one ahead sideways. Again, this travels to the front, where, if the student has nothing to hang on to, a reflection of the same sign occurs. If the student hangs on to a doorway, or other solid object, the pulse is reflected with change of sign.

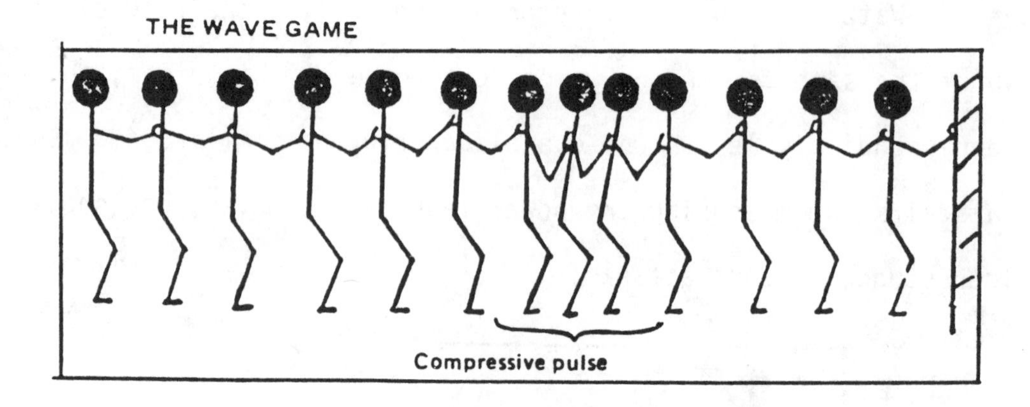

If you don't want the students to stand up, they can perform a similar experiment in their seats. Each student grabs the nearest hand of the student on either side, so that a chain is formed. On command, the end student squeezes the hand of the adjacent student who, feeling the squeeze, squeezes their nearest neighbor - and so on until the last student, feeling the squeeze, yells or puts his hand up. The time for this, divided by the number of students, is the reaction time. A transverse wave may be generated by raising and lowering the neighbour's hand, instead of squeezing. The neighbour in turn raises and lowers his other hand and that of his other neighbour. Watching this up-down move round the room is fascinating. One can, of course, generate a standing wave by reflection at both ends.

14.05 The Kinetic Theory Game

Since people are particles, the motion of particles represents a good opportunity for games. The aim is to demonstrate the three states of matter--solid liquid and gas--and the effects of pressure, volume and temperature. The demonstration of a gas involves a few students dashing madly about, bumping into one another, and the walls of the room. The faster they run, the more momentum delivered to the walls--the "pressure" is proportional to the "temperature". The smaller the space in which they can operate, the more often they strike the walls--the "volume" is inversely proportional to the "pressure".

Now, as more students are added to the gas, it "condenses" -- the students are effectively jostling about in a crowd. There is no "mean free path" as before, but motion still occurs . If, now, they attach to one another--hands or shoulders or otherwise--they form a "solid"--their long range order remains the same--the relationship of one to another remains fixed.

14.06 The Nuclear Reaction Game

Since nuclei contain more neutrons than protons, the sex in the majority in the class are neutrons. Elastic scattering--a neutron runs in and bounces off an alpha particle (two girls and two boys hugging hands around shoulders). The bouncing should be done using the hands only to prevent accidents.

Nuclear reaction--we can have a knockout reaction, where one child is knocked out of the "nucleus" of five or ten--or a pickup reaction, when one child is pulled off the group--or a high q reaction--children in the nucleus are hand to hand so they can rapidly push away from one another when triggered by an incoming "particle"--or a fission reaction--the "nucleus" splits into two with release of a neutron. One can let the nucleus be split on a random basis, and see what comes out. Is the product stable? Is the reaction exothermic (the students deliberately push one another away) or endothermic (only the impact of the foreign nucleon provides the energy to break up the nucleus). The multiplication which occurs when a reaction or bomb goes supercritical can be simulated by having groups of five (or more) students form a nucleus. When one "neutron" bumps into them, the nucleus fissions into two groups of two (or more) plus a neutron, which goes on to fission another nucleus. A class of thirty can provide six "nuclei", somewhat small for a reactor, but it does drive home the principle.

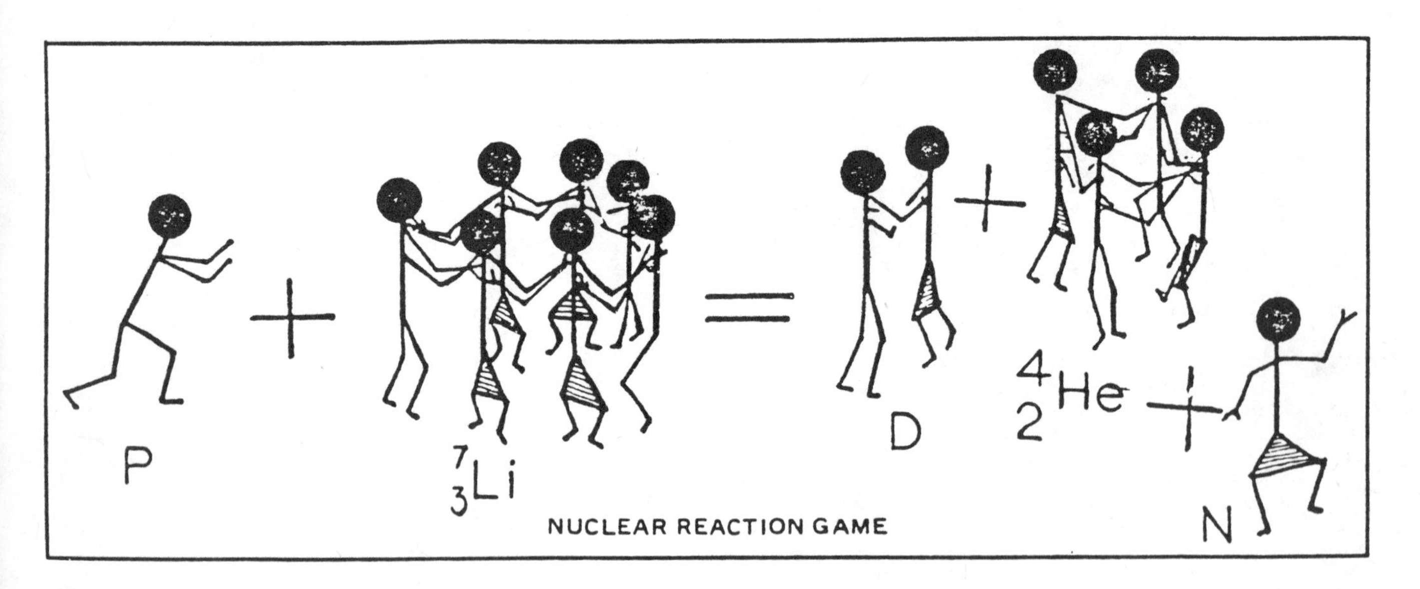

NUCLEAR REACTION GAME

14.07 The collision game

Basically two students place hands to hands, in the "pat-a-cake" position, and push one another away. Who moves fastest and farthest? Then, they rush toward one another and push away, in a similar "collision" process. See the effect of mass, with a small and large student -- and velocity.

14.08 The electron- in-a-wire game.

When we put a voltage on a wire, why don't the electrons run out the end? For this game, a small student represents the electron, to move from one side of the room to the other. Students stand, fairly close together, randomly, forming the atoms of the wire. The electron cannot move directly from one side of the room to the other, but must bump into several "atoms" in the process, coming to rest, and starting off again.

14.09. The close packing game.

This is a dilly for people who like togetherness. How close together can a group of people get in a crowd? A useful aquisite is a tape measure, or piece of string, to be placed round the outside of the group. Different groups pack differently. For example, if you have large and small students, they should alternate, like atoms in sodium chloride. If they are all the same size, and very round, they pack hexagonally as do metals. This two-dimensional close packing is the same as one layer of either face centered cubic or hexagonal close packing in three dimensions. Raising hands makes the students rounder, and gives tighter packing. Hands down, we are smaller front to back than side to side and pack most closely as shown in figure (C) in a layered structure.

It seems the change in perimeter for people-shaped objects packed differently is much less than for circular objects. (a) in the figure has the largest perimeter, but (b) and (c), the random and close packed arrangements are very little different--only an inch or two. Nevertheless, it does prove easier to explain the concept of close packing in solids after the experience.

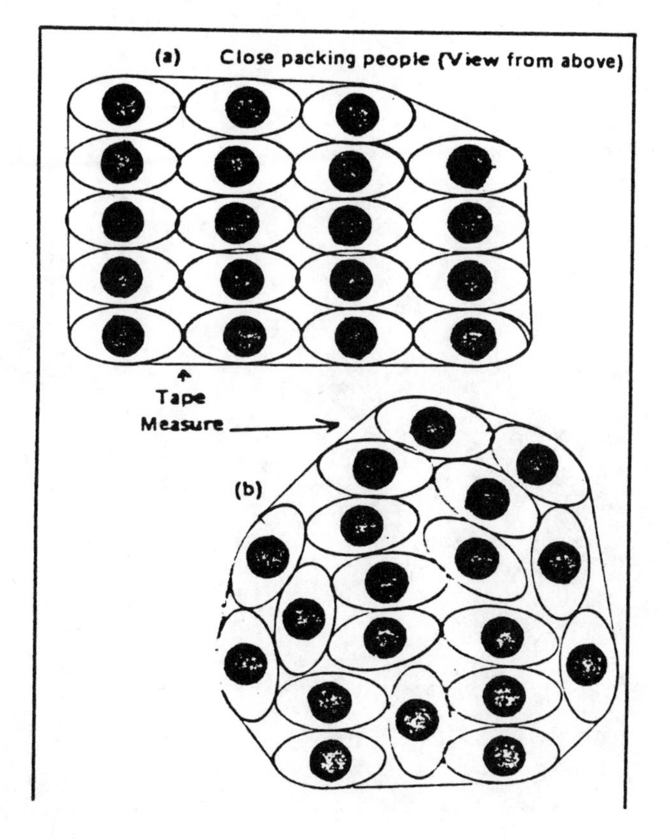

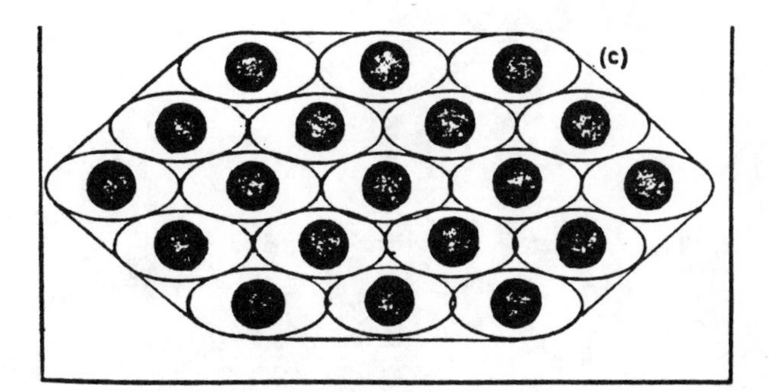

14.10 Lifting the body - Force

One student lies prone on the ground. The others gather round, and lift him (or her) by putting one hand under the body to lift up. If enough students are around, then one finger from each is enough. Divide the weight by the number of students lifting. The student can then be raised head high, and passed from hand to hand along the two supporting rows of students.

14.11 Reaction Kinematics

Have you ever had a problem trying to explain statistical equilibrium in a thermo class? Try Vampire, a game from Transylvania, of course (New Games p. 123). Students close their eyes and mill around. The vampire keeps her eyes closed, but when she bumps into someone else, there is a difference. She snatches him and lets out a blood curdling scream. The victim becomes a vampire as well, on the prowl for new victims. Now for the physics (not in the reference), let us suppose there are N students, who each collide, on an average, every τ seconds. There are N/τ collisions per second. If we have n vampires, the probability of a collision between a vampire and a non-vampire is $(N-n)/(N-1)$ (we subtract the 1 for the vampire doing the colliding). So the rate of vampire production is (number of vampire collisions/sec)(probability of a collision with a non-vampire)

$$= (n/\tau)\ (N-n)\ /\ (N-1) = dn/dt$$

where t is the time

This means the rate of vampire production increases very rapidly as time goes on--exponentially in fact, since, if N is large $(N-n)\ /\ (N-1) \simeq 1$ at first , and so $dn/dt = n/\tau$, and $dn/n = dt/\tau$ Integrating, $\log n = t/\tau + \text{constant}$, and $n \propto \exp\ (t/\tau)$ This is a dramatic effect. Every few seconds the referee stops the game and counts the vampires. Plotting vampires versus time gives a marked exponential increase.

This exponential growth is very typical of a number of different chain reactions--the best known being that occuring in a nuclear reactor, where the reaction grows exponentially once criticality is reached.

As n approaches N, $n\ /\ (N-1) \simeq 1$ so $(N - n)\ /\ \tau = dn/dt$

$$dn/(N - n) = dt/\tau$$

$$\log (N - n) = - t/\tau + \text{constant}$$

$$(N - n) \propto \exp. (- t/\tau)$$

So, towards the end, n approaches N more and more slowly, because fewer and fewer non-vampires are left. Nevertheless, ultimately everyone becomes a vampire and the game would end.

To avoid this we insert the additional condition that when two vampires feast on each other, they transform back to normal mortals. Now, our rate of vampire production has a negative term proportional to the number of vampire-vampire collisions,

(number of vampire collisions) x (probability of a vampire collision with a vampire)

so

$$\frac{dn}{dt} = \frac{n}{\tau} \times \frac{(N-n)}{(N-1)} - \frac{n}{\tau} \frac{(n-1)}{(N-1)}$$

We shall get statistical equilibrium when the rate of vampire production = rate of non-vampire production, i.e. when dn/dt = 0 which will occur when

$$(N-n) = (n - 1) \quad \text{or} \quad N + 1 = 2n$$

i.e. almost half the students will be vampires.

See how long this takes, and, by counting vampires again, how accurately the statistical equilibrium is maintained. Chemical reactions, of course, are of this nature, where, when two molecules collide they may react to give a product, but when the product molecules collide, they can reverse the process and give the original constituents. The rates of reaction are proportional to the product of the density of the two constituents reacting, as in our game. The nice feature of the game is that statistical equilibrium comes at about 50%, so it is easy to picture what is going on.

A living experiment in statics (p 57 New Games Book), Four stout hearted hulks on hands and knees form the bottom layer; three mid sized inner directed individuals climb on their backs to form the next level, and two small courageous acrobats above them. Top it off with one light, expendable child. This two dimensional force pyramid is shown in the figure 2, to demonstrate the distribution of weight. Knowing the mass of the people involved one can easily calculate the forces. With a large group, a round base, like a rugby scrum can be used, with succeedingly higher levels--a Spanish specialty which can reach great heights (calculate the forces--several hundred pounds)

From the Figure:

$$F_1 = m_1\, g/2$$

$$F_2 = (F_1 + m_2\, g)/2 = (m_1/2 + m_2)\, g/2$$

$$F_3 = (m_1/2 + m_3)\, g/2$$

$$F_4 = (F_2 + m_4 g)/2 = ((m_1/2 + m_2)/2 + m_4)\, g/2$$

$$F_5 = (F_2 + F_3 + m_5 g)/2 = ((m_1/2+m_2)/2 + (m_1/2 + m_3)/2 + m_5)\, g/2$$

The weight carried by the eighth participant is

$$F_4 + F_5 + m_8 g = (3\, m_1/8 + m_2/2 + m_4/\, 2 + m_5/2 + m_3/4 + m_8)g$$

so, if all participants weighed the same, this would be

$3\,{}^1\!/_8\, Mg$

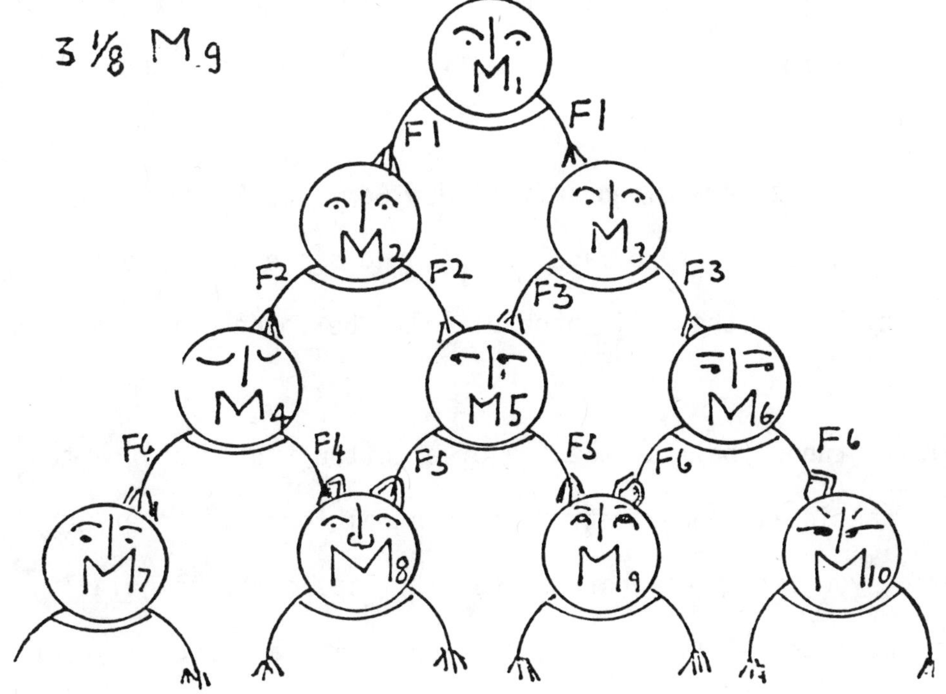

EXPERIMENTS INVOLVING MINIMUM EQUIPMENT

So far we have been looking at games involving no equipment and many students. Let us extend this to minimal equipment and two or more students. This would help involve and interest those students who believe physics is a dull subject. It is true, there is much hard labor in a study of physics--but too often students are put off before they really get started because the concepts involved are unfamiliar. If they can familiarize themselves with these concepts, then they are more likely to take the trouble to understand the mathematics and principles involved.

An important objection to games of this nature is that their content is low for the time spent; on the other hand, however, the impact on the student is high. It is quite surprising how much more firmly the concepts stick in the mind when associated with a pleasurable experience. Games such as these exemplify physics as an attitude, or a way of looking at things--an analytical approach--rather than a bunch of equations which must be memorized. Too often, we forget we look at life through different eyes from non physicists. We picture an ocean wave in terms of the motion of the water particles, the energy transfer--hydrodynamics. A "normal" person sees only the aesthetic side of it

Several such games are given in the "New Games Book" mentioned earlier. These appeal to a wide variety of age groups, and since their motto is "play hard, Play fair, nobody hurt", they will probably also be found suitable by the teacher.

Schmerlz (a rubber ball at the toe of a cotton tube sock), a toss and catch game, is the ideal demonstration for central forces. After whirling it round your head, you let go--and of course, it flies off tangentially to be caught by your partner. A fascinating game--but also a lesson in central motion and trajectories, if you think about them.

14.13 STAND UP

Stand-Up is a good game for two (or more) to show the principles of resolving forces. Sit on the ground back to back with your partner, knees bent and elbows linked. Now simply stand up together. You get a good intuitive feel for what is going on force wise. In addition to the vertical force you normally use to stand, participants exert equal and opposite horizontal forces on one another. Hence, the forces through the legs are as shown in the figure. Although more force is required than normal, it is in a direction which helps the pair to stand. When more than two participate, the force problem becomes a little more difficult, nevertheless, the forces between participants are basically horizontal. (p 65 New Games Book)

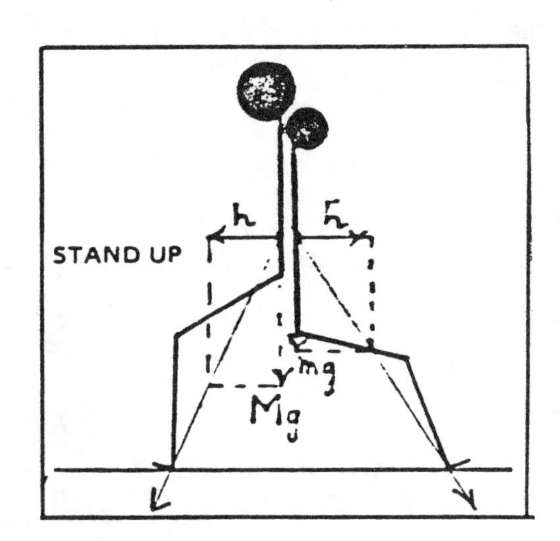

14.14 HUNKER HAWSER

Hunker Hawser, another game for two , is a very practical demonstration of Newton's third law. The players hunker down on pedestals, which can be a block of wood, an inverted trash can or flower pot, set six feet apart, each holding on to the end of a rope about one inch in diameter and at least fifteen feet long. The excess rope lies coiled between them. At the starting signal players reel in. The object is to unbalance your oponent by tightening or slackening the rope. This is not as simple as it sounds, because if you give a good tug, your opponent may just let the rope slide, and over you go. However, it is clear that an impulse on the rope is transmitted to your opponent if he doesn't let go, and an equal and opposite impulse acts upon you. Instead of an impulse, a steady force can be applied if both lean back--but this is even more dangerous. In any case, the two factors--equal and opposite--are really brought home with a vengeance in this game.

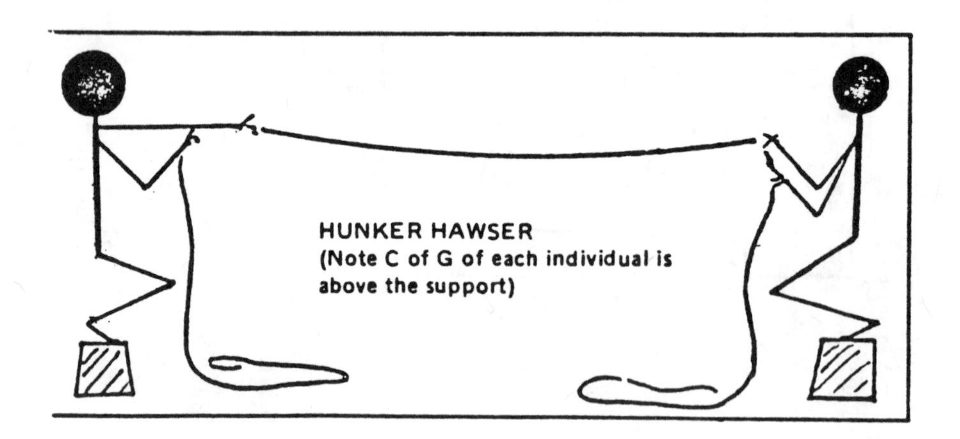
HUNKER HAWSER
(Note C of G of each individual is above the support)

14 CONCLUSION

INTRODUCING PHYSICS TO THE YOUNGEST STUDENTS

Particular attention to games should be given for younger students. Older students find physics difficult because the concepts are novel and unfamiliar. It is well known that, unlike languages or history, the first physics course is the most difficult. For this reason, the younger a student is when he is introduced to physics, the better. An excellent example of the way in which this can be done is given in Peter Werner and Elsie Burton's book, "Learning through movement"[3]. This is geared to primary grade children, and in chapter eight, for example, they show how children on a seesaw can be introduced to the idea of levers. It is a question of the teacher asking the simple questions how a heavy child can balance a light one--and the explanation given by the teacher and confirmed experimentally by the students. A concept is being taught which can be applied to opening doors, lifting objects with our muscles, using a wedge to split wood--once the concept is understood, its applications are obvious--specifically in a tug of war, each child must get his body close to the ground by lowering his center of mass to obtain maximum pulling force. Chapter nine of the same reference concerns Newton's laws--the concept of inertia is particularly relevant in football--both in starting and stopping the ball, and stopping and starting the players--but it can be exemplified in many other ways also. They go on to talk about factors affecting the human body--mass, force and work as we move about, and how these can be incorporated in learning activities The association of ideas--work as force times distance, the effects of friction, air resistance, water resistance in swimming--can all be introduced at a very early age. It is merely the way in which a child thinks about these things which differs--a stimulation of the inate curiosity in all of us.

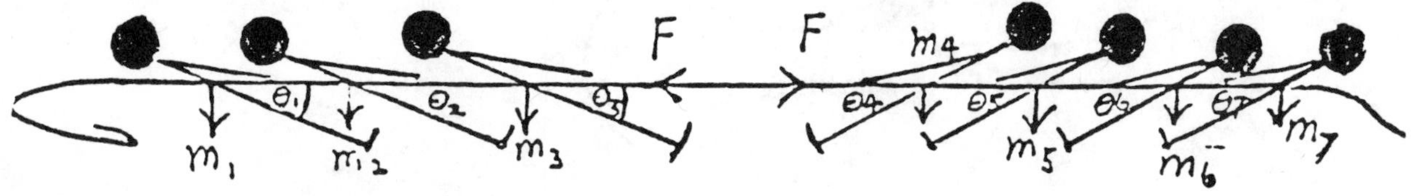

$$m_1 g/\tan\theta_1 + m_2 g/\tan\theta_2 + m_3 g/\tan\theta_3 = F$$
$$= m_4 g/\tan\theta_4 + m_5 g/\tan\theta_5 + m_5 g/\tan\theta_6$$

TUG OF WAR

Experiment 15.01 Reaction - The Rocket and Balloon

Materials: - aluminum foil - match, paper clip

Procedure:

Wrap aluminum foil around upper half of paper match. Push straight pin up under foil to head of match and remove again, leaving an exhaust channel. Place match on opened paper clip and hold lighted match to tip. Step back. The foil must cover the whole of the match head in one piece, and not be punctured. Two or three match heads may be incorporated - after all, two heads are better than one. A most unpleasant odour is given off - better do it outside.

S. J. Tweedie and F. D. Woodruff, Falls Church, Virginia.

Taken from: - The Great International Paper Airplane Book by Mander, Dippel and Gossage, Simon and Schuster- New York.

An alternative is to blow up a balloon and release it. It flies off in all directions, but a suitable tail of cardboard can insure it flies straight.

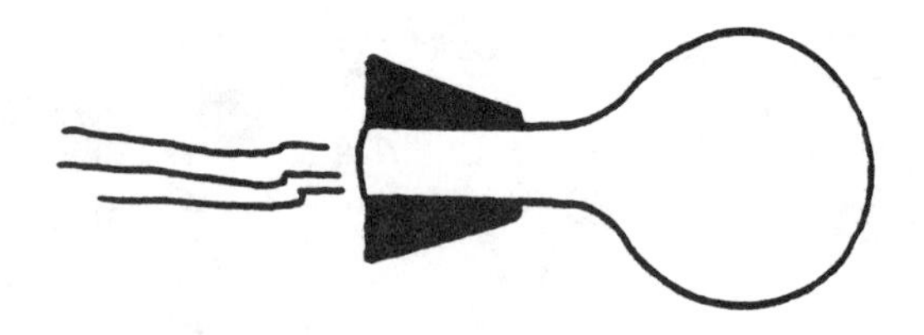

Experiment 15.02 The Platonic Solids

Materials: Straws, paper clips

Procedure: The regular solids of Plato are interesting to construct. The five solids are shown below.

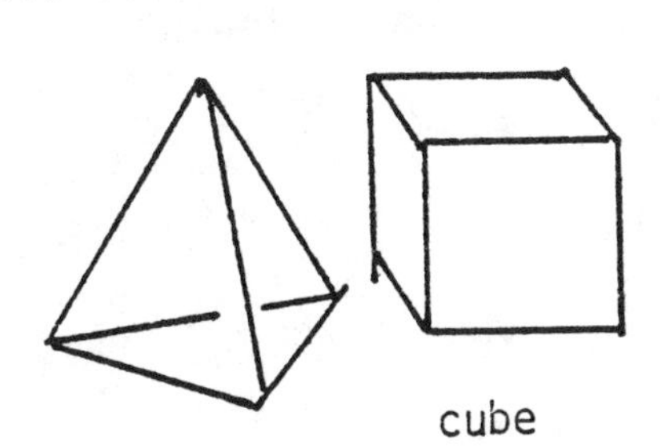
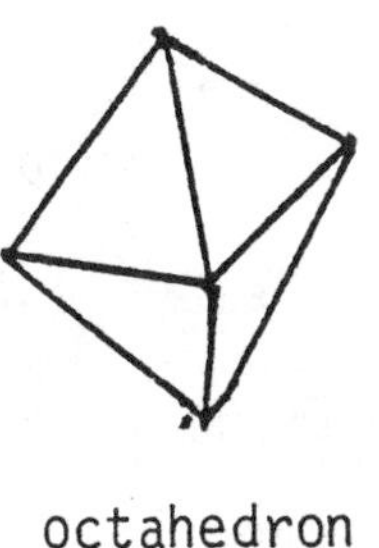
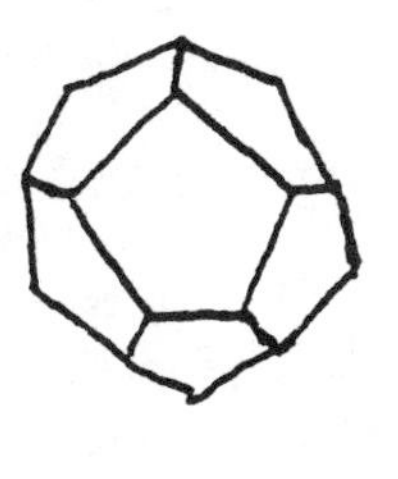
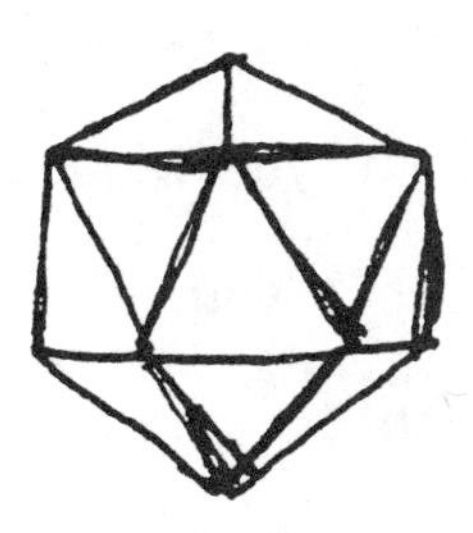

tetrahedron cube octahedron dodecahedron icosahedron

Three of the five can be constructed very easily using straws and #1 paper clips. To construct the tetrahedron, loop three paper clips together, as shown.

Now, stick the small end of each clip into a straw, so you have the ends of three straws attached together. Insert three more such looped sets of clips into the other open ends, and make up triangles by joining pieces of these sets together. In this way you will build a tetrahedron. If you loop four clips together for each corner, you will make an octahedron, and five makes an icosahedron, which has twenty sides. Six gives only a flat plane of triangles--so in a sense, it is a solid with an infinite number of sides. The cube and the pentagonal dodecahedron cannot be built in this way, since the figures are not freestanding.

Count the number of sides, edges and corners of these figures, and try to relate them.

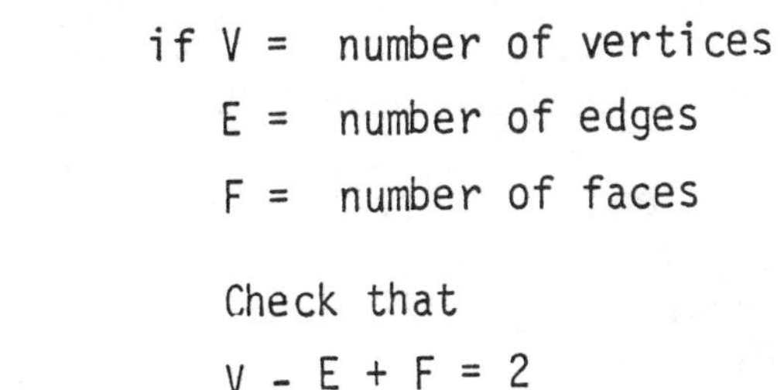

if V = number of vertices

E = number of edges

F = number of faces

Check that

$$V - E + F = 2$$

Experiment 15.03 - A string and sticky tape digital computer

Although today every student has a digital calculator, how many know what goes on inside? Perhaps they are afraid to find out! In the following experiment you will learn how to build a digital computer which will (at least until you understand it) beat you at the ancient Chinese game of NIM. One of the very first digital computers in England was displayed at the Festival of Britain in 1951 - and it played NIM with all comers. The computer we are building plays NIM to near perfection - but unfortunately, that is all it can do - however, in the process it will inform you, pleasantly, exactly how digital computers work. Glue or tape figure 1 to a piece of card 8½" x 11" (such as the back of a writing pad). This is the baseboard of the computer. Cut two-inch pieces of drinking straw to fit over C, D, E, F and G and a longer piece over H, and tape them down where shown. Now we must make the "flip flops". Two of these are required. Cut a piece of straw 2½ inches long and another 1 inch. Take a piece of tape about 4 inches long, lay the longer piece of straw across it, and tape the shorter piece to the center, as shown in figure 2. Put another piece of tape around the shorter piece as shown to ensure the first piece sticks. Pierce the baseboard from the front with thumbtacks at the points I, J, K, L, M, N, O remove them and push them through the holes from the reverse side so that they stick out from the front. The two flip flops must be pushed over K and M as shown. Now, take them off and enlarge the pivot hole with a pencil point - the flip flop must move very easily. Cut a straw 5" long to act as the indicator Q. It is pivoted at O to move from "computer to "player" and back. Score the line RS and fold the baseboard up by 90° along this line. Stick pieces of tape to hold it at 90° at each end, as shown. The marbles will be held here so as not to run off the table. Score also TU, and fold down 90° - this forms the stand for the computer and is inclined so the marbles run downhill and work the flip flops. Your computer is now complete. Put 15 marbles into a cup and you are ready to beat the computer. Here is what you do.

1) Set the flip flops as shown in the figure.

2) The rules of the game are, that you may take one, two or three marbles. It is then the turn of the computer, who may also take one, two or three marbles - then back to you. The one who leaves the other player to take the last marble is the winner.

Set the pointer to player. Place the number of marbles you select, one at a time at "INPUT". The first marble moves the first flip flop so its vertical arm moves to the right. The second marble moves it back and goes on to the second flip flop. When you have made your selection, move the point to "computer" and continue to put in marbles until the pointer moves back from "computer" to "player". Now it is your turn again. If you give no thought to how many marbles you choose, you will almost certainly lose. A student playing the game for the first time wonders whether the computer can "think" - you will probably answer no - but then, neither does the largest computer "think" in the way that you or I do. You have constructed a true binary computer, for the "flip flops" you have built can count up to two each, so together they count to four - and this is all they can do. Each "flip flop" has two stable states - and each time a marble drops onto it, it changes from one side to the other - a logic element.

You have constructed a device which

1) Has two logic elements,
2) Is a computer which makes logical decisions based on what state the logic element is in - so they must be set correctly, the computer taking its turn.
3) The computer changes the state of its logic elements as the game progresses.

4) The computer "remembers" the state of the game between moves - i.e. it has a memory.

5) The computer can direct the marbles in the correct channels to win the game.

6) The computer is "programmed" by positioning the elements at the beginning of the game.

Counting, logical functions, altering the internal states and memory are all typical of a computer, and they combine to give the appearance of playing an intelligent game - i.e. if the computer were hidden, you would be unable to decide whether it was an intelligent person or a machine playing.

"Digital" refers here to the logic elements having only two stable positions which are commonly designated "0" and "1" - we could say 0 was to the left, and 1 to the right - the two states for each element make this a binary system.

We could write a flow chart for the computer as in Fig. 3 for each marble. Let us see how this works in a special case starting with 15 marbles. If the flip flop has it's center pillar to the left, we put a bar over it, as $\bar{A}$, if to right, it is A. We set the flip flops to start so the first flip flop A points left, $\bar{A}$ and the second flip flop B points right. This is the essential ingredient if the computer is to win. Let us play a simulated game - we play first and take one marble. The computer than takes 1 marble - and moves to pointer to player. Fig. 3 shows the flip flop orientation.

	Marble Number			
player	1.	$\bar{A}$	B	The first marble flips $\bar{A}$ to A, but does not affect B
	2.	A	B	The next marble flip A, and goes on to flip B
	Pointer Moves			
Player takes 2	3.	$\bar{A}$	$\bar{B}$	
	4.	A	$\bar{B}$	Note A flips each time, but B only every other time
	5.	$\bar{A}$	B	The flip flops return to their original position
computer	6.	A	B	
	pointer moves			
	7.	$\bar{A}$	$\bar{B}$	
player	8.	A	$\bar{B}$	
	9.	$\bar{A}$	B	
computer	10.	A	B	
	pointer moves			
	11.	$\bar{A}$	$\bar{B}$	
player	12.	A	$\bar{B}$	
	13.	$\bar{A}$	B	
computer	14.	A	B	
	pointer moves			
	15.	$\bar{A}$	B	

We Lose!

This type of calculation is called Boolean algebra, which is the basis of all digital computer operation, whether a string and sticky tape device, or the largest fastest computer. We stopped the computer (i.e. switched from the computer to player) because the pointer shifted when the first flip flop moved to $\bar{A}$ and the second $\bar{B}$. In Boolean algebra this would be written

STOP = $[\bar{A} \cdot \bar{B}]$. - i.e. the computer stops when the states are $\bar{A}$ and $\bar{B}$ at the same time. All the basic operations of a digital computer, then, can be carried out by our string and sticky tape computer.

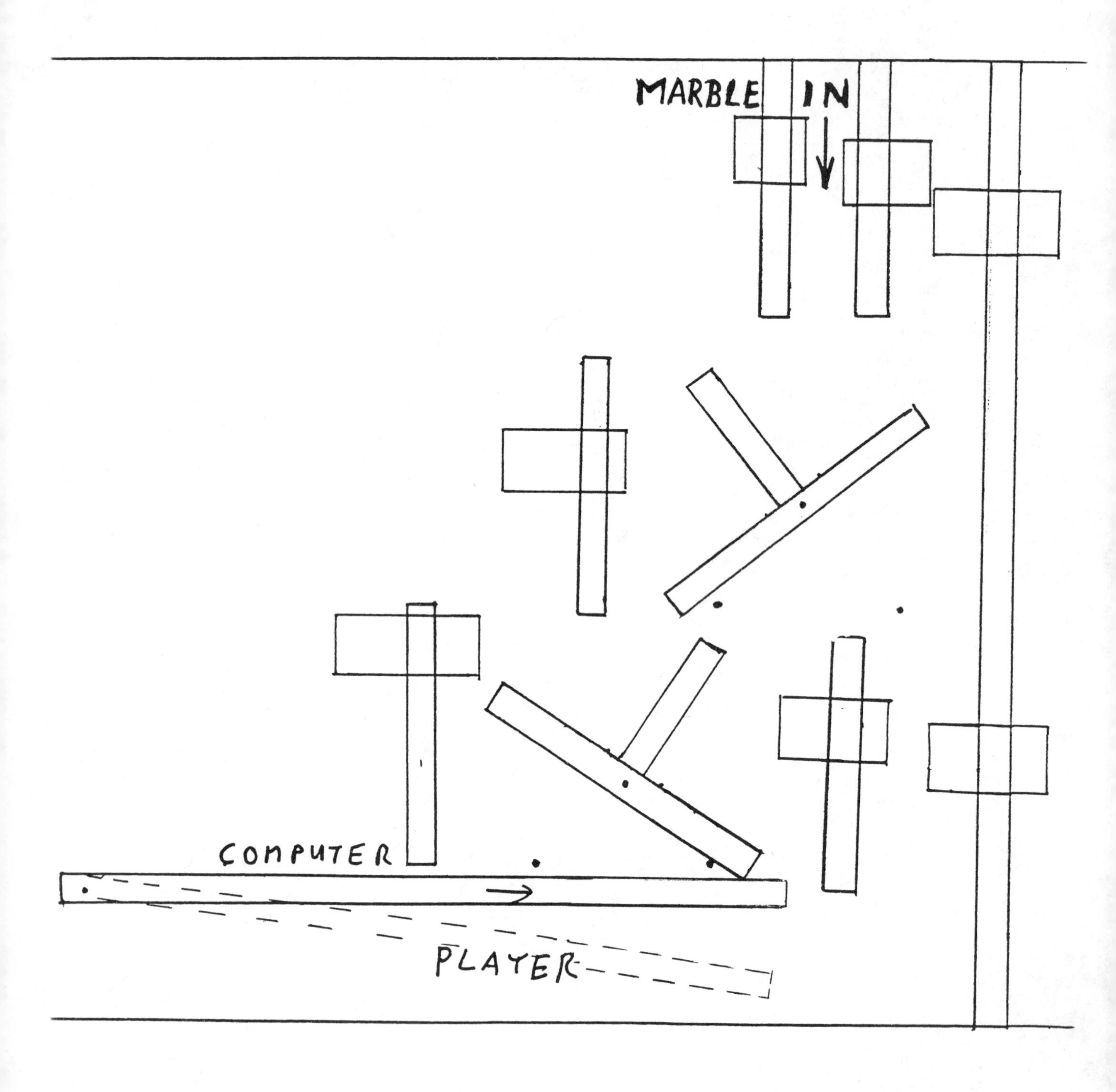
MARBLE IN
COMPUTER
PLAYER

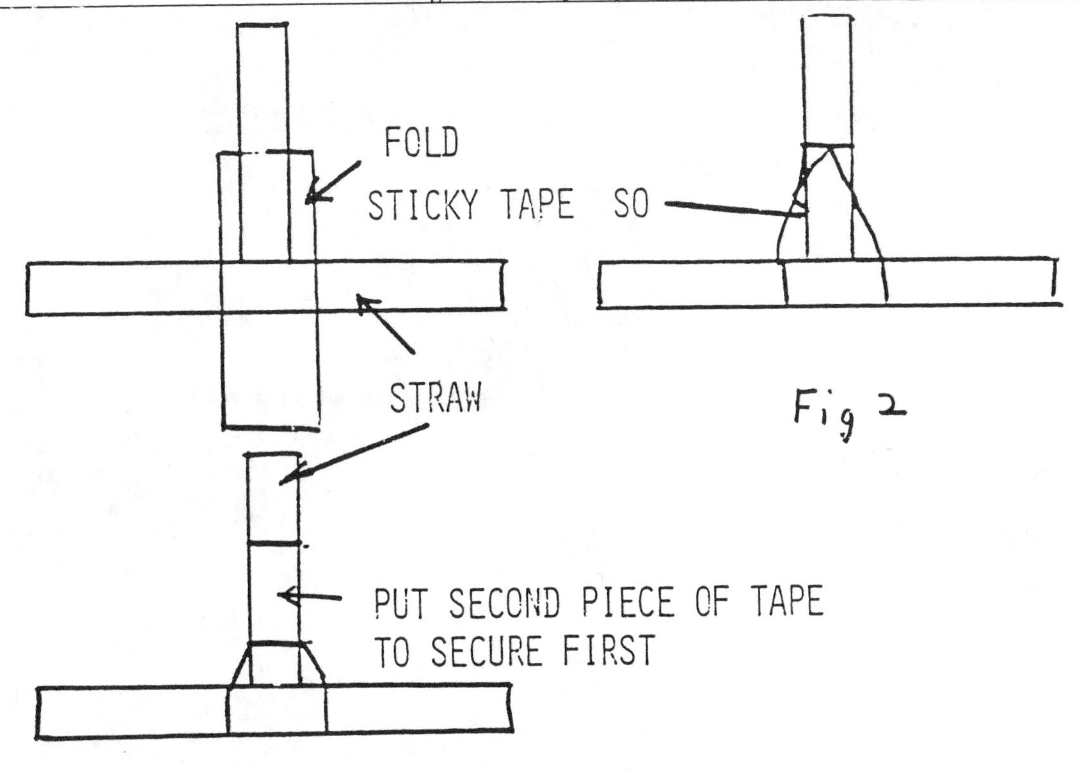
FOLD
STICKY TAPE SO
STRAW
PUT SECOND PIECE OF TAPE
TO SECURE FIRST
Fig 2

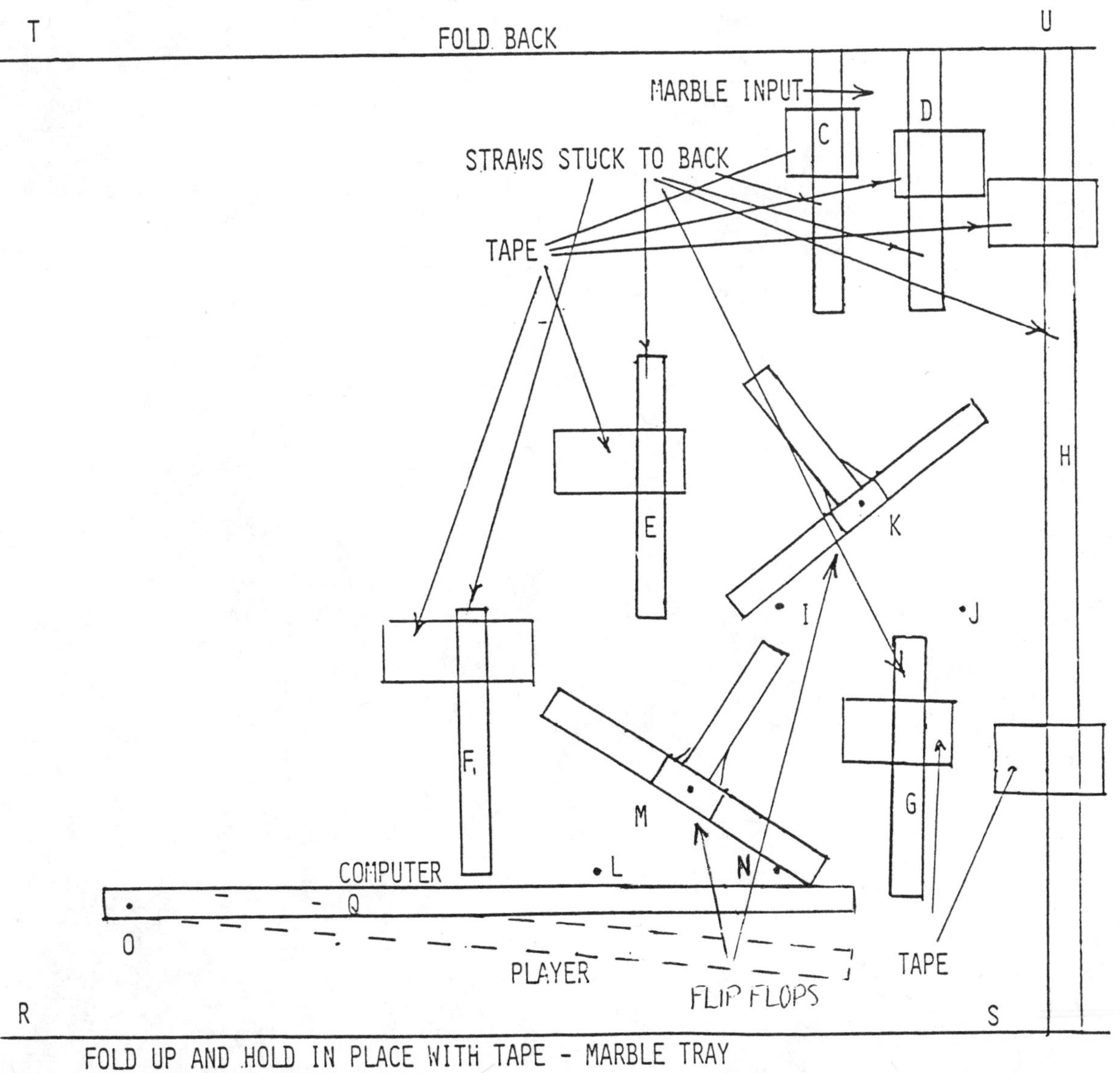
T
FOLD BACK
U
MARBLE INPUT
C
D
STRAWS STUCK TO BACK
TAPE
H
E
K
I
J
F
G
M
COMPUTER
L
N
Q
O
PLAYER
TAPE
FLIP FLOPS
R
S
FOLD UP AND HOLD IN PLACE WITH TAPE - MARBLE TRAY

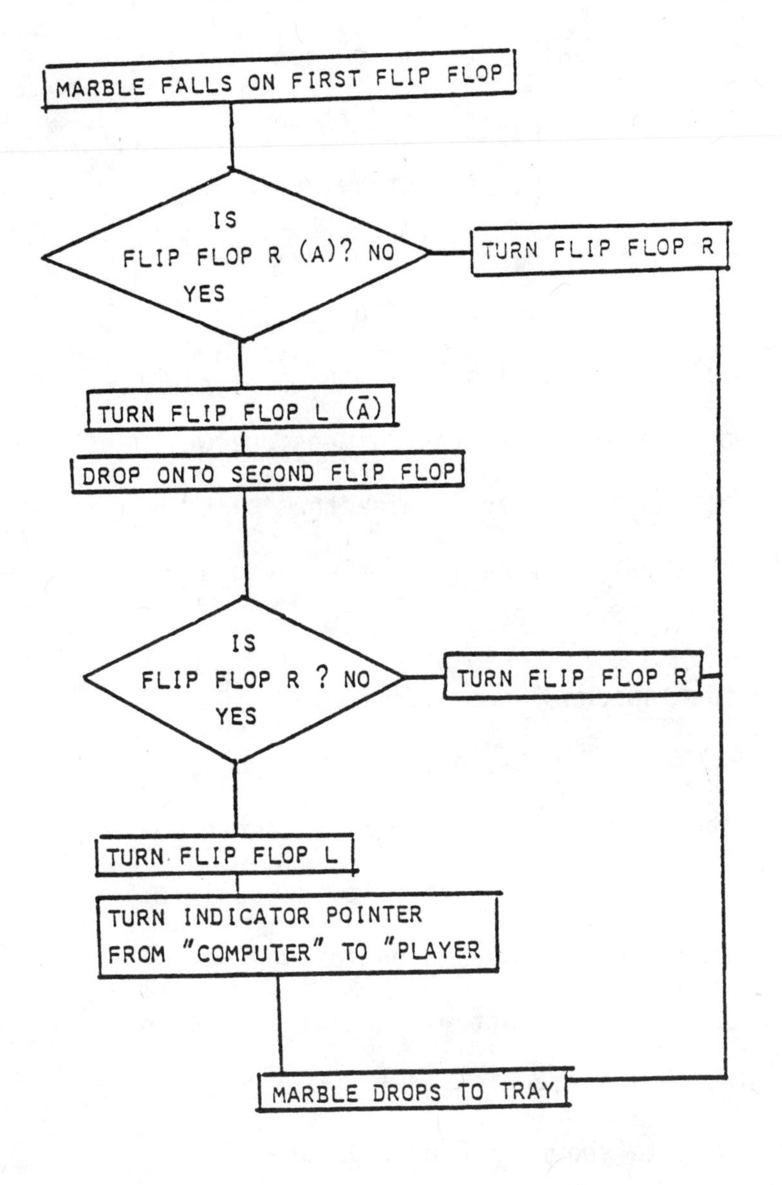

POSITION OF FLIP FLOPS & INDICATOR

FLIP FLOP B FLIP FLOP A

START $\bar{A}$

B

INDICATOR

COMPUTER

AFTER
FIRST
MARBLE

AFTER
SECOND
MARBLE

POINTER
MOVED FROM COMPUTER
TO PLAYER

PLAYER

AFTER
THIRD
MARBLE

AFTER FOURTH
MARBLE

FLIP FLOPS
BACK TO
ORIGINAL
ORIENTATION

ETC

Fig 3

Experiment 15.04 - Four Dimensional Tic-Tac-Toe

It is often difficult to explain the meaning of four-dimensional space to students. Here, as elsewhere, "understanding" four dimensions means using it. The simple game of tic-tac-toe seems to be ideally suited to demonstrating this. We start with the conventional delineation whown in the left top of Figure 1. Each player alternately puts down an X or 0, respectively, trying to obtain a line of X's or 0's. The extension of this game to three dimensions is straightforward--we have three boards laid out as shown in Figure 1, and imagine them superimposed on top of one another. Examples of winnning straight lines on the cube of three boards is shown. To extend to four dimensions we need nine boards as shown in Figure 2. One can no longer imagine them piled on top of one another, but again, a winning straight line is fairly obvious--several examples given by numbers are shown.

Tic-tac-toe may be extended to have four spaces in each dimension. Such a game in two dimensons proves useful in discussing finite, but not bounded surfaces. In Figure 3, we imagine the left hand edge of the board attached to the right hand side of the board, forming a cylinder as shown. It is possible to play on such a cylindrical board, which only has a top and bottom but no sides. We can now attach the top of the board to the bottom, so that the board has no edges at all. It is not really possible to visualize this, but it is easy to use it. a situation for such a board is shown in Figure 4, where we extend the board beyond the overlap on all edges. A sequence of four X's and 0's can continue across what was an edge of the board, as shown. A surface such as this is obviously finite, but has no boundary, which is also a property of a sphere. Note in the specific example shown, we actually have a continuous straight line of X's, much as a great circle on a sphere is a line with no end.

This can be extended to three dimensions, and proves handy in discussing finite, but not bounded, three dimensional game with four spaces is available as "Qubic" manufactured by Parker Brothers.

This game proves quite useful in helping students think in three dimensions, which is very necessary in X-ray crystallography and other subjects where one must visualize molecular and crystal structure.

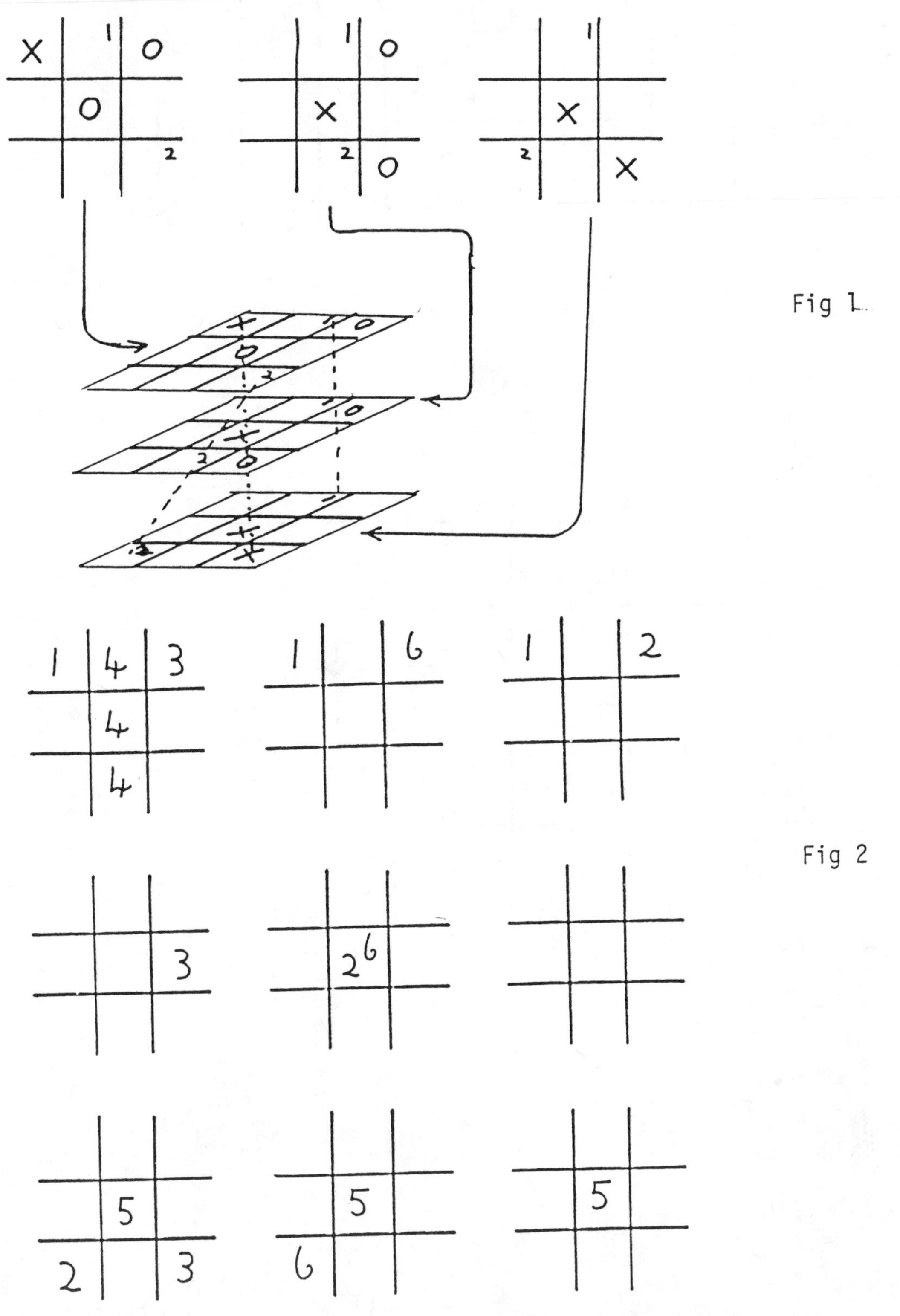

Fig 1

Fig 2

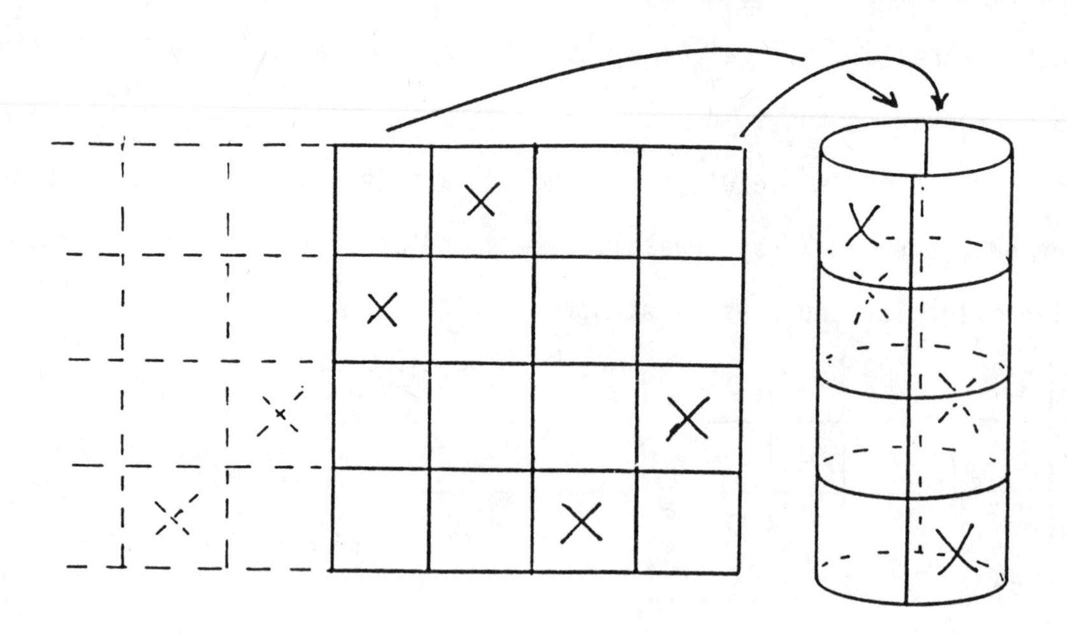

Fig 3

Fig 4

Experiment 15.05 - Simple Vacuum Experiments

We have seen how string and sticky tape may be employed to investigate mechanics, sound, optics, even heat - but what about vacuum experiments? At first sight, it would seem one would need an expensive vacuum pump, a bell jar, yards and yards of vacuum tubing, and so forth. However, there are quite a few experiments one can do simply using preserve jars, such as Mason or Ball jars. These are designed to withstand the vacuum used in canning, and the high temperatures involved in this process. The larger jars (32 oz.) are most useful - but a little dangerous.

Basically, we place a little water in the jar with the lid loosely attached. The water boils, steam fills the jar, escaping and completely displacing the air. We carefully screw the top on, holding the jar with a towel in order not to get burned. In fact, the jars are mostly self sealing, so that as they cool down the vacuum holds the top on with a force of 1.01×10^5 N/m² or Pascals (32 lb/sq in) multiplied by the area of the top (diameter 6.5 cm giving about .0033 m², or 5 sq in). This is a force of 333 N (165 lb. wt.). It follows, take great care of the evacuated jar - an implosion caused by dropping it could be dangerous. The pressure due to water vapor inside the jar at room temperature (30°C) is 32 mm Hg (4,252 Pa, .042 atm). If you put the jar in the freezer, it will fall to 4 mm Hg, (532 Pa .0052 atm) which is quite a good enough vacuum for most experiments, and, furthermore, the water will freeze on the bottom of the jar, so it will not run when the jar is shaken or moved.

What kind of experiments can be performed? The favorite is generally showing that sound will not travel through a vacuum. It would be convenient to glue a hook (a paper clip) to the lid, but unfortunately the boiling water generally prevents this holding, so a support, such as shown in fig. 1, can be bent out of a wire clothes hanger. A small bell, a bunch of keys, or anything which will jingle, is attached by a short length of string or thread to the support. A rubber band support is better, but tends to get destroyed by the boiling water. The object is to provide no path for the sound to escape via the support. Shake the jar, and convince yourself that you can hear the objects jingling. Now, evacuate the jar and repeat the process. Can you hear the keys? Pry the lid off a little to allow some air in. How much air is required before the keys become audible? In letting the air in, notice how much force is required to remove the top - 333 N (or 165 lb. wt.)

A second experiment shows how a balloon expands in a vacuum. A rubber balloon may be used, but it tends to deteriorate in boiling water. Sealable polyethylene sandwich bags work quite well, provided they are sealed quite tight. Put the bag, or balloon with just a little air in, in the jar. Boil the water and seal the jar - the balloon or bag will rapidly swell to fill the whole jar. Now let the air in slowly and note how the bag collapses.

The next experiment is to show air has weight. We need a small, sensitive balance. Poke about three holes through a soda straw, as shown, the middle one a little above the others. Support a small plastic bottle (one the heat will not warp) or piece of styrofoam from one end, using a paper clip, and attach enough paper clips to the other end just so they outweigh the ball. Put the balance in the jar, as shown, and evacuate. Whereas previously the bottle or foam was buoyed up by the atmosphere, in the jar the vacuum does not do this, and the mass of air inside the object will cause it to outweigh the clip - or, more accurately, there is no longer air being displaced by it so it weighs more heavily. This is not an easy experiment - it requires quite a lot of adjustment to get the balance just right - and the support must be reasonably free - tapping the jar helps here. The next experiment should be done immediately after removing the jar from the stove and screwing the top on. Pour cold water (ice water is best) on the top. The water inside will immediately start boiling again. Water condenses on the cold top, and because the pressure is now below equilibrium for the hot water, it boils. If one inverts the jar and puts it on ice, the water will freeze on the lid, because the freezing point of ice at low pressure is just a little less than at room temperature. (Remember the pressure of ice skates melt the ice under them? This is just the opposite.) Unfortunately, one atmosphere only raises the freezing point .007°C, so you may have difficulty with this one. A little salt on the ice will cover the freezing point so the water inside the jar freezes. Furthermore, at these low pressures, the water drops on the walls will rapidly evaporate and condense on the lid.

Another experiment is a variation of the "guinea and feather tube". Unfortunately, the feather will stick to the wall, because of the surface tension of the water in the jar. However, if you make up a little object such as that shown in fig. 3 out of thin plastic, the object will not stick because the edges and not the sides touch the walls of the jar. Put a penny and this object in the jar and note they both fall rapidly on turning the jar upside down. In air, the plastic object falls slowly. Shake the jar, and note the object moves from one end of the jar to the other because of its inertia. It will not do this if the jar is full of air.

One of the simplest and most spectacular vacuum experiments is to take an empty soft drink can, boil a little water in it until steam is emitted vigorously, then rapidly invert the can and plunge it, open end downward, in a basin of cold water. The steam immediately condenses and collapses the can.

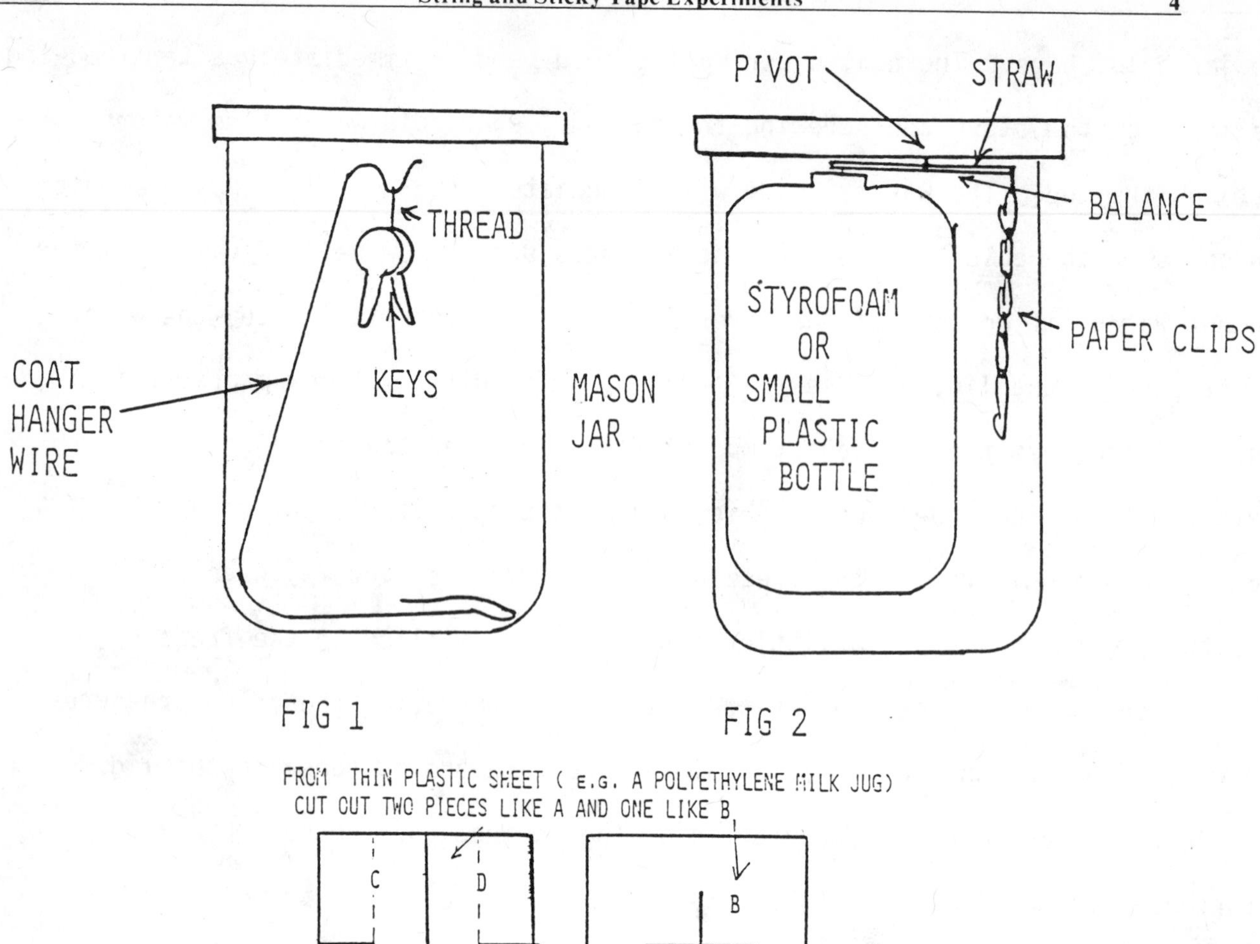
THREAD
KEYS
COAT HANGER WIRE
MASON JAR
PIVOT
STRAW
BALANCE
STYROFOAM OR SMALL PLASTIC BOTTLE
PAPER CLIPS

FIG 1

FIG 2

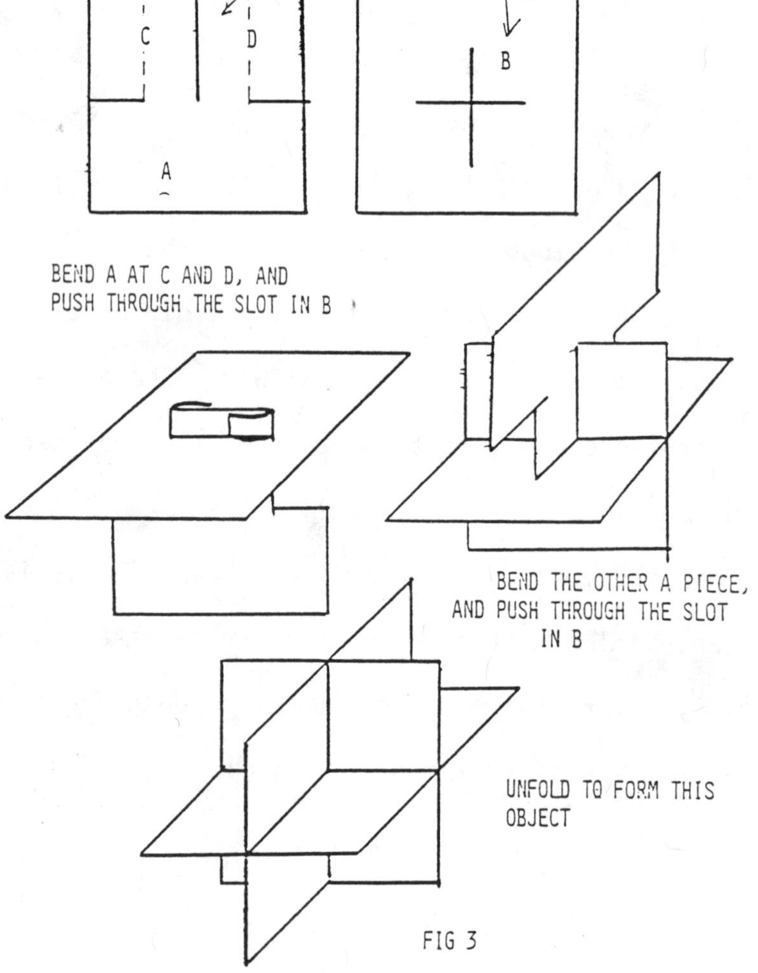
FROM THIN PLASTIC SHEET (E.G. A POLYETHYLENE MILK JUG)
CUT OUT TWO PIECES LIKE A AND ONE LIKE B
C
D
A
B
BEND A AT C AND D, AND
PUSH THROUGH THE SLOT IN B
BEND THE OTHER A PIECE,
AND PUSH THROUGH THE SLOT
IN B
UNFOLD TO FORM THIS
OBJECT

FIG 3